Exploration and Development of Unconventional Oil and Gas Resources: Latest Advances and Prospects

Exploration and Development of Unconventional Oil and Gas Resources: Latest Advances and Prospects

Editors

Shu Tao
Wei Ju
Shida Chen
Zhengguang Zhang
Jiang Han

Basel • Beijing • Wuhan • Barcelona • Belgrade • Novi Sad • Cluj • Manchester

Editors

Shu Tao
China University of
Geosciences (Beijing)
Beijing
China

Wei Ju
China University of Mining
and Technology
Xuzhou
China

Shida Chen
China University of
Geosciences (Beijing)
Beijing
China

Zhengguang Zhang
China National
Administration of Coal
Geology
Beijing
China

Jiang Han
Yanshan University
Qinhuangdao
China

Editorial Office
MDPI AG
Grosspeteranlage 5
4052 Basel, Switzerland

This is a reprint of articles from the Special Issue published online in the open access journal *Energies* (ISSN 1996-1073) (available at: https://www.mdpi.com/journal/energies/special_issues/7FJ247N9P1).

For citation purposes, cite each article independently as indicated on the article page online and as indicated below:

Lastname, A.A.; Lastname, B.B. Article Title. *Journal Name* **Year**, *Volume Number*, Page Range.

ISBN 978-3-7258-1969-0 (Hbk)
ISBN 978-3-7258-1970-6 (PDF)
doi.org/10.3390/books978-3-7258-1970-6

© 2024 by the authors. Articles in this book are Open Access and distributed under the Creative Commons Attribution (CC BY) license. The book as a whole is distributed by MDPI under the terms and conditions of the Creative Commons Attribution-NonCommercial-NoDerivs (CC BY-NC-ND) license.

Contents

Dishu Chen, Jinxi Wang, Xuesong Tian, Dongxin Guo, Yuelei Zhang and Chunlin Zeng
Geological Constraints on the Gas-Bearing Properties in High-Rank Coal: A Case Study of the Upper Permian Longtan Formation from the Songzao Coalfield, Chongqing, Southwest China
Reprinted from: *Energies* 2024, 17, 1262, doi:10.3390/en17051262 1

Shutong Wang, Yanhai Chang, Zefan Wang and Xiaoxiao Sun
Evaluation of Grain Size Effects on Porosity, Permeability, and Pore Size Distribution of Carbonate Rocks Using Nuclear Magnetic Resonance Technology
Reprinted from: *Energies* 2024, 17, 1370, doi:10.3390/en17061370 23

Hanwen Zheng, Zhansong Zhang, Jianhong Guo, Sinan Fang and Can Wang
Numerical Simulation Study on the Influence of Cracks in a Full-Size Core on the Resistivity Measurement Response
Reprinted from: *Energies* 2024, 17, 1386, doi:10.3390/en17061386 38

Tao Liu, Zongbao Liu, Kejia Zhang, Chunsheng Li, Yan Zhang, Zihao Mu, et al.
Intelligent Identification Method for the Diagenetic Facies of Tight Oil Reservoirs Based on Hybrid Intelligence—A Case Study of Fuyu Reservoir in Sanzhao Sag of Songliao Basin
Reprinted from: *Energies* 2024, 17, 1708, doi:10.3390/en17071708 56

Hanyu Zhang, Yang Wang, Haoran Chen, Yanming Zhu, Jinghui Yang, Yunsheng Zhang, et al.
Study on Sedimentary Environment and Organic Matter Enrichment Model of Carboniferous–Permian Marine–Continental Transitional Shale in Northern Margin of North China Basin
Reprinted from: *Energies* 2024, 17, 1780, doi:10.3390/en17071780 76

Shailee Bhattacharya, Shikha Sharma, Vikas Agrawal, Michael C. Dix, Giovanni Zanoni, Justin E. Birdwell, et al.
Influence of Organic Matter Thermal Maturity on Rare Earth Element Distribution: A Study of Middle Devonian Black Shales from the Appalachian Basin, USA
Reprinted from: *Energies* 2024, 17, 2107, doi:10.3390/en17092107 98

Wei Zou and Yongan Gu
Solvent Exsolution and Liberation from Different Heavy Oil–Solvent Systems in Bulk Phases and Porous Media: A Comparison Study
Reprinted from: *Energies* 2024, 17, 2287, doi:10.3390/en17102287 121

Guangjie Zhao, Fujie Jiang, Qiang Zhang, Hong Pang, Shipeng Zhang, Xingzhou Liu and Di Chen
Hydrocarbon Accumulation Process and Mode in Proterozoic Reservoir of Western Depression in Liaohe Basin, Northeast China: A Case Study of the Shuguang Oil Reservoir
Reprinted from: *Energies* 2024, 17, 2583, doi:10.3390/en17112583 141

Sławomir Kędzior and Lesław Teper
Occurrence and Potential for Coalbed Methane Extraction in the Depocenter Area of the Upper Silesian Coal Basin (Poland) in the Context of Selected Geological Factors
Reprinted from: *Energies* 2024, 17, 2592, doi:10.3390/en17112592 157

Xiaoshan Li, Liu Yang, Dezhi Sun, Bingjian Ling and Suling Wang
Experimental Study of Forced Imbibition in Tight Reservoirs Based on Nuclear Magnetic Resonance under High-Pressure Conditions
Reprinted from: *Energies* 2024, 17, 2993, doi:10.3390/en17122993 177

Jiaping Tao, Siwei Meng, Dongxu Li, Lihao Liang and He Liu
Experimental Evaluation of Enhanced Oil Recovery in Shale Reservoirs Using Different Media
Reprinted from: *Energies* **2024**, *17*, 3410, doi:10.3390/en17143410 . . . . **195**

Shihu Zhao, Yanbin Wang, Yali Liu, Zengqin Liu, Xiang Wu, Xinjun Chen and Jiaqi Zhang
Evaluation of Favorable Fracture Area of Deep Coal Reservoirs Using a Combination of Field Joint Observation and Paleostress Numerical Simulation: A Case Study in the Linxing Area
Reprinted from: *Energies* **2024**, *17*, 3424, doi:10.3390/en17143424 . . . . **207**

Article

Geological Constraints on the Gas-Bearing Properties in High-Rank Coal: A Case Study of the Upper Permian Longtan Formation from the Songzao Coalfield, Chongqing, Southwest China

Dishu Chen [1,2,3], Jinxi Wang [1,2], Xuesong Tian [1,2,*], Dongxin Guo [1,2], Yuelei Zhang [1,2] and Chunlin Zeng [1,2]

[1] Key Laboratory of Shale Gas Exploration, Ministry of Natural Resources, Chongqing Institute of Geology and Mineral Resources, Chongqing 401120, China; cds1003chengdu@163.com (D.C.); m17347807272@163.com (J.W.); shmilydongxin@163.com (D.G.); zyl7049@163.com (Y.Z.); ceng8623@189.cn (C.Z.)
[2] National and Local Joint Engineering Research Center of Shale Gas Exploration and Development, Chongqing Institute of Geology and Mineral Resources, Chongqing 401120, China
[3] School of Earth Resources, China University of Geosciences (Wuhan), Wuhan 430074, China
* Correspondence: xuesongtian@outlook.com; Tel.: +86-15205155670

Abstract: The Permian Longtan Formation in the Songzao coalfield, Southwest China, has abundant coalbed methane (CBM) stored in high-rank coals. However, few studies have been performed on the mechanism underlying the differences in CBM gas content in high-rank coal. This study focuses on the characterization of coal geochemical, reservoir physical, and gas-bearing properties in the coal seams M_6, M_7, M_8, and M_{12} based on the CBM wells and coal exploration boreholes, discusses the effects of depositional environment, tectono-thermal evolution, and regional geological structure associated with CBM, and identifies major geological constraints on the gas-bearing properties in high-rank coal. The results show that high-rank coals are characterized by high TOC contents (31.49~51.32 wt%), high T_{max} and R_0 values (averaging 539 °C and 2.17%), low HI values (averaging 15.21 mg of HC/g TOC), high porosity and low permeability, and high gas-bearing contents, indicating a post-thermal maturity and a good CBM production potential. Changes in the shallow bay–tidal flat–lagoon environment triggered coal formation and provided the material basis for CBM generation. Multistage tectono-thermal evolution caused by the Emeishan mantle plume activity guaranteed the temperature and time for overmaturation and thermal metamorphism and added massive pyrolytic CBM, which improved the gas production potential. Good geological structural conditions, like enclosed fold regions, were shown to directly control CBM accumulation.

Keywords: coalbed methane enrichment; Permian Longtan Formation; high-rank coal; depositional environment; tectono-thermal evolution; regional geological structure

1. Introduction

Abundant coalbed methane (CBM) resources are stored in high-rank coals in China, representing a geological resource of 1.044×10^{13} m^3, which accounts for approximately one-third of the total CBM resources [1–3]. Realizing the development and utilization of high-rank CBM plays an important and reliable role in guaranteeing national green energy security, reducing the hazards of coal mine gas, and decreasing carbon dioxide emissions [4–7]. To date, high-rank CBM has garnered extensive attention, with major breakthroughs and commercial developments in the Jincheng and Shouyang–Yangquan areas (Qingshui basin) in North China, the southern part of the Hancheng–Yanchuan area (Ordos basin), and several areas of the Qianbei–Qianxi–Chuannan area in Southwest China [8–13].

However, to efficiently increase the supply of green energy and successfully achieve the carbon peak and carbon neutrality in China by 2030 and 2060, high-rank CBM exploration and development should be carried out immediately. Some areas that have not been extensively explored for high-rank CBM in Southwest China will become a major focus of research, such as typical coalfields distributed in Chongqing city. Chongqing city is rich in high-rank CBM resources, with a conservative value of 2×10^{11} m^3. In particular, the CBM resources from the Songzao coalfield account for 65.7% of the total resources, with 2×10^8 m^3/km^2, indicating a great resource potential. In addition, the Songzao coalfield is also an important anthracite production base [14] and one of the coalfields with the most serious coal and gas outburst accidents in China. Therefore, the Songzao coalfield is an ideal area for further exploration, development, and utilization of high-rank CBM resources in Chongqing city, Southwest China.

Previous studies effectively summarized the systematic geologic theory of high-rank CBM formation in the carboniferous Taiyuan Formation and Permian Shanxi Formation of the Jincheng area and Shouyang–Yangquan areas within the Qinshui basin, including their geochemistry, reservoir physical, and gas-bearing properties, accumulation mechanism, enrichment pattern, main controlling factors, a geological model, and a resource prospect [9,12,15–22]. High-rank CBM reservoirs are highly diverse, complex, and heterogeneous, with limited permeability, undersaturation, low pressure, overmaturation, and high gas contents. However, there are distinct geological variables impacting high-rank CBM accumulation in different regions of China, posing hurdles to improving the CBM production potential [2,15,23,24]. Compared with the great progress regarding high-rank CBM within the Qinshui basin in North China, although some geological investigations on high-rank CBM in the Permian Longtan Formation in the northern and western parts of Guizhou province in Southwest China were conducted [8,10,25–29], the geological characteristics of high-rank CBM in the Songzao coalfield in Chongqing are still lacking in pertinence and validity, and CBM exploration has not yet achieved a major breakthrough. A detailed study of the geological constraints on the CBM gas content in the high-rank coals of the Longtan Formation from the coalfield is thus indispensable.

In this study, coal geochemical, coal reservoir physical, and gas-bearing properties of the main high-rank coal seams in the Longtan Formation in the Songzao coalfield of Chongqing city were analyzed. The effects of depositional environment, tectono-thermal evolution, and regional geological structural conditions on coal formation, CBM gas production potential, and gas accumulation in high-rank coals are comprehensively discussed, and major geological constraints on the gas-bearing properties of high-rank coal from the Longtan Formation in the Songzao Coalfield are identified.

2. Geological Setting

The Songzao coalfield is situated in the Qijiang District in the southwestern part of Chongqing city in Southwest China and has a total area of approximately 235.5 km^2 (Figure 1a,b). It mainly consists of twelve key coal mines, i.e., the Songzao, Tonghua, Guanyinqiao, Yangchatan, Yuyang, Datong, Shihao, Zhangshiba, Liyuanba, Daluo, Xiaoyutuo, and Macun mines (Figure 1b).

2.1. Regional Structural Features

The coalfield is located in the secondary fold belt on the western flank of the Jiudianya, Jiulongshan, and Sangmuchang anticlines (Figure 1b). Its structural pattern presents a radial shape that converges to the northeast and spreads to the southwest. The Lianghekou syncline, Yangchatan anticline, Damushu syncline, and Yutiao anticline from east to west in the coalfield form a "bulge-shaped structure" to the northwestward rise. This structure is distinguished by wide, low anticlines and compact synclines with a gradual weakening of fold amplitude from east to west. Surface fracture phenomena are relatively insignificant and minor, and only those associated with the four folds affect the mining conditions. In addition, the stratigraphic denudation in the anticline cores is more serious than that in the syncline cores, the extension direction of the fracture zones is nearly parallel or perpendicular to the anticline axis, and the fracture extensions are not far.

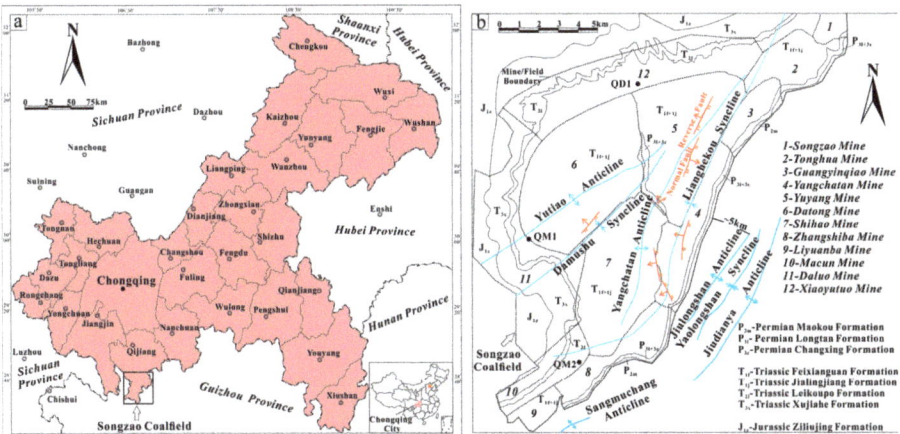

Figure 1. (a) Location of the Songzao coalfield in Chongqing city, Southwest China; (b) regional structural and lithostratigraphic divisions of the Songzao coalfield.

2.2. Regional Structural Features

The coalfield is located in the secondary fold belt on the western flank of the Jiudianya, Jiulongshan, and Sangmuchang anticlines (Figure 1b). Its structural pattern presents a radial shape that converges to the northeast and spreads to the southwest. The Lianghekou syncline, Yangchatan anticline, Damushu syncline, and Yutiao anticline from east to west in the coalfield form a "bulge-shaped structure" to the northwestward rise. This structure is distinguished by wide, low anticlines and compact synclines with a gradual weakening of fold amplitude from east to west. Surface fracture phenomena are relatively insignificant and minor, and only those associated with the four folds affect the mining conditions. In addition, the stratigraphic denudation in the anticline cores is more serious than that in the syncline cores, the extension direction of the fracture zones is nearly parallel or perpendicular to the anticline axis, and the fracture extensions are not far.

2.3. Coal-Bearing Stratigraphic Characteristics

The strata exposed in the coalfield mainly include the Paleozoic Permian series and the Mesozoic Triassic and Jurassic series (Figure 1b). The Jurassic and Triassic strata are widely distributed in the synclines in the western, eastern, and southeastern parts of the coalfield, while the Permian strata are mainly exposed along the anticline axis or near the axis in the eastern and southeastern parts of the coalfield. By the latest Permian integrative stratigraphy and timescale of China [30], the upper Permian series in the coalfield include the Wuchiapingian and Changhsingian stages. The Changhsingian stage includes the Changxing Formation, and the Wuchiapingian stage includes the Longtan Formation.

The coal measure strata of the coalfield are exposed in the Longtan Formation, which mainly consists of coal seams, bioclastic and siliceous limestone, sandstone, siltstone, silty mudstone, calcareous mudstone, argillaceous shale, and tuffaceous sediments, and belong to the shallow bay–tidal flat–lagoon mixed deposits of alternating marine–continental transitional environments along the western margin of a shallow carbonate platform within an epicontinental sea ([14,31–37] and Figure 1). The Kangdian Oldland is the dominant terrestrial source for the coalfield. The total thickness of the coal measure strata is generally approximately 66 to 80 m, containing 5 to 13 coal seams with a high metamorphic degree (type III kerogen) and an average maximum vitrinite reflectance (R_0, max) greater than 2.0%, among which, the main coal seams include M_6, M_7, M_8, and M_{12} throughout the whole coalfield, which are important targets of CBM exploration and development (Figure 2). The middle Permian Maokou Formation disconformably underlies the Longtan coal measure strata, which consists of medium-to-thick-bedded and massive bioclastic limestones that are rich in marine fossils, mostly including fusulinids, corals, brachiopods, ammonites, conodonts, benthic foraminifera, and calcareous algae [38,39]. The Longtan Formation overlies the Changxing Formation, which is composed of medium-to thick-bedded bioclastic limestone containing less dolomite and less banded and nodular cherts dominated by marine fusulinid, coral, brachiopod, and ammonite fossils in a shallow carbonate platform environment.

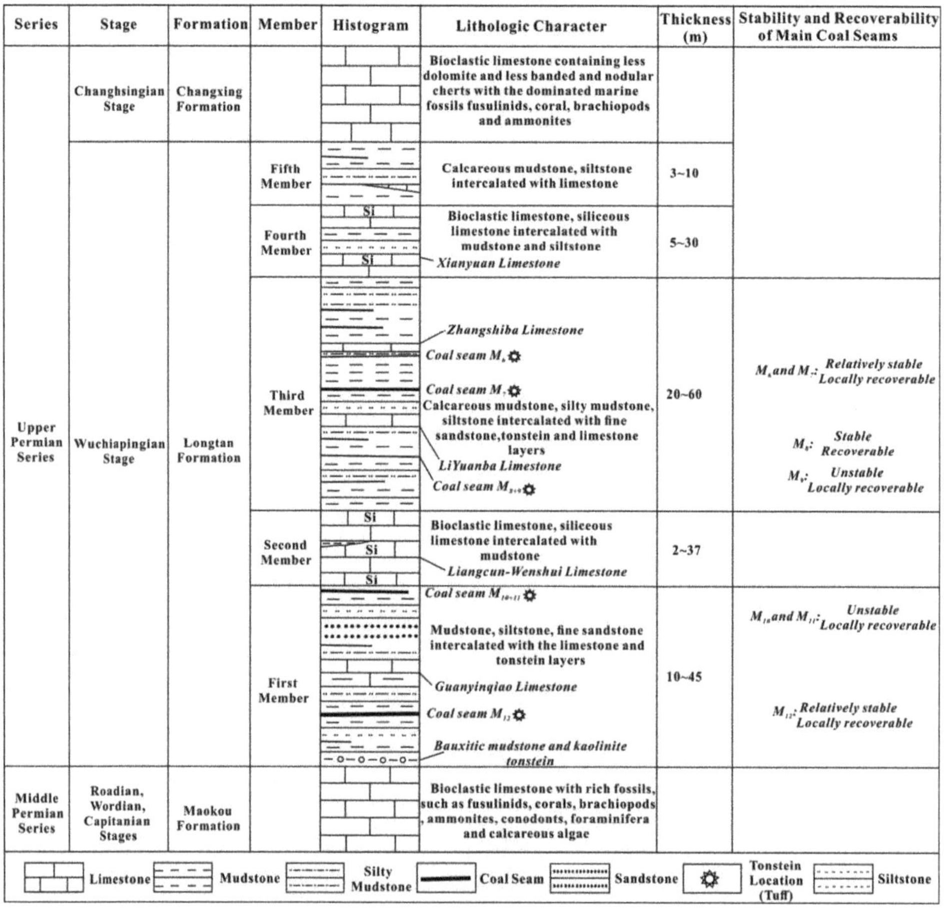

Series	Stage	Formation	Member	Histogram	Lithologic Character	Thickness (m)	Stability and Recoverability of Main Coal Seams
Upper Permian Series	Changhsingian Stage	Changxing Formation			Bioclastic limestone containing less dolomite and less banded and nodular cherts with the dominated marine fossils fusulinids, coral, brachiopods and ammonites		
	Wuchiapingian Stage	Longtan Formation	Fifth Member		Calcareous mudstone, siltstone intercalated with limestone	3~10	
			Fourth Member		Bioclastic limestone, siliceous limestone intercalated with mudstone and siltstone — Xianyuan Limestone	5~30	
			Third Member		— Zhangshiba Limestone Coal seam M_6 Coal seam M_7 Calcareous mudstone, silty mudstone, siltstone intercalated with fine sandstone, tonstein and limestone layers — LiYuanba Limestone Coal seam $M_{8,9}$	20~60	M_6 and M_7: Relatively stable Locally recoverable M_8: Stable Recoverable M_9: Unstable Locally recoverable
			Second Member		Bioclastic limestone, siliceous limestone intercalated with mudstone — Liangcun-Wenshui Limestone	2~37	
			First Member		Coal seam M_{10-11} Mudstone, siltstone, fine sandstone intercalated with the limestone and tonstein layers — Guanyinqiao Limestone Coal seam M_{12} Bauxitic mudstone and kaolinite tonstein	10~45	M_{10} and M_{11}: Unstable Locally recoverable M_{12}: Relatively stable Locally recoverable
Middle Permian Series	Roadian, Wordian, Capitanian Stages	Maokou Formation			Bioclastic limestone with rich fossils, such as fusulinids, corals, brachiopods, ammonites, conodonts, foraminifera and calcareous algae		

Legend: Limestone, Mudstone, Silty Mudstone, Coal Seam, Sandstone, Tonstein Location (Tuff), Siltstone

Figure 2. Typical lithologic stratigraphic framework of the Permian Longtan Formation in the Songzao coalfield.

2.4. Thickness and Distribution of the Main Coal Seams

The thickness and distribution of the main coal seams M_6, M_7, M_8, and M_{12} in the Longtan Formation in the Songzao coalfield are relatively stable (Figures 2 and 3 and Table 1). The coal seams are usually buried at a depth of 400~1700 m, and their total thickness is 4.01~9.88 m, with an average of 7.58 m and a total recoverable value of 5.45 m. There is a thinning zone with a range of 2 km^2 in the northwestern part of the coalfield, which is situated at the junction of the Xiaoyutuo and Daluo mines. The total thickness of the four coal seams averages approximately 3 m, but their largest area, distributed in the Shihao, Datong, and Daluo mines located on the southeastern flank of the Damushu syncline, has a total thickness of approximately 8 m. But in the monoclinal structure in the southwestern area of the coalfield, the thickness of the four coal seams is generally small.

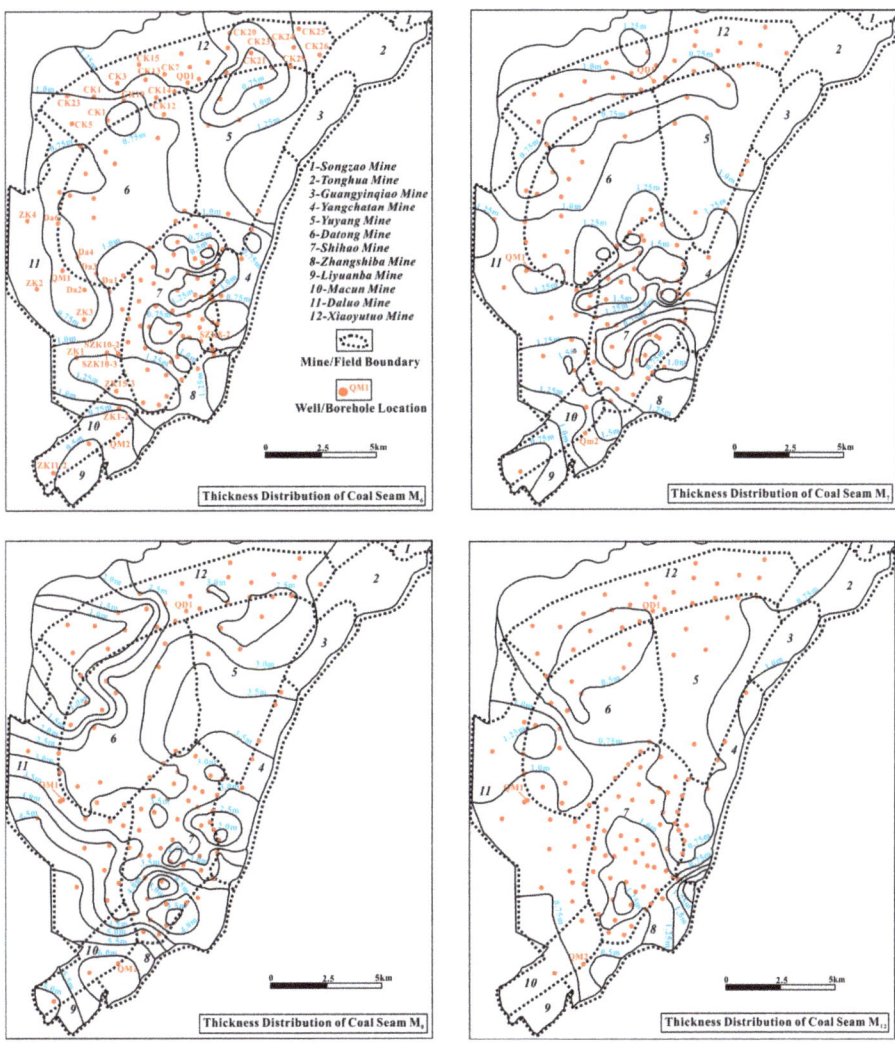

Figure 3. Thickness variation in the coal seams M_6, M_7, M_8, and M_{12}.

Table 1. General parameters of the coal seams M_6, M_7, M_8, and M_{12}.

Coal Seam Number	Depth of Coal Seams (m)	Thickness (m) Minimum–Maximum Average	Coal Seam Interlayer Spacings (m)	Tonstein Thickness (m) Minimum–Maximum Average	Lithological Characters of Coal Seam Roof and Floor		Stable and Recoverable
					Coal Seam Floor	Coal Seam Roof	
M_6		0.4~1.47 0.94	7.1	0.03~0.24 0.11	mudstone–siltstone	mudstone–siltstone	relatively stable locally recoverable
M_7	400~1700	0.71~1.62 1.11	6.6	0.01~0.57 0.26	mudstone–siltstone	mudstone	relatively stable locally recoverable
M_8		0.83~6.43 3.04		0.03~0.58 0.24	sandstone–siltstone	mudstone–siltstone	stable recoverable
M_{12}		0.31~3.33 0.86	22.6	0.01~0.1 0.04	siltstone–sandy mudstone	siltstone–sandy mudstone	relatively stable locally recoverable

The coal seams M_6, M_7, and M_8 are located in the middle part of the Longtan coal measure strata, with interlayer spacings of 7.1 m and 6.6 m. The thickness of the coal seams

M_7 and M_8 revealed a trend of gradual increase as the depth increased. The coal seam M_{12} is located in the lower part of the coal measure strata, with an interlayer separation of approximately 22.6 m from the coal seam M_8. The total thickness of the coal seam M_6 is 0.4~1.47 m, with an average of 0.94 m, and it serves as an unstable coal seam, containing 0~1 layer of tonstein and 2 local layers. The total thickness of these tonsteins is 0.03~0.24 m, with an average of 0.11 m. The coal seam is thinner at the junction of the Xiaoyutuo and Daluo mines, with a thickness of less than 0.75 m. The overall thickness of the coal seam M_7 is 0.71~1.62 m, with an average of 1.11 m. It is a thin coal seam with a simple structure and a stable thickness. In certain regions, the middle part of the coal seam is interspersed with a 0.01~0.57 m thick layer of argillaceous tonstein. The most important recoverable coal seam, M_8, is 0.83~6.43 m thick, with an average thickness of 3.04 m, and is a medium-thickness and stable coal seam. The structure of the coal seam is simple, and the tonsteins are generally located in its upper part, with a total thickness of 0.03~0.58 m and an average thickness of 0.24 m. The thickness of the zone delimiting the lower part of the coal seam is generally 6~9 times that of the upper part, leading to the formation of a three-layer tonstein structure with two coal layers and one tonstein layer. The coal seam M_{12}, situated in the plunging crown of the Yutiao anticline, is located in the area including the coal seams M_{11} and M_{12}. It is directly overlain by aluminum mudstone, of which the east side is the independent stratification area of the coal seams M_{11} and M_{12}. The coal seam M_{11} is located above the coal seam M_{12}, with a thickness of 0.41~3.41 m. The coal seam M_{11} is partially recoverable, while the coal seam M_{12} does not have, generally, a recoverable thickness, belonging to a nonrecoverable coal seam. In this study, the two coal seams are jointly referred to as coal seam M_{12}. The total thickness of the coal seam M_{12} is 0.31~3.33 m, with an average of 0.86 m, indicating that it is a thin coal seam; its general thickness is between recoverable and critically recoverable, making it a relatively stable coal seam.

3. Sampling and Methods

3.1. Evaluation of the Samples

Samples from three CBM wells (QD1, QM1, and QM2 wells) and more than eighty coal exploration boreholes of the Upper Permian Longtan Formation in the Songzao coalfield were collected, with burial depths ranging from approximately 400 to 1700 m, and the distribution of CBM gas in the main coal seams is described. Then, 99 experimental samples of high-rank coal from 3 drilling cores in the QD1, QM1, and QM2 wells and 5 coal exploration boreholes (ZK1, ZK4, SZK8-2, SZK10-3, and SZK10-2) in the coalfield were analyzed in depth. The experimental materials were systematically extracted from the coal seams M_6, M_7, M_8, and M_{12} of the Longtan Formation to determine geological parameters such as macerals, vitrinite reflectance (R_0), total organic carbon (TOC), amount of free hydrocarbons plus yield of residual hydrocarbons ($S_1 + S_2$), maximum pyrolysis temperature (T_{max}) in rock pyrolysis, pore structure, porosity, and permeability, gas content, and components. All the materials were sealed with desiccators and then measured in the laboratory of the Chongqing Mineral Resources Supervision and Testing Center, Chinese Ministry of Land and Resources. Some data for the coal seams M_6, M_7, M_8, and M_{12} of the Songzao coalfield, such as thickness, CBM gas content, and macerals, were primarily obtained from previous studies [14,31,33–35,37,40,41] and geological reports on the detailed investigation of coal resources in various coal mines of the coalfield.

3.2. Analytical Methods

In this study, macerals and R_0 were measured using a Leica DM4500P light microscope (Leica, Wetzlar, Germany) with a 40× objective to analyze the volume percentages of macerals and evaluate the thermal maturity of organic matter based on reflectance spectrometry, fluorescence, and transmission spectrometry. The analytical methods referred to the Chinese oil and gas industry standards SY/T 6414-2014 [42] and SY/T 5124-2012 [43]. Five thermal evolution stages could be generally identified, i.e., immature ($R_0 < 0.5\%$), lowly mature (R_0, 0.5~0.7%), mature (R_0, 0.7~1.3%), highly mature (R_0, 1.3~2.0%), and overmature ($R_0 > 2.0\%$) stages.

To determine the original parent organic matter material in the hydrocarbon source rock, coal samples were pyrolyzed using a China Haicheng Rock-Eval VIII instrument (Haicheng Petrochemical Instrument Factory, Haicheng, China) with a flame ionization detector. The program was carried out in accordance with the national standard GB/T 18602-2012 [44].

The pore structure was observed, and the pores were counted by scanning electron microscopy (SEM) and high-pressure mercury intrusion porosimetry. SEM imaging was performed using an American Thermo Fisher Scientific Apreo SHiVac-Type field-emission scanning electron microscope (FE-SEM) (Thermo Fisher Scientific, Waltham, MA, USA) and an American Gatan 697 Ilion II Argon ion-polishing mill (AMETEK, Berwyn, PA, USA) to determine the pore characteristics in the coal samples based on the Chinese oil and gas industry standard SY/T 5162-2014 [45]. The classification of the pore type referred to a previous work, which described organic matter pores, interparticle mineral pores, intra-particle mineral pores, and fracture pores [46]. Based on the guidelines of the International Union of Pure and Applied Chemistry (IUPAC), pores in coal can be classified into three categories, i.e., "micropores", with a diameter between 0 and 0.002 μm (0~2 nm), "meso-pores" with a diameter between 0.002 μm and 0.05 μm (2~50 nm), and "macropores" with a diameter greater than 0.05 μm (>50 nm) [47]. The mercury instrusion porosimetry was conducted via an American Mike Autopore IV 9500 mercury porosimeter (Micromeritics, Atlanta, GA, USA) to determine different pore volumes under the national standard of GB/T 21650.1-2008 [48].

The porosity and permeability were determined via an American CORETEST SYSTEMS Inc. AP-609 porosity–permeability tester (CoreTest, Atlanta, GA, USA) with the analyzed porosity ranging from 0.1 to 40%, and permeability ranging from 0.001 to 10,000 mD on the basis of the national standard SY/T 6385-2016 [49].

The gas content and its components were determined by in situ gas desorption and isothermal adsorption experiments and gas composition determination. The contour map of CBM gas concentration was established by referring to two geostatistical methods, kriging and triangulation, based on previous research and geological data on coal resources in the Songzao coalfield. The in situ CBM desorption was measured using a self-developed in situ gas-bearing test instrument from the Chongqing Institute of Geology and Mineral Resources to evaluate CBM potential and sweet spot prediction. The in situ test methods referred to the national standard GB/T 19559-2008 [50]. The isothermal adsorption experiment was performed using a ZJ466 Rubotherm IsoSORP HP StaticIII-Type magnetic suspension balance gravimetric high-pressure isothermal adsorption–desorption instrument (Rubotherm, Bochum, Germany). The Langmuir volume pressure is referred to as the Langmuir adsorption isothermal [51]. The gas composition was evaluated by using an ITQ 900 gas chromatographer (GC) (Thermo Fisher Scientific, Waltham, MA, USA) equipped with a thermal conductivity detector and a flame ionization detector based on the national standard GB/T 13610-2014 [52].

4. Results

4.1. Coal Geochemical Characterization

The coal geochemical parameters were directly obtained from the maceral, rock pyrolysis, and R_0 experiments. The coal maceral analysis of samples from the Datong mine and the QD1 well revealed that the average content of organic components in the coal seams M_7, M_8, and M_{12} was 81.5 to 88.0%, with the coal seam M_8 having the highest content (Table 2). The content of typical inorganic components varied from 12.0 to 18.5%, with the coal seam M_{12} possessing the highest content. On average, vitrinite was found to contribute 60.1~69.0% of the organic components, whereas inertinite accounted for 14.3~25.3% of them. Clay minerals appeared to be the most abundant inorganic component, followed by sulfide minerals, while oxide and carbonate minerals were less prevalent.

Table 2. Coal macerals from the Datong mine and the QD1 well.

Coal Seam Number	Coal Macerals from Main Coal Seams $\left(\frac{\text{Minimum} \sim \text{maximum}}{\text{Average (quantity)}}\right)$							
	Organic Component			Inorganic Component				
	Vitrinite (%)	Inertinite (%)	Subtotal (%)	Clay Mineral (%)	Sulfide Mineral (%)	Oxide Mineral (%)	Carbonate Mineral (%)	Subtotal (%)
M_7	$\frac{55.2 \sim 64.3}{60.1\ (5)}$	$\frac{19.3 \sim 34.2}{25.3\ (5)}$	$\frac{81.8 \sim 89.4}{85.4\ (5)}$	$\frac{5.6 \sim 13.8}{9.7\ (5)}$	$\frac{1.1 \sim 4.9}{2.5\ (5)}$	$\frac{0.7 \sim 1.8}{1.1\ (5)}$	$\frac{0.4 \sim 2.2}{1.3\ (5)}$	$\frac{13.9 \sim 18.2}{14.6\ (5)}$
M_8	$\frac{60.3 \sim 77.1}{69.0\ (7)}$	$\frac{10.5 \sim 25.6}{19.0\ (7)}$	$\frac{85.6 \sim 90.3}{88.0\ (7)}$	$\frac{3.3 \sim 9.8}{7.6\ (7)}$	$\frac{1.2 \sim 5.2}{2.5\ (7)}$	$\frac{0.1 \sim 5.8}{1.6\ (7)}$	$\frac{0.1 \sim 0.6}{0.3\ (7)}$	$\frac{9.7 \sim 14.4}{12.0\ (7)}$
M_{12}	$\frac{62.3 \sim 72.2}{67.2\ (5)}$	$\frac{11.0 \sim 18.3}{14.3\ (5)}$	$\frac{75.6 \sim 85.5}{81.5\ (5)}$	$\frac{8.7 \sim 16.3}{12.5\ (5)}$	$\frac{2.0 \sim 7.1}{4.1\ (5)}$	$\frac{0.1 \sim 1.5}{0.4\ (5)}$	$\frac{0.1 \sim 3.7}{1.5\ (5)}$	$\frac{14.5 \sim 24.4}{18.5\ (5)}$

The four coal samples from the coal seams M_6, M_7, M_8, and M_{12} in the QD1 well had high TOC contents ranging from 31.49 to 51.32 wt% (Table 3). These studied coals presented low S_1 and S_2 values in the ranges of 0.0916~0.12 mg/g and 4.3565~8.4797 mg/g, respectively. The TOC and $S_1 + S_2$ values indicated that the coals are overmature, as discussed below, and have a fair hydrocarbon generation potential. The T_{max} values ranged from 534 to 549 °C, with an average of 539 °C, suggesting that the coals experienced thermal evolution to overmaturation. The hydrogen index (HI) values ranged from 13.83 to 16.52 mg HC/g TOC, with a mean value of 15.21 mg HC/g TOC, indicating that type III kerogen (less than 200 mg HC/g TOC) is dominant in the coal seams M_6, M_7, M_8, and M_{12} of the Longtan Formation. The R_0 values varied from 2.09 to 2.24%, averaging 2.17%. In addition, the coals in the Longtan coal measure strata from Chongqing city contain mostly semianthracite and anthracite, with R_0 values of 1.88~2.6% on average, according to previous studies [14,29,31], and underwent a highly thermal evolution process, leading to high-rank coal with a post-thermal maturity and good potential for CBM accumulation.

Table 3. Geochemical parameters of the high-rank coal in the QD1 well.

Coal Seam Number	S_1 (mg/g)	S_2 (mg/g)	T_{max} (°C)	HI (mg/g)	TOC (wt%)	R_0 (%)
M_6	0.0975	7.183	534	15.66	45.86	2.13
M_7	0.12	8.4797	535	16.52	51.32	2.09
M_8	0.0916	7.3502	535	14.81	49.63	2.24
M_{12}	0.1047	4.3565	549	13.83	31.49	2.2

Remarks: HI = $S_2 \times 100$/TOC, mg HC/g TOC.

4.2. Coal Reservoir Characterization

4.2.1. Pore Structure

The coal samples from the coal seams M_6, M_7, M_8, and M_{12} analyzed by SEM showed that the coal pores were mainly gas holes and erosion pores, which were distributed inside the massive organic components. These pores' diameters were generally 0.13~3.45 μm, with a maximum of 10.69 μm (Figure 4). Coal fissures had not developed, and only a few of them were visible. The width of these fissures was generally 1.0~7.18 μm. The fissures were mainly shell-like and step-shaped. The organic components were distributed in flatter blocks and strips, with clastic, agglomerate, and granular clay minerals dominating the mineral composition of the coal.

The coal pore volumes in the study area varied from 1.48 to 48.40 × 10^{-4} cm^3/g, with an average of 10.86 × 10^{-4} cm^3/g, and the average volume ratio was 33.33% (Table 4). Meanwhile, the volume ratio of the coal seam M_6 was more than 40%, and micropores were predominant. The variation range of the micropore volumes was 2.34~48.40 × 10^{-4} cm^3/g, with an average of 11.78 × 10^{-4} cm^3/g, and the average volume ratio of the micropores was 32.28%. The variation range of the mesopore volume was 1.48~11.40 × 10^{-4} cm^3/g, with an average of 4.56 × 10^{-4} cm^3/g. The volume ratio of the mesopores was 7.22~20.11%, with an average of 13.59%. The mesopore volume in these coal seams was much smaller. Moreover, the variation range of the macropore volumes was 7.87~27.24 × 10^{-4} cm^3/g, with an average of 16.22 × 10^{-4} cm^3/g, and the volume ratio of the macropores ranged from 28.12 to 81.37%, with an average of 54.14%. In summary, the coal in the coalfield appeared to contain mainly macropores and micropores, and their total proportion was greater than 80%. Among them, the coal seam M_6 revealed a prevalence of micropores, and the other coal seams presented a prevalence of macropores.

4.2.2. Porosity and Permeability

Depending on the porosity and permeability data of 19 coal samples from different mines within the coalfield, it was determined that the coal porosity ranged from 2.36% to 5.26%, with an average of 4.29%. The permeability varied from 0.0029 to 0.0221 mD, with the majority of the samples having a permeability below 0.01 mD and an average permeability of 0.0069 mD and thus placed in the ultralow-permeability coal seam group (Table 5). Except for the QM1 well, the coal permeability in the Daluo and Shihao mines was found to be extremely low, with the highest permeability not exceeding 0.01 mD, and the average being 0.006 mD, which is related to the fact that the tested coal samples were taken from deep coal seams (buried more than 1000 m).

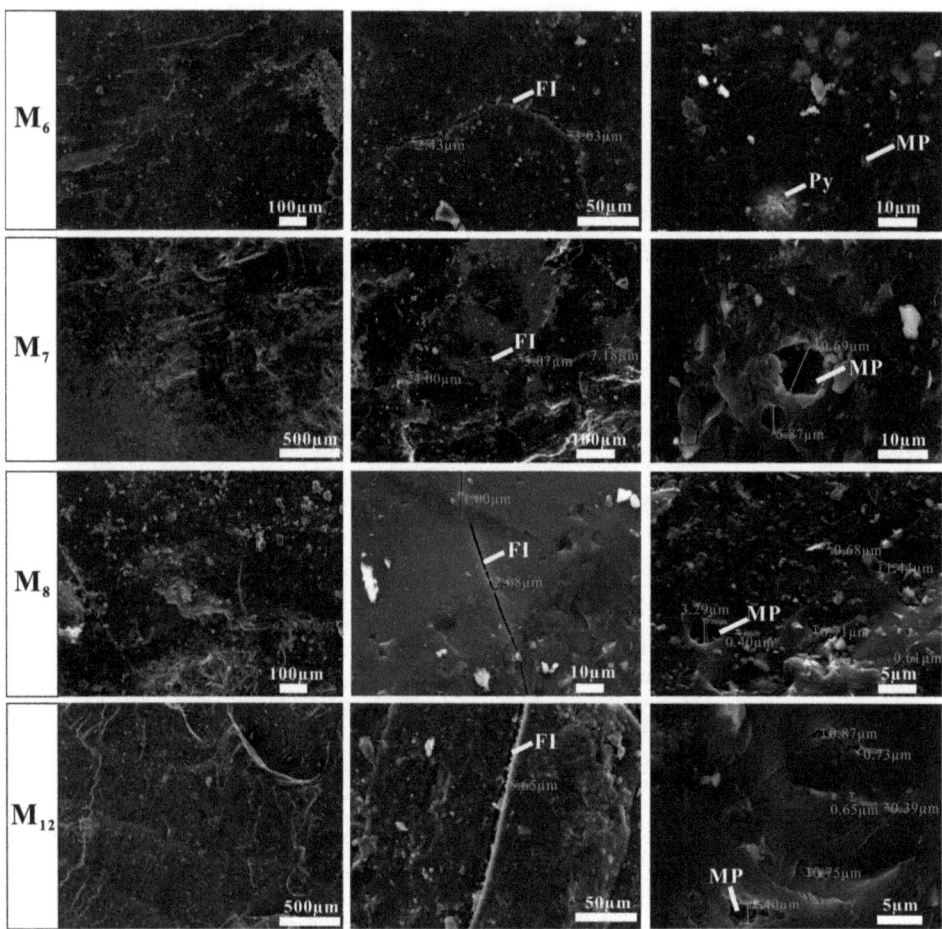

Figure 4. Microstructural photographs of coal from the coal seams M_6, M_7, M_8, and M_{12}. FI, fissure; MP, macroscopic pore; Py, pyrite.

Table 4. Coal pore structure and volume parameters in the coal seams M_6, M_7, M_8, and M_{12} determined by mercury injection porosimetry.

Coal Seam Number	Buried Depth (m)	Pore Volume (10^{-4} cm^3/g)				Pore Volume Ratio (%)			Well/Borehole Number
		V1	V2	V3	Vt	V1/Vt	V2/Vt	V3/Vt	
M_6	886.20	7.87	3.47	8.56	19.90	39.55	17.44	43.02	QM1 Well
	1661.85	11.12	6.43	22.00	39.55	28.12	16.26	55.56	Daluo Mine, ZK1
	1381.45	15.29	6.99	20.30	42.58	35.91	16.42	47.67	Daluo Mine, ZK4
	912.41	15.32	5.16	13.70	34.18	44.82	15.10	40.08	Shihao Mine, SZK8-2
	1074.34	25.22	11.40	48.40	85.02	29.66	13.41	56.93	Shihao Mine, SZK10-2
	1444.85	19.75	6.89	15.50	42.14	46.87	16.35	36.78	Shihao Mine, SZK10-3

Table 4. Cont.

Coal Seam Number	Buried Depth (m)	Pore Volume (10^{-4} cm^3/g)				Pore Volume Ratio (%)			Well/Borehole Number
		V1	V2	V3	Vt	V1/Vt	V2/Vt	V3/Vt	
M$_7$	1672.09	15.84	2.23	3.24	21.31	74.33	10.46	15.20	Daluo Mine, ZK1
	1393.31	15.73	2.50	5.98	24.21	64.97	10.33	24.70	Daluo Mine, ZK4
	917.76	14.77	3.32	6.52	24.61	60.01	13.49	26.49	Shihao Mine, SZK8-2
	1452.42	14.39	5.27	6.55	26.20	54.92	20.11	25.00	Shihao Mine, SZK10-3
	1081.12	14.64	3.12	10.00	27.76	52.74	11.24	36.02	Shihao Mine, SZK10-2
	898.70	14.20	3.76	8.05	26.01	54.59	14.46	30.95	QM1 Well
M$_8$	1685.02	18.97	2.19	3.54	24.70	76.80	8.87	14.33	Daluo Mine, ZK1
	1400.17	14.54	3.16	9.11	26.81	54.23	11.79	33.98	Daluo Mine, ZK4
	929.95	14.36	3.13	5.37	22.86	62.82	13.69	23.49	Shihao Mine, SZK8-2
	1461.11	16.48	5.19	14.60	36.27	45.44	14.31	40.25	Shihao Mine, SZK10-3
	1091.12	18.45	5.45	14.00	37.90	48.68	14.38	36.94	Shihao Mine, SZK10-2
	905.90	16.68	1.48	2.34	20.50	81.37	7.22	11.41	QM1 Well
M$_{12}$	1704.07	10.89	3.50	7.51	21.90	49.73	15.98	34.29	Daluo Mine, ZK1
	1431.64	27.24	2.74	6.32	36.30	75.04	7.55	17.41	Daluo Mine, ZK4
	1114.23	20.42	10.90	24.50	55.82	36.58	19.53	43.89	Shihao Mine, SZK10-2
	934.30	14.76	2.10	3.14	20.00	73.80	10.50	15.70	QM1 Well

Remarks: V1, V2, and V3 are the pore volumes of macropores, mesopores, and micropores, respectively. Vt is the total pore volume.

Table 5. Coal porosity and permeability data of samples from the coal seams M$_6$, M$_7$, M$_8$, and M$_{12}$.

Coal Seam Number	Burying Depth (m)	Porosity (%)	Permeability (mD)	Well/Borehole Number
M$_6$	1662	4.25	0.0063	Daluo Mine, ZK1
	1381	4.32	0.0065	Daluo Mine, ZK4
	1445	3.82	0.0077	Shihao Mine, SZK10-3
	912	4.57	0.0050	Shihao Mine, SZK8-2
	1074	4.67	0.0063	Shihao Mine, SZK10-2
	899	2.36	0.0221	QM1 Well
M$_7$	1672	5.01	0.0062	Daluo Mine, ZK1
	1393	3.95	0.0054	Daluo Mine, ZK4
	918	5.08	0.0059	Shihao Mine, SZK8-2
	1452	4.68	0.0086	Shihao Mine, SZK10-3
	1081	3.54	0.0072	Shihao Mine, SZK10-2
M$_8$	1685	3.95	0.0043	Daluo Mine, ZK1
	930	5.26	0.0068	Shihao Mine, SZK8-2
	1461	4.16	0.0094	Shihao Mine, SZK10-3
	1091	4.19	0.0031	Shihao Mine, SZK10-2
	1400	3.94	0.0050	Daluo Mine, ZK4
M$_{12}$	1704	4.31	0.0075	Daluo Mine, ZK1
	1432	4.21	0.0042	Daluo Mine, ZK4
	1114	5.18	0.0029	Shihao Mine, SZK10-2

4.3. Coal Gas-Bearing Properties

4.3.1. Composition of CBM

According to the gas component data of the 15 coal samples from the QD1, QM1, and QM2 wells (Table 6), the concentration of desorbed CH_4 in the coal seams M$_6$, M$_7$, M$_8$, and M$_{12}$ ranged from 88.62 to 99.41%, with an average of 94.45%. The content of C^{2+} was 0~0.18%, while the inorganic component comprised minor amounts of CO_2 and N_2. The CO_2 content ranged from 0.48 to 1.55%, while the N_2 content was typically less than 9.71%.

4.3.2. Distribution of the CBM Gas Contents

Based on the in situ desorption analysis of the CBM gas content in the QM1, QM2, and QD1 wells, the in situ desorption gas contents in the coal seams M$_6$, M$_7$, M$_8$, and M$_{12}$ were 12.5~15.3 m^3/t, 21.4~25.8 m^3/t, 15.9~25.6 m^3/t, and 12.1~21.1 m^3/t, respectively. The gas contents in the main coal seams from the three CBM wells were more than 8.0 m^3/t, indicating a good material foundation for gas generation. Meanwhile, vertically, the gas contents in the coal seams M_7 and M_8 were relatively higher than those in the coal seams M_6 and M_{12}.

Table 6. Gas component data of samples from the coal seams M_6, M_7, M_8, and M_{12} in the QD1, QM1 and QM2 wells.

Well Number	Coal Seam Number	Content without Air of Components (Volume)/%			
		N_2	CO_2	CH_4	C^{2+}
QD1 Well	M_6	6.90	0.88	92.18	0.04
QM1 Well		2.56	1.12	96.21	0.11
QM2 Well		7.81	0.90	91.12	0.18
QM2 Well	M_7	1.51	0.90	97.44	0.16
QM1 Well		9.54	0.65	89.71	0.10
QD1 Well		6.26	0.89	92.84	0.01
QM1 Well		6.72	1.05	92.15	0.08
QM1 Well	M_8	0.00	0.50	99.41	0.09
QM2 Well		9.71	1.55	88.62	0.13
QM2 Well		8.69	1.24	89.92	0.15
QM2 Well		2.08	1.24	96.50	0.18
QD1 Well		3.72	1.23	95.05	0.00
QM2 Well	M_{12}	1.21	0.83	97.74	0.23
QM2 Well		0.45	0.53	98.85	0.17
QM1 Well		0.35	0.48	99.03	0.13

As can be seen from the distribution of the CBM gas contents in the coal seams M_6, M_7, M_8, and M_{12} from different mines (Figure 5 and Table 7), the average CBM gas content in these coal seams in the Xiaoyutuo mine ranged from 12.47 to 21.45 m^3/t, with the highest content was found in the coal seam M_8. The average CBM gas contents in the coal seams of the Datong and Shihao mines showed a very similar variation trend and were only 11.99~16.98 m^3/t and 11.15~17.42 m^3/t, respectively. The average CBM gas content in the coal seams of the Daluo mine ranged from 26.14 to more than 30 m^3/t and was the highest in the study area.

Table 7. Average CBM gas contents of the coal seams M_6, M_7, M_8, and M_{12} in different coal mines of the Songzao coalfield.

Coal Mine	Coal Seam	Average Depth (m)	Average Gas Content (m^3/t)
Xiaoyutuo	M_6	776.98	12.47
	M_7	850.29	15.42
	M_8	918.95	21.45
	M_{12}	1057.1	18.14
Datong	M_6	533.39	11.99
	M_7	640.27	15.92
	M_8	690.34	16.98
Daluo	M_7	1152.59	26.14
	M_8	1549.19	26.25
	M_{12}	1587.06	28.18
Shihao	M_6	885.48	11.17
	M_7	1025.49	17.42
	M_8	1076.72	17.02
	M_{12}	1079.65	11.51

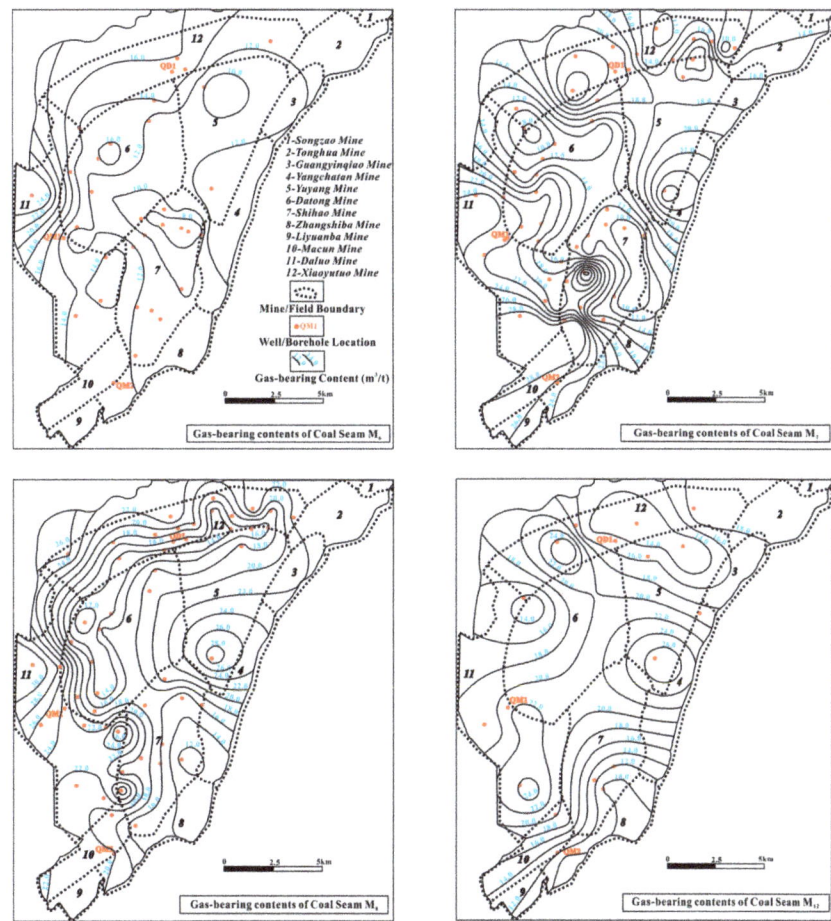

Figure 5. Regional distribution of the CBM gas contents in high-rank coal from the coal seams M_6, M_7, M_8, and M_{12}. Typical CBM gas content data are reported in Table S1.

The average CBM gas content in the coal seam M_8 in the coalfield was usually higher than 16 m^3/t, and some areas with lower gas contents were found only at the junction of the Datong, Xiaoyutuo, and Daluo mines and at the junction of the Datong and Shihao mines. The highest CBM gas content in regional coal seams was above 30 m^3/t, and these high-content sites are mainly distributed in deep areas of the Xiaoyutuo, Daluo, and Shihao mines along the Yutiao anticline. The total CBM gas content in these coal seams appeared to increase from east to west and as the elevation of the coal seam floor decreased.

4.3.3. Adsorption–Desorption Characteristics

The isothermal adsorption results of CBM gas analysis revealed that the Langmuir volume and the Langmuir pressure in the coal seam M_6 were 6.40 cm^3/g and 1.30 MPa, respectively (Figure 6). The coal seam M_8 was characterized by the largest Langmuir volume of 24.88 cm^3/g and a Langmuir pressure of 1.04 MPa. The Langmuir volume of the coal seam M_{12} was 15.24 cm^3/g, and its Langmuir pressure was only 0.91 MPa.

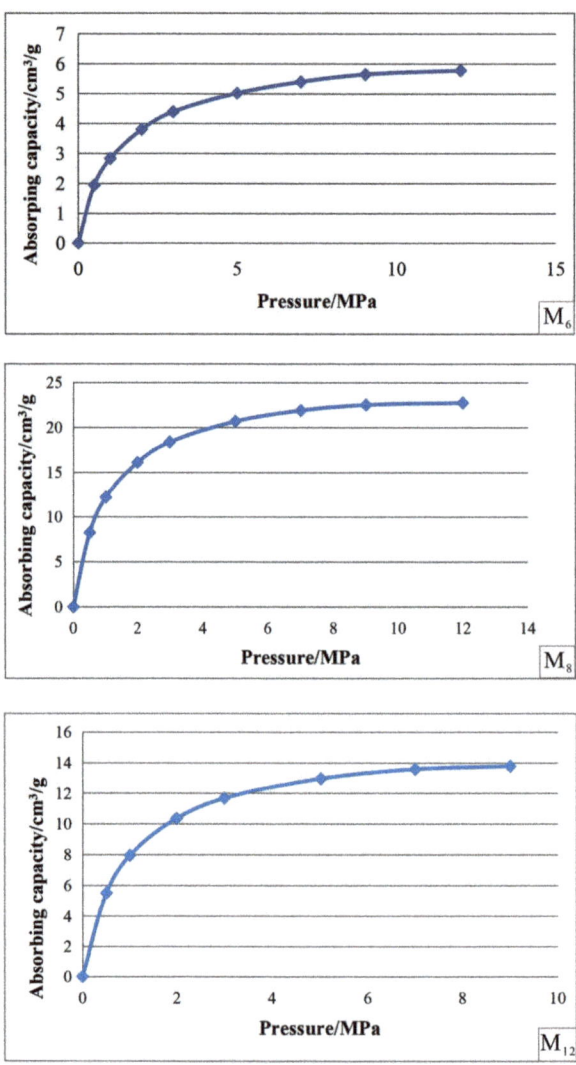

Figure 6. Isothermal adsorption data of CBM gas in the coal seams M_6, M_8, and M_{12} from the QD1 well.

The central depths (vertical depths) of the coal seams M_6, M_8, and M_{12} in the QD1 well were 888.1 m, 904.7 m, and 933.6 m, respectively. Based on the formation pressure coefficient of 1.0 for the Xiaoyutuo mine, the formation pressures of the coal seams M_6, M_8, and M_{12} were 8.88 MPa, 9.05 MPa, and 9.34 MPa, respectively. Combining these data with the isothermal adsorption curve of the coal samples and the Langmuir isothermal adsorption equation, the theoretical CBM gas contents of the coal seams M_6 and M_{12} were 5.58 m^3/t and 13.88 m^3/t, respectively. However, the coal seam M_8 showed the highest theoretical gas content of 22.32 m^3/t, appearing as the most promising candidate for CBM exploration and development efforts.

5. Discussion

5.1. Constraint of the Depositional Environment on Coal Formation

The depositional environment constrains the characteristics of coal accumulation, the petrographic composition, and the spatial combinations of coal seams [14,31,53–55], which largely provide the material basis for CBM generation. When the depositional conditions are good, the coal seam thickness is large, and its distribution is stable, leading to a significant possibility of CBM gas production. In contrast, when the subsidence amplitude is not obvious, and the depositional conditions are poor, the coal seam thickness is unevenly distributed, and CBM gas production may also be relatively small.

A set of lowland residual plain deposits, dominated by bauxitic mudstone and kaolinite tonstein, developed steadily and were widely distributed throughout the weathering and denudation substrate at the top of the Maokou Formation during the early Wuchiapingian Period in the Songzao coalfield, as the crust started to sink slowly, and a large-scale sea recession stopped (Figures 2 and 7a,b).

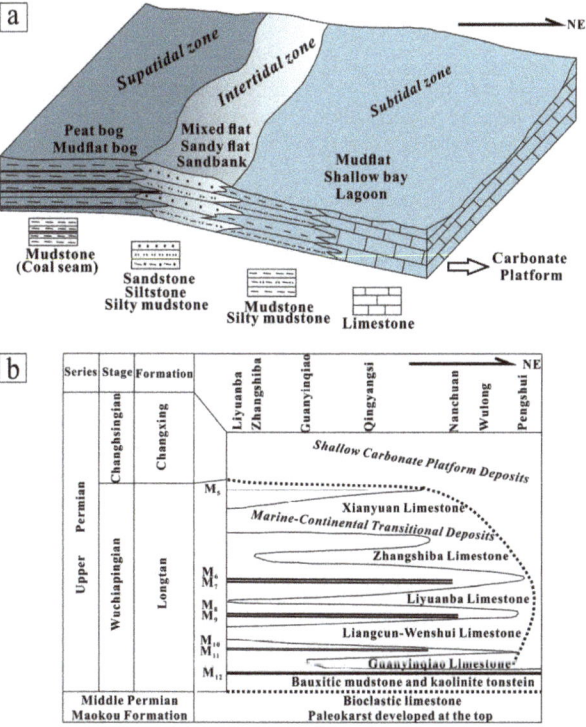

Figure 7. (a) Horizontal Wuchiapingian depositional pattern in the Songzao coalfield; (b) longitudinal Wuchiapingian depositional system of the Songzao coalfield.

As the crust continued to sink, and the first transgression invaded the area from the northeast to the southwest of Chongqing, the range of the marine–continental transitional zone gradually expanded, and large-scale coal accumulation occurred throughout the coalfield, resulting in the formation of the stably developed coal seam M_{12}, which is the product of regional transgression and is commonly presented during the initial stage of the early Wuchiapingian Period.

After that, seawater continued to rise slowly, the effect of coal gathering ended, the littoral tidal flat environment began to develop in a large region, and a set of fine clastic sediments such as siltstone and silty mudstone generally formed. Meanwhile, under the dual effects of further crust sinking and seawater rising, marine carbonate sediments appeared locally in the coalfield, and a thicker layer of limestone formed, represented by the Guanyinqiao limestone, which was deposited in a shallow bay with varying amounts of siliceous clastics. As the transgression stopped briefly, and the crust rose slowly, tidal flat and lagoon deposits developed on the top of the Guanyinqiao limestone, consisting primarily of sandstone, shale, and mudstone, locally interspersed with thin layers of limestone and unstable coal seams such as the coal seams M_{10} and M_{11}. The largest transgression of the Wuchiapingian Period occurred after the deposition of the coal seams M_{10} and M_{11}, forming the Wenshui–Liangcun limestone of the shallow bay environment throughout the whole coalfield. The Wenshui–Liangcun limestone represents the highest position of the transgression during the Wuchiapingian Period, and then the depositional sequence of lagoon and tidal flat redeveloped due to a seawater falling trend toward the east side of the coalfield and a crustal basement imbalance, forming the coal seam M_7, which also reflects the fluctuating in and out movement of seawater. During this stage, the coal seams M_8 and M_9 with regional spreading also formed, among which the coal seam M_8 is the best developed.

With the beginning of a new transgression, the crust sank, and seawater rose, and the depositional sequence shallow bay–lagoon–tidal flat manifested again over a wide range, forming fine clastic sediments dominated by siltstone and silty mudstone, thin coal seams (the relatively stable coal seams M_6 and M_7), and thin marker limestone layers (the Liyuanba, Zhangshiba, and Xianyuan limestone layers). During this stage, there were several short periods of regression, and the coal measure strata better developed to the west side of the coalfield. By the Changhsingian Period, a long-term and stable shallow carbonate platform had emerged in the Songzao field, implying the end of the Wuchiapingian marine–continental transitional environment.

However, the interpretation of the attributes of sparse vertical and horizontal sections and of borehole data using a geologic model, due to the heterogeneity and the inability to explain their spatial distribution, is difficult [56]. The traditional geostatistical interpolation approaches identified unhandled uncertainty in the Wuchiapingian marine–continental transitional environment pattern. This issue can be overcome by incorporating supplemental testing data to obtain more accurate inference results using hybrid techniques, such as a hybrid ensemble-based automated deep learning methodology [56]. In conclusion, the Songzao coalfield experienced repeated transgression and regression events from northeast to southwest throughout the Late Permian Wuchiapingian Period, with shallow bay–tidal flat–lagoon deposits dominating the depositional system (Figure 7a,b). Large-scale and stable coal accumulation mainly occurred in the early and middle Wuchiapingian. After the progressive rising of seawater and the variable fluctuation of the crust, no favorable coal-forming environment developed; hence, few coal seams formed in the coalfield in the middle to late Wuchiapingian.

5.2. Tectono-Thermal Evolution Constraining the CBM Production Potential

The Emeishan mantle plume activity was a large-scale tectono-thermal evolution event in Southwest China that constructed the Emeishan large igneous province in the latest middle Permian [57–59], triggering multistage intermediate-acidic volcanic eruptions during the late Permian Wuchiapingian and Changhsingian [60–62]. This event, with different development stages (emplacement, doming, and erosion of the Emeishan mantle plume and continued volcanism), deeply impacted the marine sedimentary strata in this time interval, forming a high geothermal field [63–67]. The Dongwu movement between the middle and the late Permian was a rapid differential uplift of the crust caused by mantle plume activity, and the top of the middle Permian Maokou Formation exposed at the surface underwent weathering and denudation [38]. The resulting tectonic fractures, such as the Huayingshan and Qiyueshan fault belts, provided migration channels for magmatic upwelling, intrusion, and volcanic activity throughout some regions of Sichuan province and Chongqing city in Southwest China (Figure 8a,b).

As the crust subsided again, marine–continental transitional deposits began to develop during the late Permian Wuchiapingian Period, forming the coal measure strata of the Longtan Formation. In addition, some tonstein (or tuff) layers were found near or within the coal seams of the Longtan Formation in southern Sichuan province, southern Chongqing city, western Guizhou province, and eastern Yunnan province ([33,35,36,68–70] and Figure 2), which belong to the outer zone of the Emeishan large igneous province and resulted from the waning activity of the mantle plume (Figure 8a,b). The tonsteins (or tuffs), originating from various partial melting conditions, indicated that the volcanic activities were characterized by multiple eruptions, relatively short time intervals, and small scales during peat accumulation. These geological conditions ensured the required temperature and time for the overmaturation and the achievement of the corresponding thermal metamorphic degrees of the whole coal seams in the Longtan Formation. Multistage volcanic eruptions during the late Permian could have resulted in pronounced increases in the geothermal gradient and heat flow [62,64,67,71,72], promoting the thermal metamorphism of the coal seams and accelerating CBM gas formation.

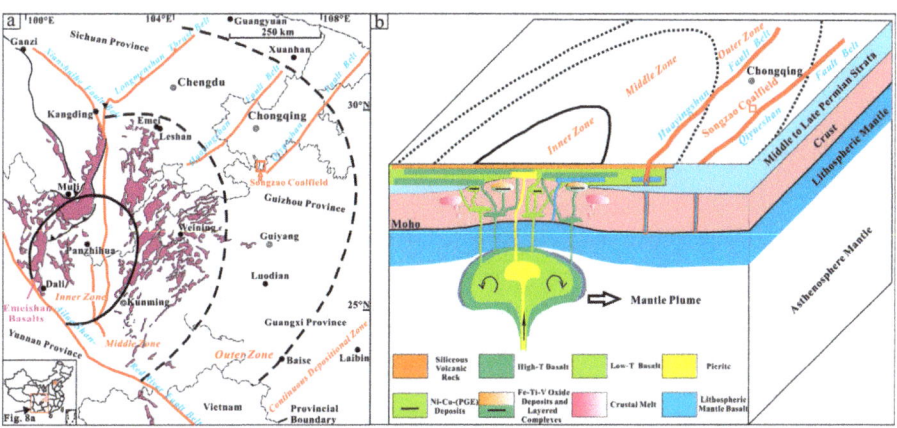

Figure 8. (a) Distribution of the Emeishan large igneous province showing the location of the Songzao coalfield, modified after [73]; (b) middle–late Permian magmatic and tectonic-thermal responses to the Emeishan large igneous province, modified after [59,65].

The reconstructed heat flow history modeling of the Emeishan large igneous province region based on multiple paleogeothermal parameters indicated a high heat flow of 80~110 mW/m^2 in the late Permian. The Longtan coal measure strata of the coalfield are located within the region ([63,66] and Figure 8a,b), which could be viewed as direct evidence of temperature anomalies related to mantle plume activity. Furthermore, thermal metamorphism at higher temperatures could also change the molecular composition of coal, resulting in an increase in the degree of coal metamorphism (high-rank coal) and vitrinite reflectivity (1.88~2.14%). Thermal metamorphism may also lead to increases in the local CBM pore volume (including gas hole and erosion pore numbers inside organic components) and gas content, producing a huge amount of pyrolytic methane adsorbed in the coal seams and further enhancing the gas production potential in high-rank coals.

5.3. Effect of the Regional Geological Structure on CBM Accumulation

The influence of regional geological structures on the CBM gas content is generally significant. The structure may not only influence the folding, twisting, shape change, fracture dislocation, and interbed sliding of coal seams [11,74,75], but also cause the escape and redistribution of CBM stored in coal reservoirs, affecting the gas content in coal seams in different structural sites [12,76,77]. In the Yuyang and Yangchatan mines of the Songzao coalfield, several tiny reverse and normal faults have developed underground (Figure 1).

Meanwhile, there is not much variation between the thickness of the coal seams M_6, M_7, M_8, and M_{12} of the Permian Longtan Formation in the two mines (Figure 3). However, the CBM gas contents in the main coal seams in the Yuyang mine are significantly lower than those in the Yangchatan mine (Figure 5). This implies that CBM is easily discharged, and the gas content in the coal seams frequently decreases significantly in underground tension fractures in the Yuyang mine. Nevertheless, in the Yangchatan mine, underground compression fractures can effectively close and collect CBM, and the gas content in the coal seams increases as the formation pressure increases. Few underground faults developed in the other 10 mines of the Songzao coalfield, although the distribution of the CBM gas content in the coal seams M_6, M_7, M_8, and M_{12} has nearly always a certain regularity. For instance, the coal seam M_8 has a thickness of only 0.33 m at the plunging crown of the Yutiao anticline, yet abundant tectonic coals developed inside it, with a CBM gas content of up to 32.77 m^3/t. Furthermore, the CBM gas content of the coal seam M_8 was found to be more than 15 m^3/t in the trap areas of the Yutiao and Yangchatan anticlines, such as the Datong and Shihao mines. The coal seams of the adjacent Zhangshiba and Liyuanba mines are monoclinic, with thickness ranging from 3.5 to 6.0 m. The gas content in the coal seam M_8 normally fell between 12 and 20 m^3/t. Moreover, the gas content in the coal seam M_8 was found to increase as the formation pressure rose in the northern and southern compound structural areas at the plunging convergence site between the Lianghekou syncline and the Yangchatan anticline, i.e., in the northern part of the Guanyinqiao mine, the southern part of the Tonghua mine, and the southern parts of the Yangchatan and Shihao mines (Figure 6). This case illustrates that whether the fold structure is enclosed is the most critical element influencing variances in the lateral distribution of the CBM gas content. Tectonic coals are very developed in the Songzao coalfield, with a high CBM gas content in strongly folded areas or tightly bonded areas, such as the plunging crown of an anticline and the trap area between two anticlines, destroying the original pores and fissures of coal seams and increasing the CBM gas content to a high degree in certain enclosed areas.

6. Conclusions

(1) The high-rank coals in the coal seams M_6, M_7, M_8, and M_{12} of the Permian Longtan Formation from the Songzao coalfield have high vitrinite and TOC contents (60.1~69.0%, 31.49~51.32 wt%), high T_{max} and R_0 values (averaging 539 °C, 2.17%), low HI values (averaging 15.21 mg HC/g TOC), high porosity and low permeability, and comparatively high gas contents.

(2) The frequent changes among shallow bay, tidal flat, and lagoon depositional environments triggered the formation of multiple coal seams and furnished the material basis for CBM generation. The multistage tectono-thermal evolution caused by the Emeishan mantle plume activity provided favorable temperatures and the necessary time for the overmaturation and thermal metamorphism of the coal seams and the acceleration of pyrolytic CBM formation.

(3) The effective regional structures, such as the enclosed fold regions like the plunging crown of the anticline and the trap area between two anticlines, directly optimized the conditions for CBM enrichment in the high-rank coals.

Supplementary Materials: The following supporting information can be downloaded at: https://www.mdpi.com/article/10.3390/en17051262/s1, Table S1: Typical average CBM gas contents in different coal seams from boreholes in the Songzao coalfield.

Author Contributions: Conceptualization, D.C., X.T. and J.W.; methodology and data curation, Y.Z. and C.Z.; formal analysis and investigation, D.G.; writing—original draft preparation, D.C.; writing—review and editing, X.T.; funding acquisition, D.C. and X.T. All authors have read and agreed to the published version of the manuscript.

Funding: This study was supported by the Natural Science Foundation of Chongqing (CSTB2022NSCQ-MSX1221), the National Natural Science Foundation of China (42302027), and the Research Project of SINOPEC East China Company (34600000-23-ZC0611-0003).

Data Availability Statement: All data are contained within the manuscript.

Acknowledgments: We would like to thank the two anonymous reviewers for their valuable comments and constructive suggestions that helped to improve the original quality of this paper.

Conflicts of Interest: The authors declare no conflicts of interest.

References

1. Zhu, Q.Z.; Yang, Y.H.; Zuo, Y.Q.; Song, Y.; Guo, W.; Tang, F.; Ren, J.; Wang, G. On the scientific exploitation of high-rank CBM resources. *Nat. Gas Ind. B* **2020**, *7*, 403–409. [CrossRef]
2. Sang, S.X.; Han, S.J.; Liu, S.Q.; Zhou, X.Z.; Li, M.X.; Hu, Q.J.; Zhang, C. Comprehensive study on the enrichment mechanism of coalbed methane in high rank reservoirs. *J. China Coal Soc.* **2022**, *47*, 388–403, (In Chinese with English Abstract). [CrossRef]
3. Zhu, Q.Z. Improving the production efficiency of high rank coal bed methane in the Qinshui Basin. *Nat. Gas Ind. B* **2022**, *9*, 477–486. [CrossRef]
4. Shao, L.Y.; Hou, H.H.; Tang, Y.; Lu, J.; Qiu, H.J.; Wang, X.T.; Zhang, J.Q. Selection of strategic replacement areas for CBM exploration and development in China. *Nat. Gas Ind. B* **2015**, *2*, 211–221. [CrossRef]
5. Zou, C.N.; Yang, Z.; Huang, S.P.; Ma, F.; Sun, Q.P.; Li, F.H.; Pan, S.Q.; Tian, W.G. Resource types, formation, distribution and prospects of coal-measure gas. *Pet. Explor. Dev.* **2019**, *46*, 433–442. [CrossRef]
6. Qin, Y. Strategic thinking on research of coal measure gas accumulation system and development geology. *J. China Coal Soc.* **2021**, *46*, 2387–2399, (In Chinese with English Abstract). [CrossRef]
7. Li, Y.; Pan, S.P.; Ning, S.Z.; Shao, L.Y.; Jing, Z.H.; Wang, Z.S. Coal measure metallogeny: Metallogenic system and implication for resource and environment. *Sci. China Earth Sci.* **2022**, *65*, 1211–1228. [CrossRef]
8. Tang, S.L.; Tang, D.Z.; Xu, H.; Tao, S.; Li, S.; Geng, Y.G. Geological mechanisms of the accumulation of coalbed methane induced by hydrothermal fluids in the western Guizhou and eastern Yunnan regions. *J. Nat. Gas Sci. Eng.* **2016**, *33*, 644–656. [CrossRef]
9. Qin, Y.; Moore, T.A.; Shen, J.; Yang, Z.B.; Shen, Y.L.; Wang, G. Resources and geology of coalbed methane in China: A review. *Int. Geol. Rev.* **2017**, *60*, 777–812. [CrossRef]
10. Bi, C.Q.; Zhang, J.Q.; Shan, Y.S.; Hu, Z.F.; Wang, F.G.; Chi, H.P.; Tang, Y.; Yuan, Y.; Liu, Y.R. Geological characteristics and co-exploration and co-production methods of Upper Permian Longtan coal measure gas in Yangmeishu syncline, western Guizhou Province, China. *China Geol.* **2020**, *3*, 38–51, (In Chinese with English Abstract). [CrossRef]
11. Cao, D.Y.; Ning, S.Z.; Guo, A.J.; Li, H.T.; Chen, L.M.; Liu, K.; Tan, J.Q.; Zheng, Z.H. Coalfield structure and structural controls on coal in China. *Int. J. Coal Sci. Technol.* **2020**, *7*, 220–239. [CrossRef]
12. Liu, D.M.; Jia, Q.F.; Cai, Y.D.; Gao, C.J.; Qiu, F.; Zhao, Z.; Chen, S.Y. A new insight into coalbed methane occurrence and accumulation in the Qinshui Basin, China. *Gondwana Res.* **2022**, *111*, 280–297. [CrossRef]
13. Sang, S.X.; Zheng, S.J.; Yi, T.S.; Zhao, F.P.; Han, S.J.; Jia, J.L.; Zhou, X.Z. Coal measures superimposed gas reservoir and its exploration and development technology modes. *Coal Geol. Explor.* **2022**, *50*, 13–21, (In Chinese with English Abstract). [CrossRef]
14. Cheng, J.; Li, D.H.; Liu, D.; Yao, G.H.; Ren, S.C.; Li, C.L.; Zhang, L.H.; Tang, B.F. *Coal-Forming Pattern and Quantitative Forecast of the Coal Resources in the Chongqing City*; China University of Geosciences Press: Wuhan, China, 2015; 282p. (In Chinese)
15. Zhu, Q.Z.; Zuo, Y.Q.; Yang, Y.H. How to solve the technical problems in the CBM development: A case study of a CMB gas reservoir in the southern Qinshui Basin. *Nat. Gas Ind. B* **2015**, *35*, 106–109. [CrossRef]
16. Li, Z.T.; Liu, D.M.; Ranjith, P.G.; Cai, Y.D. Geological controls on variable gas concentrations: A case study of the northern Gujiao Block, northwestern Qinshui Basin, China. *Mar. Pet. Geol.* **2018**, *92*, 582–596. [CrossRef]
17. Liang, J.T.; Huang, W.H.; Wang, H.L.; Blum, M.J.; Chen, J.; Wei, X.L.; Yang, G.Q. Organic geochemical and petrophysical characteristics of transitional coalmeasure shale gas reservoirs and their relationships with sedimentary environments: A case study from the Carboniferous-Permian Qinshui Basin, China. *J. Pet. Sci. Eng.* **2020**, *184*, 106510. [CrossRef]
18. Zhu, Q.Z. *New Technologies and Practice of Exploration and Development for High-Rank Coal Bed Methane*; Petroleum Industry Press: Beijing, China, 2021; 268p. (In Chinese)
19. Jiang, W.P.; Zhang, P.H.; Li, D.D.; Li, Z.C.; Wang, J.; Duan, Y.N. Reservoir characteristics and gas production potential of deep coalbed methane: Insights from the no. 15 coal seam in shouyang block, Qinshui Basin, China. *Unconv. Resour.* **2022**, *2*, 12–20. [CrossRef]
20. Cao, L.; Yao, Y.; Cui, C.; Sun, Q. Characteristics of in-situ stress and its controls on coalbed methane development in the southeastern Qinshui Basin, North China. *Energy Geosci.* **2020**, *1*, 69–80. [CrossRef]
21. Zhang, P.; Ya, M.; Liu, C.; Guo, Y.; Yan, X.; Cai, L.M.; Cheng, Z. In-situ stress of coal reservoirs in the Zhengzhuang area of the southern Qinshui Basin and its effects on coalbed methane development. *Energy Geosci.* **2023**, *4*, 100144. [CrossRef]
22. Yang, Y.; Li, X.; Zhang, Y.; Mei, Y.; Ding, R. Insights into moisture content in coals of different ranks by low field nuclear resonance. *Energy Geosci.* **2020**, *1*, 93–99. [CrossRef]

23. Lv, Y.M.; Tang, D.Z.; Xu, H.; Luo, H.H. Production characteristics and the key factors in high-rank coalbed methane fields: A case study on the Fanzhuang Block, Southern Qinshui Basin, China. *Int. J. Coal Geol.* **2012**, *96-97*, 93–108. [CrossRef]
24. Yao, H.F.; Kang, Z.Q.; Li, W. Deformation and reservoir properties of tectonically deformed coals. *Pet. Explor. Dev.* **2014**, *41*, 460–467. [CrossRef]
25. Gui, B.L. Geological characteristics and enrichment controlling factors of coalbed methane in Liupanshui region. *Acta Petrol. Sin.* **1999**, *20*, 31–37, (In Chinese with English Abstract). [CrossRef]
26. Li, S.; Tang, D.Z.; Pan, Z.J.; Xu, H.; Guo, L.L. Evaluation of coalbed methane potential of different reservoirs in western Guizhou and eastern Yunnan, China. *Fuel* **2015**, *139*, 257–267. [CrossRef]
27. Shen, Y.L.; Qin, Y.; Guo, Y.H.; Yi, T.S.; Yuan, X.X.; Shao, Y.B. Characteristics and sedimentary control of a coalbed methane-bearing system in Lopingian (late Permian) coal-bearing strata of western Guizhou Province. *J. Nat. Gas Sci. Eng.* **2016**, *33*, 8–17. [CrossRef]
28. Luo, W.; Hou, M.C.; Liu, X.C.; Huang, S.G.; Chao, H.; Zhang, R.; Deng, X. Geological and geochemical characteristics of marine-continental transitional shale from the Upper Permian Longtan formation, Northwestern Guizhou, China. *Mar. Pet. Geol.* **2018**, *89 Pt 1*, 58–67. [CrossRef]
29. Luo, Q.Y.; Xiao, Z.H.; Dong, G.Y.; Ye, X.Z.; Li, H.J.; Zhang, Y.; Ma, Y.; Ma, L.; Xu, Y.H. The geochemical characteristics and gas potential of the Longtan formation in the eastern Sichuan Basin, China. *J. Pet. Sci. Eng.* **2019**, *179*, 1102–1113. [CrossRef]
30. Shen, S.Z.; Zhang, H.; Zhang, Y.C.; Yuan, D.X.; Chen, B.; He, W.H.; Mu, L.; Lin, W.; Wang, W.Q.; Chen, J.; et al. Permian integrative stratigraphy and timescale of China. *Sci. China Earth Sci.* **2019**, *62*, 154–188. [CrossRef]
31. Zhang, Y.C.; Li, C.L.; Hong, X.F.; Yuan, Y.C.; Wang, X.H.; Huang, Y.A.; Tang, D.Y.; Zhu, C.S.; Gu, K.S.; Yuan, P.S. *Sedimentary Environments and Coal Accumulation of Late Permian Coal Formation in Southern Sichuan, China*; Guizhou Science and Technology Press: Guiyang, China, 1993; 204p, (In Chinese with English Abstract).
32. China National Administration of Coal Geology. *Sedimentary Environments and Coal Accumulation of Late Permian Coal Formation in Western Guizhou, Southern Sichuan and Eastern Yunnan, China*; Chongqing University Press: Chongqing, China, 1996; 277p, (In Chinese with English Abstract).
33. Dai, S.F.; Zhou, Y.P.; Ren, D.Y.; Wang, X.B.; Li, D.; Zhao, L. Geochemistry and mineralogy of the Late Permian coals from the Songzao Coalfield, Chongqing, southwestern China. *Sci. China-Earth Sci.* **2007**, *50*, 678–688. [CrossRef]
34. Dai, S.F.; Wang, X.B.; Chen, W.M.; Li, D.H.; Chou, C.L.; Zhou, Y.P.; Zhu, C.S.; Li, H.; Zhu, X.W.; Xing, Y.W.; et al. A high-pyrite semianthracite of Late Permian age in the Songzao Coalfield, southwestern China: Mineralogical and geochemical relations with underlying mafic tuffs. *Int. J. Coal Geol.* **2010**, *83*, 430–445. [CrossRef]
35. Dai, S.F.; Wang, X.B.; Zhou, Y.P.; Hower, J.C.; Li, D.H.; Chen, W.M.; Zhu, X.W.; Zou, J.H. Chemical and mineralogical compositions of silicic, mafic, and alkali tonsteins in the late Permian coals from the Songzao coalfield, Chongqing, Southwest China. *Chem. Geol.* **2011**, *282*, 29–44. [CrossRef]
36. Dai, S.F.; Li, T.; Seredin, V.V.; Ward, C.R.; Hower, J.C.; Zhou, Y.P.; Zhang, M.Q.; Song, X.L.; Zhao, C.L. Origin of minerals and elements in the Late Permian coals, tonsteins, and host rocks of the Xinde Mine, Xuanwei, eastern Yunnan, China. *Int. J. Coal Geol.* **2014**, *121*, 53–78. [CrossRef]
37. Zhao, L.; Ward, C.R.; French, D.; Graham, I.T. Mineralogical composition of Late Permian coal seams in the Songzao Coalfield, southwestern China. *Int. J. Coal Geol.* **2013**, *116–117*, 208–226. [CrossRef]
38. Tian, X.S.; Shi, Z.J.; Yin, G.; Long, H.Y.; Wang, K. A correlation between the Large Igneous Provinces and mass extinctions: Constraint on the end-Guadalupian mass extinction and the Emeishan LIP in South China, eastern Tethys. *Int. Geol. Rev.* **2016**, *58*, 1215–1233. [CrossRef]
39. Tian, X.S.; Shi, Z.J.; Yin, G.; Wang, Y.; Tan, Q. Carbonate diagenetic products and processes from various diagenetic environments in Permian paleokarst reservoirs: A case study of the limestone strata of Maokou formation in Sichuan Basin, South China. *Carbonates Evaporites* **2017**, *32*, 215–230. [CrossRef]
40. Wu, X.J.; Li, Z.F.; Sun, D.F. Coal seam M_8 stoping face water gushing characteristics and countermeasures in Songzao Mining Area. *Coal Geol. China* **2015**, *27*, 35–38, (In Chinese with English Abstract). [CrossRef]
41. Wu, G.D.; Zeng, C.L.; Cheng, J.; Guo, D.X.; Wang, J.; Wang, D.; Xie, Q.M. Characteristics of groundwater dynamic field and its effect on coalbed methane accumulation in Songzao mining area. *Coal Geol. Explor.* **2018**, *46*, 55–60, (In Chinese with English Abstract). [CrossRef]
42. *SY/T 6414-2014*; Maceral Identification and Statistical Methods on Polished Surface of Whole Rocks. Petroleum Geology Exploration Standardization Committee: Beijing, China, 2014. (In Chinese)
43. *SY/T 5124-2012*; Method of Determining Microscopically the Reflectance of Vitrinite in Sedimentary. Petroleum Geology Exploration Standardization Committee: Beijing, China, 2012. (In Chinese)
44. *GB/T 18602-2012*; Rock Pyrolysis Analysis. Chinese Standard: Beijing, China, 2012. (In Chinese)
45. *SY/T 5162-2014*; Analytical Method of Rock Sample by Scanning Electron Microscope. Petroleum Geology Exploration Standardization Committee: Beijing, China, 2014. (In Chinese)
46. Loucks, R.G.; Reed, R.M.; Ruppel, S.C.; Hammes, U. Spectrum of pore types and networks in mudrocks and a descriptive classification for matrix-related mudrock pores. *AAPG Bull.* **2012**, *96*, 1071–1098. [CrossRef]
47. Kenneth, S.W.S. Characterization of porous solids: An introductory survey. *Stud. Surf. Sci. Catal.* **1991**, *62*, 1–9. [CrossRef]

48. *GB/T 21650.1-2008*; Pore Size Distribution and Porosity of Solid Materials by Mercury Porosimetry and Gas Adsorption-Part 1: Mercury Porosimetry. Chinese Standard: Beijing, China, 2008. (In Chinese)
49. *SY/T 6385-2016*; Porosity and Permeability Measurement under Overburden Pressure. Petroleum Geology Exploration Standardization Committee: Beijing, China, 2016. (In Chinese)
50. *GB/T 19559-2008*; Method of Determining Coalbed Gas Content. Chinese Standard: Beijing, China, 2008. (In Chinese)
51. Alafnan, S.; Awotunde, A.; Glatz, G.; Adjei, S.; Alrumaih, I.; Gowida, A. Langmuir adsorption isotherm in unconventional resources: Applicability and limitations. *J. Pet. Sci. Eng.* **2021**, *207*, 109172. [CrossRef]
52. *GB/T 13610-2014*; Analysis of Natural Gas Composition-Gas Chromatography. Chinese Standard: Beijing, China, 2014. (In Chinese)
53. Dai, S.F.; Bechtel, A.; Eble, C.F.; Flores, R.M.; French, D.; Graham, I.T.; Hood, M.M.; Hower, J.C.; Korasidis, V.A.; Moore, T.A.; et al. Recognition of peat depositional environments in coal: A review. *Int. J. Coal Geol.* **2020**, *219*, 103383. [CrossRef]
54. Damoulianou, M.E.; Kalaitzidis, S.; Pasadakis, N. Turonian-Senonian organic-rich sedimentary strata and coal facies in Parnassos-Ghiona Unit, Central Greece: An assessment of palaeoenvironmental setting and hydrocarbon generation potential. *Int. J. Coal Geol.* **2022**, *258*, 104029. [CrossRef]
55. Wang, E.Z.; Guo, T.L.; Liu, B.; Li, M.W.; Xiong, L.; Dong, X.X.; Zhang, N.X.; Wang, T. Lithofacies and pore features of marine-continental transitional shale and gas enrichment conditions of favorable lithofacies: A case study of Permian Longtan Formation in the Lintanchang area, southeast of Sichuan Basin, SW China. *Pet. Explor. Dev.* **2022**, *49*, 1310–1322. [CrossRef]
56. Abbaszadeh Shahri, A.; Shan, C.L.; Larsson, S. A hybrid ensemble-based automated deep learning approach to generate 3D geo-models and uncertainty analysis. *Eng. Comput.* **2023**. [CrossRef]
57. Ali, J.R.; Thompson, G.M.; Zhou, M.F.; Song, X.Y. Emeishan large igneous province, SW China. *Lithos* **2005**, *79*, 475–489. [CrossRef]
58. He, B.; Xu, Y.G.; Wang, Y.M.; Luo, Z.Y. Sedimentation and lithofacies paleogeography in Southwestern China before and after the Emeishan flood volcanism: New Insights into surface response to mantle plume activity. *J. Geol.* **2006**, *114*, 117–132. [CrossRef]
59. Shellnutt, J.G. The Emeishan large igneous province: A synthesis. *Geosci. Front.* **2014**, *5*, 369–394. [CrossRef]
60. Xu, Y.G.; Chung, S.L.; Jahn, B.M.; Wu, G.Y. Petrologic and geochemical constraints on the petrogenesis of Permian-Triassic Emeishan flood basalts in southwestern China. *Lithos* **2001**, *58*, 145–168. [CrossRef]
61. Yang, J.H.; Cawood, P.A.; Du, Y.S.; Huang, H.; Huang, H.W.; Tao, P. Large Igneous Province and magmatic arc sourced Permian-Triassic volcanogenic sediments in China. *Sediment. Geol.* **2012**, *261–262*, 120–131. [CrossRef]
62. Dai, S.F.; Ward, C.R.; Graham, I.T.; French, D.; Hower, J.C.; Zhao, L.; Wang, X.B. Altered volcanic ashes in coal and coal-bearing sequences: A review of their nature and significance. *Earth-Sci. Rev.* **2017**, *175*, 44–74. [CrossRef]
63. He, L.J.; Xu, H.H.; Wang, J.Y. Thermal evolution and dynamic mechanism of the Sichuan Basin during the Early Permian-Middle Triassic. *Sci. China-Earth Sci.* **2011**, *54*, 1948–1954. [CrossRef]
64. Zhu, C.Q.; Hu, S.B.; Qiu, N.S.; Rao, S.; Yuan, Y.S. The thermal history of the Sichuan Basin, SW China: Evidence from the deep boreholes. *Sci. China Earth Sci.* **2016**, *59*, 70–82. [CrossRef]
65. Zhu, C.Q.; Hu, S.B.; Qiu, N.S.; Jiang, Q.; Rao, S.; Liu, S. Geothermal constraints on Emeishan mantle plume magmatism: Paleotemperature reconstruction of the Sichuan Basin, SW China. *Int. J. Earth Sci.* **2018**, *107*, 71–88. [CrossRef]
66. Jiang, Q.; Qiu, N.S.; Zhu, C.Q. Heat flow study of the Emeishan large igneous province region: Implications for the geodynamics of the Emeishan mantle plume. *Tectonophysics* **2018**, *724–725*, 11–27. [CrossRef]
67. He, L.J. Emeishan mantle plume and its potential impact on the Sichuan Basin: Insights from numerical modeling. *Phys. Earth Planet. Inter.* **2022**, *323*, 106841. [CrossRef]
68. Zhou, Y.P.; Ren, Y.L.; Bohor, B.F. Origin and distribution of tonsteins in late Permian coal seams of Southwestern China. *Int. J. Coal Geol.* **1982**, *2*, 49–77. [CrossRef]
69. Zhou, Y.P.; Bohor, B.F.; Ren, Y.L. Trace element geochemistry of altered volcanic ash layers (tonsteins) in Late Permian coal-bearing formations of eastern Yunnan and western Guizhou Provinces, China. *Int. J. Coal Geol.* **2000**, *44*, 305–324. [CrossRef]
70. Shen, M.L.; Dai, S.F.; Rechaev, V.P.; French, D.; Graham, I.T.; Liu, S.D.; Chekryzhov, I.Y.; Tarasenko, I.A.; Zhao, S.W. Provenance changes for mineral matter in the latest Permian coals from western Guizhou, southwestern China, relative to tectonic and volcanic activity in the Emeishan Large Igneous Province and Paleo-Tethys region. *Gondwana Res.* **2023**, *113*, 71–88. [CrossRef]
71. Chen, Z.S.; Wu, Y.D. Late Permian Emeishan basalt and coal-bearing formation in southern Sichuan area. *Coal Geol. China* **2010**, *22*, 14–18, (In Chinese with English Abstract). [CrossRef]
72. Feng, Q.Q.; Qiu, N.S.; Fu, X.D.; Li, W.Z.; Xu, Q.; Li, X.; Wang, J.S. Permian geothermal units in the Sichuan Basin: Implications for the thermal effect of the Emeishan mantle plume. *Mar. Pet. Geol.* **2021**, *132*, 105226. [CrossRef]
73. He, B.; Xu, Y.G.; Chung, S.L.; Xiao, L.; Wang, Y.M. Sedimentary evidence for a rapid, kilometer-scale crustal doming prior to the eruption of the Emeishan food basalts. *Earth Planet. Sci. Lett.* **2003**, *213*, 391–405. [CrossRef]
74. Tang, S.L.; Tang, D.Z.; Li, S.; Xu, H.; Tao, S.; Geng, Y.G.; Ma, L.; Zhu, X.G. Fracture system identification of coal reservoir and the productivity differences of CBM wells with different coal structures: A case in the Yanchuannan Block, Ordos Basin. *J. Pet. Sci. Eng.* **2018**, *161*, 175–189. [CrossRef]
75. Ju, Y.W.; Qiao, F.; Wei, M.M.; Li, X.; Xu, F.Y.; Feng, G.R.; Li, Y.; Wu, C.F.; Cao, Y.X.; Li, G.F.; et al. Typical coalbed methane (CBM) enrichment and production modes under the control of regional structure and evolution. *Coal Geol. Explor.* **2022**, *50*, 2, (In Chinese with English Abstract). [CrossRef]

76. Song, Y.; Liu, H.L.; Feng, H.; Qin, S.F.; Liu, S.B.; Li, G.Z.; Zhao, M.J. Syncline reservoir pooling as a general model for coalbed methane (CBM) accumulations: Mechanisms and case studies. *J. Nat. Gas Sci. Eng.* **2012**, *88–89*, 5–12. [CrossRef]
77. Chen, Y.; Tang, D.Z.; Xu, H.; Li, Y.; Meng, Y.J. Structural controls on coalbed methane accumulation and high production models in the eastern margin of Ordos Basin, China. *J. Nat. Gas Sci. Eng.* **2015**, *23*, 524–537. [CrossRef]

Disclaimer/Publisher's Note: The statements, opinions and data contained in all publications are solely those of the individual author(s) and contributor(s) and not of MDPI and/or the editor(s). MDPI and/or the editor(s) disclaim responsibility for any injury to people or property resulting from any ideas, methods, instructions or products referred to in the content.

Article

Evaluation of Grain Size Effects on Porosity, Permeability, and Pore Size Distribution of Carbonate Rocks Using Nuclear Magnetic Resonance Technology

Shutong Wang [1,2], Yanhai Chang [3], Zefan Wang [1,2] and Xiaoxiao Sun [1,2,*]

1. School of Energy Resources, China University of Geosciences, Beijing 100083, China; wstylyh@163.com (S.W.); wzf290516@163.com (Z.W.)
2. Beijing Key Laboratory of Unconventional Natural Gas Geological Evaluation and Development Engineering, China University of Geosciences, Beijing 100083, China
3. State Key Laboratory of Mining Response and Disaster Prevention and Control in Deep Coal Mines, Anhui University of Science and Technology, Huainan 232001, China; yhchang@aust.edu.cn
* Correspondence: xsun90@cugb.edu

Abstract: Core analysis is an accurate and direct method for finding the physical properties of oil and natural gas reservoirs. However, in some cases coring is time consuming and difficult, and only cuttings with the drilling fluid can be obtained. It is important to determine whether cuttings can adequately represent formation properties such as porosity, permeability, and pore size distribution (PSD). In this study, seven limestone samples with different sizes were selected (Cubes: $4 \times 4 \times 4$ cm, $4 \times 4 \times 2$ cm, $4 \times 2 \times 2$ cm and $2 \times 2 \times 2$ cm, Core: diameter of 2.5 cm and a length of 5 cm, Cuttings: 1–1.7 mm and 4.7–6.75 mm in diameter), and low-field nuclear magnetic resonance (NMR) measurements were performed on these samples to obtain porosity, PSD, and permeability. The results showed that the porosity of cubes and cuttings with different sizes are consistent with cores, which is about 1%. Whereas the PSDs and permeabilities of the two cutting samples (less than in size 6.75 mm) differ significantly within cores. It is suggested that interparticle voids and mechanical pulverization during sample preparation have a negligible effect on porosity and a larger effect on PSD and permeability. Combined with factors such as wellbore collapse and mud contamination suffered in the field, it is not recommended to use cuttings with a particle size of less than 6.75 mm to characterize actual extra-low porosity and extra-low permeability formation properties.

Keywords: NMR; carbonate cuttings; porosity; pore size distribution; permeability

Citation: Wang, S.; Chang, Y.; Wang, Z.; Sun, X. Evaluation of Grain Size Effects on Porosity, Permeability, and Pore Size Distribution of Carbonate Rocks Using Nuclear Magnetic Resonance Technology. *Energies* **2024**, *17*, 1370. https://doi.org/10.3390/en17061370

Academic Editor: Reza Rezaee

Received: 23 January 2024
Revised: 29 February 2024
Accepted: 11 March 2024
Published: 13 March 2024

Copyright: © 2024 by the authors. Licensee MDPI, Basel, Switzerland. This article is an open access article distributed under the terms and conditions of the Creative Commons Attribution (CC BY) license (https://creativecommons.org/licenses/by/4.0/).

1. Introduction

Carbonate rocks are sedimentary rocks composed of authigenic carbonate minerals such as calcite and dolomite. According to Roehl et al. [1], the global oil and gas reserves in carbonate reservoirs account for about 40% of the total oil and gas reserves and their production accounts for about 60% of the total oil and gas production. The high heterogeneity and complex pore system of carbonate reservoirs lead to low accuracy of reservoir physical property evaluation, which restricts exploration efficiency and reserve evaluation.

Cores, obtained by drilling core or sidewall coring, are generally used to analyze reservoirs' physical characteristics. At present, core analysis is an accurate and direct measurement method. However, in many cases, such as loose formations, horizontal wells, etc., coring is expensive and difficult. In these situations, drilling cuttings become the only source of the physical properties of the reservoir. Thus, cuttings are always used to characterize the mineralogy and lithology of the reservoir.

In recent decades, various methods for evaluating petrophysical characteristics using cuttings have been proposed. Santarelli et al. [2] found that the feasibility of cuttings represents core properties in terms of acoustic, mechanical, and petrophysical properties. For sandstone, it is suggested that the minimum size of cuttings used to characterize porosity is dependent on the porous medium structure, and that cuttings with diameter ≥ 3 mm can provide an accurate porosity [3]. Yang et al. [4] analyzed four sets of cuttings ranging from 0.42–4.0 mm in diameter and suggested that the error in porosity for cuttings with larger grain size was smaller than that of cuttings with smaller particle size. With NMR and μCT methods, Hübner et al. [5] found that cuttings with diameter of 1–10 mm can provide an accurate porosity of sandstone. Lenormand and Fonta [6] believed that cuttings from 0.5 mm to 5 mm can accurately provide porosity with an improved method for removing interparticle moisture. Moreover, for tight sandstone, Solano et al. [7] believed that cuttings with diameter > 1 mm provide accurate porosity and permeability. In addition, Ortega and Aguilera [8] defined the minimum cutting size of tight sandstone that can be used for porosity measurement to be >1 mm in diameter. For shale, Fellah et al. [9] found that shale cuttings with 1–5 mm in diameter provide accurate porosity. For coal, Chang et al. [10] found that the porosity and PSD of coal can be found using coal cuttings with diameter > 1 mm. For both conventional reservoir (sandstone) and unconventional reservoir (tight sandstone, shale, and coal), cuttings can provide porosity and PSD, but there are minimum size limitations. For carbonates, Siddiqui et al. [11] measured the porosity of cuttings with a diameter > 2.5 mm using CT scanning. Meanwhile, Egemann et al. [12] measured the permeability of cuttings and found that the permeability of cuttings with size between 2 mm and 3 mm was consistent with that of cores. In general, research on smaller-sized cuttings is limited. Therefore, the study on low-porosity and low-permeability carbonate rocks is still weak.

NMR is a non-invasive technique that can provide information about the porosity, PSD and permeability of the rock, with no restrictions on the size and shape of the sample [13,14]. Therefore, in this study, low-field NMR was used to analyze the influence of scale effect on the physical parameters of carbonate rocks, and to provide suggestions on particle size of cuttings for field application.

2. Geological Setting

The Bohai Bay Basin has a huge resource potential as one of the important oil and gas basins in eastern China, with an area of about 20×10^4 km^2 (Figure 1a) [15]. The Bozhong Depression is located in the offshore part of the east-central Bohai Bay Basin and is the main sedimentation and subsidence center of the Bohai Bay Basin. The study area has undergone the superposed complex tectonic evolution of multi-stages faulting and neotectonic movement; the overlapping structure of fault depression is developed. The basement of the depression is mainly composed of Archean metamorphic rocks, overlying Paleozoic carbonate rocks, middle to lower Jurassic clastic rocks, and upper Jurassic to Lower Cretaceous andesite, limestone, and clastic rocks. The dark mudstone of Shahejie Formation and Dongying Formation in Tertiary is the main source rock and cap rock in the study area. The Majiagou Formation of Ordovician is one of the reservoirs in the study area, which is the target seam of this study. Affected by Caledonian movement and Yanshan and Himalayan movements, the upper boundary of Majiagou Formation is a regional unconformity. The Lower Majiagou Formation is mainly composed of mudstone and dolomite interbeds, and the Upper Majiagou Formation is mainly composed of limestone (Figure 2) [16].

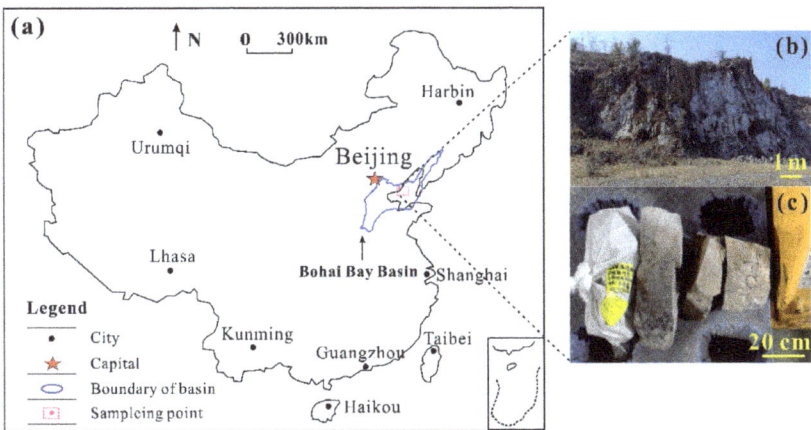

Figure 1. Map showing the location of bohai bay basin in China. (**a**) Sampling location; (**b**) Outcrop of Majiagou formation; (**c**) Photographs of limestone samples.

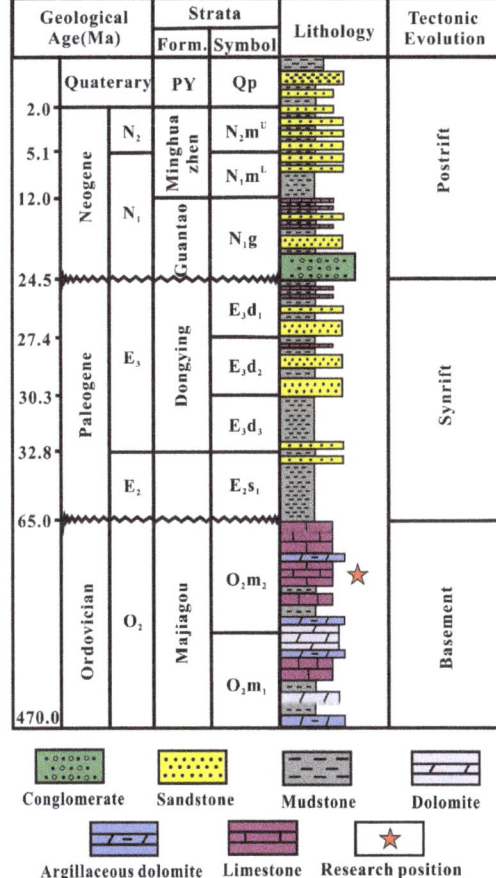

Figure 2. Stratigraphic lithology of the Bohong Depression Region [16].

3. Experiments

3.1. Samples

Limestone samples were collected from the outcrop of the Upper Majiagou Formation of the Bohai Bay Basin, China (Figure 1b,c). Eight sets of standard cylindrical core plug (C1, C2, C3, C4, C5) were drilled with a diameter of 2.5 cm and a length of 5 cm. The limestone samples were also cut into cube shapes with different size, Cube–4 × 4 × 4 cm, Cube–4 × 4 × 2 cm, Cube–4 × 2 × 2 cm and Cube–2 × 2 × 2 cm. The remaining parts were crushed and sieved into two groups of cuttings: Powder-a (4.7–6.75 mm in diameter), and Powder−b (1–1.7 mm in diameter) (Figure 3). All samples were measured by 100% water saturation NMR. For five core samples, the pulse attenuation method was used to measure permeability. Moreover, the remaining parts of limestone samples were also taken for characterizing pores via SEM analyses following the Chinese standard SY/T 5162-2014 [17].

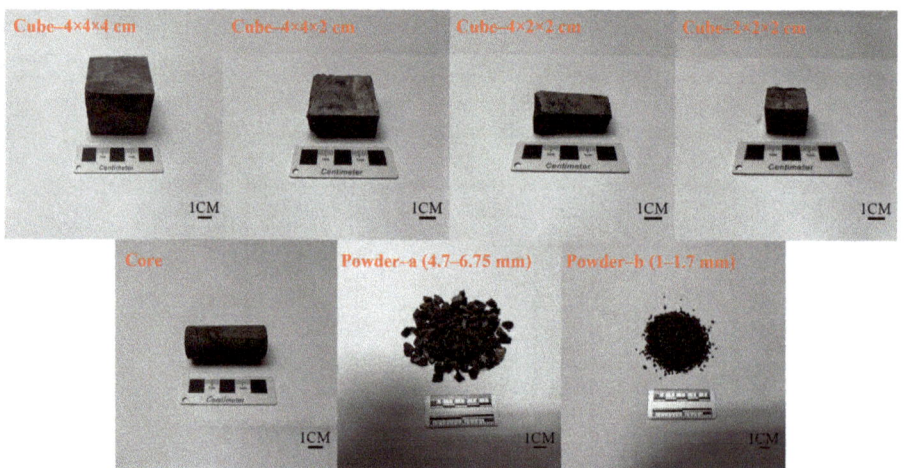

Figure 3. Limestone samples of cores, cubic, and cuttings.

3.2. NMR and Centrifugal Experiments

The water-saturated NMR technique is based on the NMR relaxation behavior of spinning hydrogen nuclei (^1H) in fluids contained in rocks in the presence of uniformly distributed static magnetic and radio frequency fields. The measured transverse relaxation time (T_2) consists of three components [18]:

$$\frac{1}{T_2} = \frac{1}{T_{2S}} + \frac{1}{T_{2D}} + \frac{1}{T_{2B}} \quad (1)$$

where T_{2S}, T_{2D}, and T_{2B} are the surface relaxation time, diffusion relaxation time and bulk relaxation time, respectively. Diffusion relaxation is the self-diffusion of fluid molecules, which is formed by proton spin diffusion via a strong internal field gradient, and can be expressed by the equation [19]:

$$\frac{1}{T_{2D}} = \frac{DG^2\gamma^2 T_{CP}^2}{12} \quad (2)$$

where D is the free diffusion constant for the relevant molecule, G is the magnetic field gradient, γ is the magnetogyric ratio, and T_{CP} is the Carr–Purcell spacing (0.5 of the echo spacing). In a uniform magnetic field, there is no magnetic field gradient, and thus the diffusion relaxation time is not considered during the experiment. The bulk relaxation time depends on the properties of fluid (chemical composition, viscosity, etc.), which is much larger than T_{2S}. Therefore, for ^1H bearing fluids in porous media, T_2 is mainly affected by surface relaxation, which is determined by pore size [19]. Thus, T_2 is expressed as follows:

$$\frac{1}{T_2} \approx \frac{1}{T_{2S}} = \rho_2 \left(\frac{S}{V}\right) = F_S \frac{\rho_2}{r_c} \tag{3}$$

where ρ_2 is the surface relaxivity; S is the specific area of the pore; V is the volume of the pore; F_S is the pore geometry morphologic factor, with $Fs = 3$ for spherical pores [20]; and r_c is the diameter of the pores. Meanwhile, for the T_2 spectra of a 100% saturated water sample, the NMR signal intensity can be converted to porosity using a standard calibration method [21]:

$$\varnothing = \frac{M_p}{M} \times \frac{V_p}{V} \times 100\% \tag{4}$$

where V_p is water volume in 100% water-saturated sample in cm^3, V is water volume in the standard sample in cm^3, M_p is the total signal intensity of a 100% water-saturated sample, and M is the total signal intensity of the standard sample.

To analyze the porosity characteristics and to determine the T_2 cutoff values (T_{2C}) of the limestone samples, all samples at 100% water saturation (S_w) were set for NMR measurement. Four core samples (C1, C2, C3, C4) were then centrifuged using a CSC-10 Super Core Centrifuge to obtain an irreducible water condition with a 6000 RPM corresponding to centrifuge pressure of 143 psi [22], after which NMR measurements were performed again.

Notably, for water-saturated cuttings samples, some water is distributed in the intergranular spaces between individual cuttings, which may cause errors in porosity measurements. Therefore, water on the surface of the saturated sample cuttings was carefully removed by placing the saturated cuttings in a moist sponge and then pressing the sponge for a few seconds to remove intergranular water between the cuttings [6].

In this study, NMR analysis was performed using an Oxford GeoSpec 12/53 core analyzer with a uniform magnetic field of 12 MHz. All NMR measurements were performed at 30 °C. the parameters of CPMG pulse sequence were set as follows: echo spacing of 0.1 ms, echo numbers of 13,158, recycle delay of 300 ms, and scan number of 16.

3.3. Permeability Measurement

The evaluation of rock permeability involves both direct and indirect methods. For the five core samples (C1–C5), the absolute permeability (K_p) was obtained using pulse attenuation method by flowing air through the core sample until the pressure at both ends is equilibrated. For cubes and cuttings samples, it is difficult to measure permeability using the direct method. Therefore, the Coates and SDR empirical models, based on NMR-T_2 spectra, are used [23,24].

For the Coates model, a relationship between porosity, irreducible water condition, and permeability is established as follows:

$$K_c = C\varnothing^4 \left(\frac{1-S_{ir}}{S_{ir}}\right)^2 \tag{5}$$

where $\varnothing$ is the porosity, %; C is a constant; and S_{ir} is the irreducible water saturation, %:

$$S_{ir} = \frac{\int_{T_{2min}}^{T_{2c}} A(T_{2i}) dT_2}{\int_{T_{2min}}^{T_{2max}} A(T_{2i}) dT_2} \tag{6}$$

where T_{2min} is the minimum T_2 relaxation time, ms, T_{2max} is the maximum T_2 relaxation time, ms; and $A(T_{2i})$ is the signal amplitude corresponding to any relaxation time T_{2i}. The T_{2C} value varies for different types of carbonates, which is related to different types of pore space, pore structure, and mineral compositions [25]. Generally, T_{2C} values are determined using the centrifugal method [26],

The SDR model takes pore size distribution into account to estimate permeability:

$$K_s = C_1 \varnothing^4 (T_{2g})^2 \qquad (7)$$

where $\varnothing$ is the nuclear magnetic porosity in %, C_1 is a constant, and T_{2g} is the geometric mean of the T_2 distribution.

4. Results

4.1. NMR Spectra Result for Cubes and Cuttings

Figure 4 shows the NMR-T_2 spectra for different sizes of samples at 100% water saturation. It is worth noting that the samples used in this experiment were of different masses, thus the T_2 spectra of all samples were normalized. For five core samples, the NMR-T_2 spectra show a single peak-like shape, overlapping well and centered at approximately 10 ms. According to Equation (2), T_2 relaxation time was converted to pore diameter size. The surface relaxivity is 37.33 for limestone [27]. Moreover, pores are classified into micropore (<100 nm), mesopore (100–1000 nm), and macropores (>1000 nm) [28]. The results show that the limestone samples are dominated by mesopores and macropores, and the proportion of micropores is only about 10%. For cube samples, the NMR spectra were generally consistent with the core samples. However, for two groups of cuttings, the NMR spectra were shifted to the right (Figure 4b), which indicates that some changes have occurred in the pore-size characteristics of this cuttings sample [11].

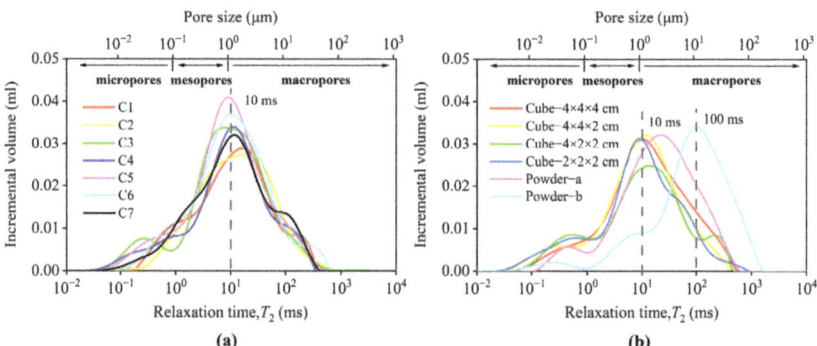

Figure 4. T_2 spectra of water-saturated limestone samples: (**a**) core samples, (**b**) cubic and cuttings samples.

4.2. NMR Results for Core Plugs at S_w and S_{ir}

The NMR results of the cores under 100% water saturation (S_w) and irreducible water saturation (S_{ir}) are shown in Figure 5. After the high-speed centrifugation experiment, the signal amplitude of the single peak of the sample decreases with the loss of the movable moisture. The decrease or disappearance of the signals provides information about the movable water in the limestone pores, and the remaining signals provides information on the irreducible water in the limestone pores [29]. For the four core samples C1, C2, C3, and C4, the S_{ir}s were 95.72%, 93.60%, 94.06%, and 95.18%, respectively. This indicates that for tight carbonate rocks, the irreducible water content is high and difficult to displace. It can be inferred that a large amount of irreducible oil is difficult to extract due to low porosity and low permeability conditions for this limestone reservoir.

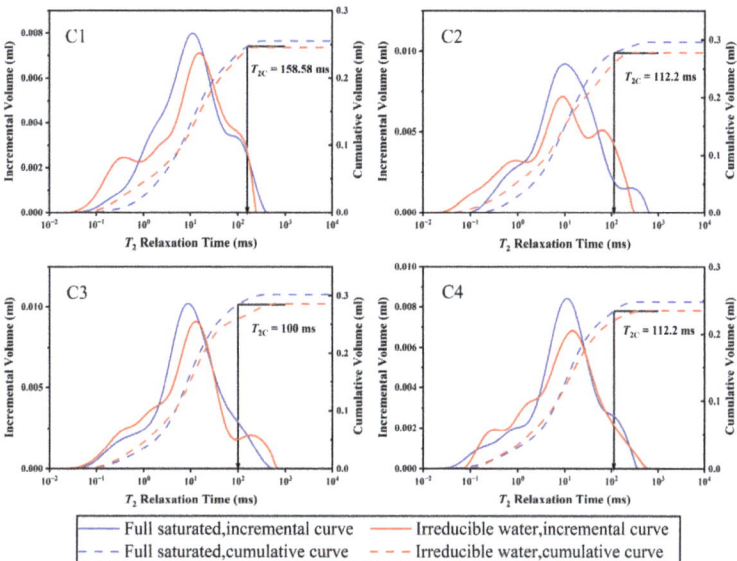

Figure 5. T_2 distributions of core samples at 100% water saturation and after centrifuging to 143 psi, showing the method to calculate a T_2 cutoff value T_{2C}.

The method of determining the T_{2C} based on NMR centrifugal experiments is described by Yao et al. [26]. First, the NMR T_2 distribution of core samples in 100% water saturation was transformed into a cumulative porosity curve (blue dashed line in Figure 5), and the NMR T_2 distribution of cores under the optimal centrifugal force of 143 psi was transformed into a irreducible water cumulative porosity curve (red dashed line in Figure 5). Then, based on the maximum cumulative porosity value under irreducible water condition, a horizontal line parallel to the transverse T_2 time was plotted to intersect the accumulation curve for 100% water saturation at one point. Finally, based on this intersection point, a line perpendicular to the transverse coordinate T_2 time is plotted, and T_{2c} is the intersection of this line with the transverse axis. In this study, the centrifugation-T_{2C} for the selected sample cores ranged from 100 ms to 158.58 ms, which is consistent with the previous studies [25]. The average value of the T_{2C} of the four samples was chosen as the T_{2C} for this study, i.e., T_{2C} = 120.75 ms.

4.3. Pore and Permeability Characterizations of Limestone Samples

The mineral composition of the limestone of the Ordovician Upper Majiagou Formation is mainly calcite and dolomite. As shown in the SEM microscopy results (Figure 6), intergranular pores, dissolution pores, intercrystalline pores, and microfractures are developed in the limestone.

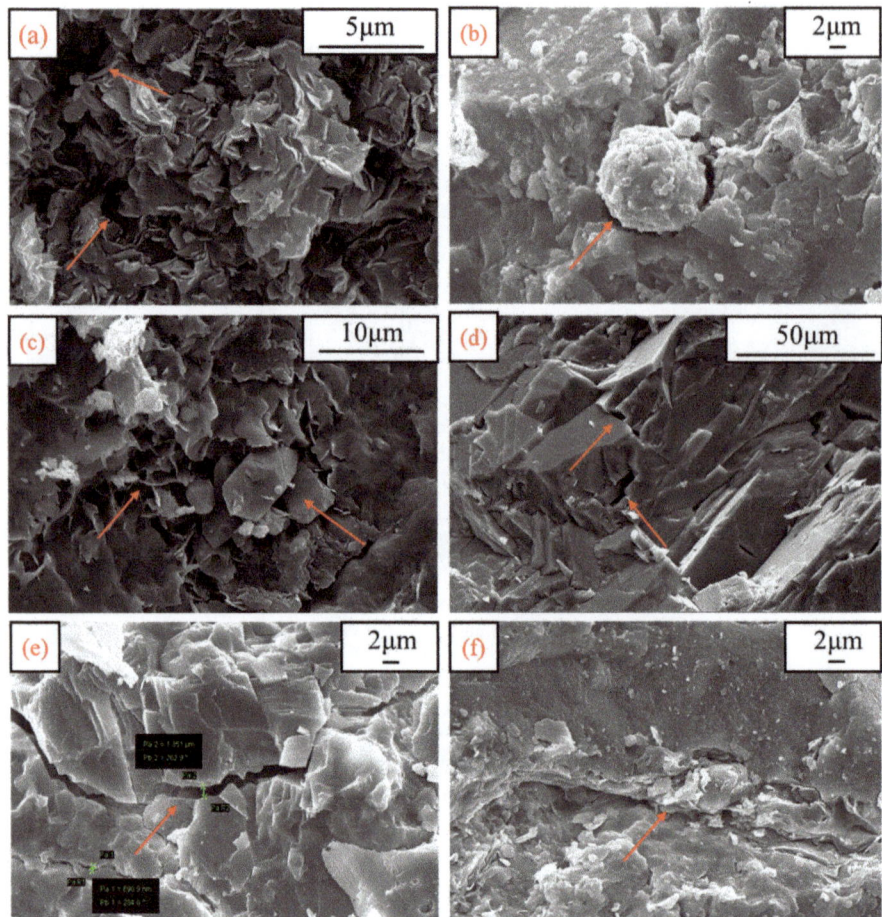

Figure 6. SEM images of intergranular pores (**a**,**b**), dissolution pores (**c**), intercrystalline pores (**d**), microfractures (**e**,**f**).

Intergranular pores included the residual intergranular pores between the deposited particles and the intergranular dissolved pores formed by the residual intergranular pores further expanding outwardly. Most of the intergranular pores observed under the microscope were residual intergranular pores, some of which are filled with authigenic illite, which shows unevenly foliated grains with irregular grain boundaries under the microscope [30] (Figure 6a), and some of these intergranular pores are filled with pyrite, surrounded by calcite (Figure 6b). Under the microscope, the pyrite mainly appears framboidal and spherical [31] and the calcite crystals mainly appear rhombohedra [32]. Dissolved pores are secondary pores formed via the dissolution of fragmented particles. The dissolved pores in the study area are mainly calcite dissolved pores, and some of them are filled with autogenous quartz and silky illite (Figure 6c). Intercrystalline pores are micropores between matrix grains and between intercrystals such as authigenic clay minerals. Microscopically observed intercrystalline pores of calcite are usually triangular or polygonal in shape with smooth, straight edges (Figure 6d). The microfractures in the study area are mainly related to tectonic movement [33]. There are many stages of fractures in the study area; a large number of microfractures are unfilled (Figure 6e) and some are filled with silky illite (Figure 6f).

Moreover, Table 1 shows the results of the pulsed attenuation method for the core samples and the porosity obtained by NMR. The porosity of the five core samples ranges from 0.96% to 1.21%, with an average of 1.07%, and the permeability ranges from 0.00252×10^{-3} μm^2 to 0.00502×10^{-3} μm^2, with an average of 0.00379×10^{-3} μm^2. Carbonate reservoirs with porosity less than 4% and permeability less than 1×10^{-3} μm^2 are classified as extra-low porosity and extra-low permeability reservoirs in the China evaluation standard for physical properties of sedimentary reservoirs of SY/T 6285-2011. Compared with the experimental results, the Lower Paleozoic Ordovician carbonate reservoir is the extra-low porosity and extra-low permeability reservoirs.

Table 1. The result of core samples.

Serial	V_t (cm^3)	$\varnothing_{n1}$ (%)	S_{ir} (%)	T_{2g} (ms)	K_p ($\times 10^{-3}$ μm^2)
C1	24.943	1.03	94.51	10.155	0.00252
C2	25.013	1.19	94.05	10.243	0.00502
C3	24.943	1.21	95.70	8.478	0.00313
C4	25.057	1.14	95.62	8.441	0.00329
C5	25.575	0.96	93.84	10.605	0.00353

Notes: V_t = total volume; $\varnothing_{n1}$ = NMR porosity; K_p = permeability by pulse attenuation method.

Table 2 shows the results of the NMR porosity of the cubic and cutting samples. There was little difference in porosity between the samples of different sizes, but the S_{ir} and T_{2g} of the two sets of cutting samples was significantly different from that of the cubic and core samples. This is due to a change in the pore size of the cuttings sample [5], which is discussed in detail in Section 5.2.

Table 2. Cubic and cutting samples test results.

Serial	m (g)	$\varnothing_{n2}$ (%)	S_{ir} (%)	T_{2g} (ms)
Cube–4 × 4 × 4 cm	178.04	1.05	92.22	11.305
Cube–4 × 4 × 2 cm	98.34	1.02	95.45	8.762
Cube–4 × 2 × 2 cm	44.48	0.89	91.76	9.278
Cube–2 × 2 × 2 cm	31.66	0.93	95.48	7.902
Powder–a	102.33	1.10	89.48	18.542
Powder–b	130.48	1.00	61.95	64.468

Notes: m = completely dry mass; $\varnothing_{n2}$ = NMR porosity.

5. Discussion

5.1. Effect of Rock Sample Size on Porosity

As shown in Figure 7, the porosity of the cube and cuttings is basically equal to the average porosity of the core; the relative error is between 1.87–16.82% and the average relative error is 7.63%. In this work, the relative standard deviation (RSD) was used to evaluate the differences between the porosity of samples with different sizes. For core samples of the same size, RSD = 9.60%. In comparison, the RSD of the four groups of cube samples of different sizes is 7.71%, the RSD of the two groups of cuttings samples is 6.73%, and the RSD of the cube and cuttings samples of different sizes is 7.74%, all of which are less than 9.60%, indicating that the porosity is independent of the sample shape and size. Therefore, for carbonate rocks, cubic samples and cuttings samples larger than 1 mm porosity can be used to represent reservoir porosity.

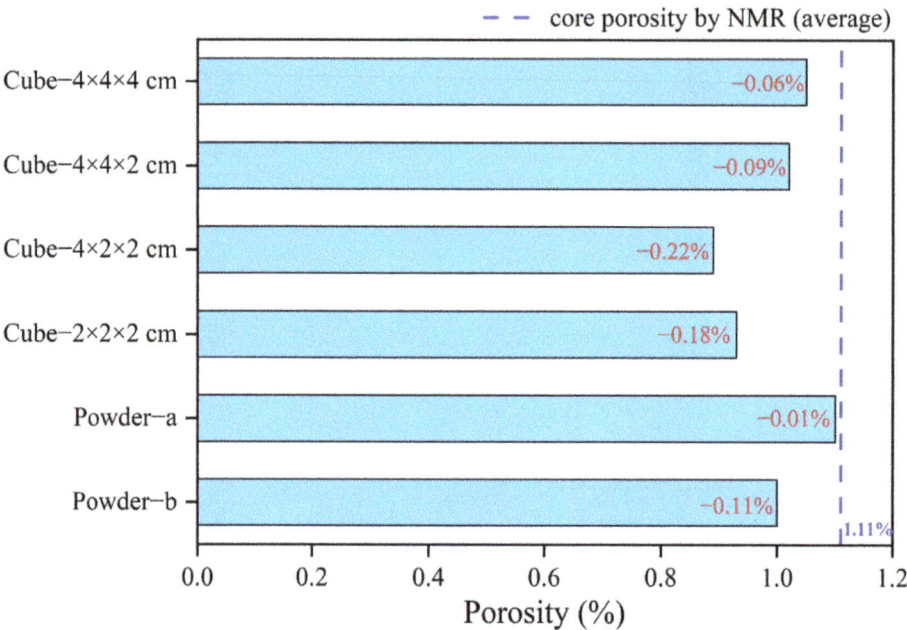

Figure 7. Porosity of limestone samples with different sizes.

5.2. Effect of Rock Sample Size on PSD

According to the transformation method of T_2 relaxation time to PSD and the classification of pores, the percentage of micropore, mesopore, and macropore of limestone samples were obtained (Figure 8). For five core samples, the average proportions of micropores, mesopores, and macropores are 9%, 38%, and 53%. For four cube samples, the proportions of the three types of pores are close, which are 12%, 34%, and 54%; For the two groups of cuttings samples, the pore proportions are quite different. For cuttings with particle size of 4.7–6.75 mm, the proportions of micropores, mesopores, and macropores are 6%, 25%, and 69%, respectively. For cuttings of 1–1.7 mm, the proportions are 4%, 10%, and 86%, respectively. The above results show that for core and cubic samples the proportion of the three types of pores is almost the same. Compared with the cuttings sample, the proportion of micropores and mesopores decreases while the proportion of macropores increases significantly as the particle size of cuttings decreases. This result is mainly related to the mechanical pulverization effect and to the interparticle void effect. The mechanical pulverization effect during the process of cutting and pulverizing samples can cause the proportion of macropore to increase and the proportion of micropore and mesopore to decrease. Arson and Pereira [34] showed that undamaged and damaged rocks have different PSD characteristics. Han et al. [35] showed that pulverization has a negligible effect on the 2 nm micropores and that the damage of the sample will form secondary pores and cracks, which are mainly macropores and mesopores. In this study, for cube samples, the mechanical pulverization effect was small and the vast majority of pores remained intact. Meanwhile, for cuttings samples, the mechanical pulverization effect was large and most of the pores were destroyed, causing some micropores or mesopores to open or form cracks. Moreover, SEM images of the samples show that most of the pores of the limestone samples are filled with minerals. During the crushing process, the discrete minerals filled in the pores are easily separated, resulting in an increased proportion of macropores. The remaining water in the interparticle void can also increase the NMR measurement of macropores. Smith and Schentrup [36] demonstrated that larger size particles contained larger interparticle pores. Lenormand and Fonta [6] suggested that for cuttings with smaller particle sizes, the amount of liquid trapped by capillarity between the cuttings is very large, and some methods can be used to reduce the influence of interparticle water. Although we performed pretreatments between NMR tests to eliminate the effect of interparticle voids, this interparticle space effect may still affect the results of macropore measurements.

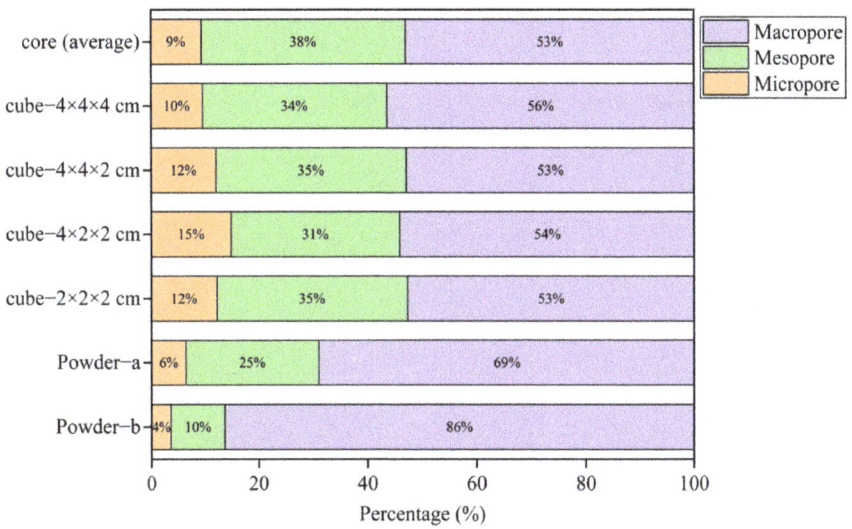

Figure 8. Distribution of pore sizes of different sized chert samples.

5.3. Effect of Rock Sample Size on Permeability

For both the Coates and SDR models, the key is to obtain the constants C and C_1 in Equations (4) and (6) from the experiment data. The constants C and C_1 vary significantly for different rocks; for example, C in the Coates model for sandstone is generally 0.0001. For carbonate, the C_1 in the SDR model is 0.1, according to Fleury et al. [37]. Moreover, according to Amabeoku et al. [38], for rocks with the same lithology, the constant C and C_1 are selected differently in different regions. Therefore, in this study, the optimal constant C and C_1 was obtained by mathematical fitting of the permeability K_p and NMR data of five cores. As shown in Figure 9a, the optimal value of the Coates model C is 0.7487, with $R^2 = 0.8655$; in Figure 9b, the optimal value of the SDR model C_1 is 0.00003, with $R^2 = 0.9146$.

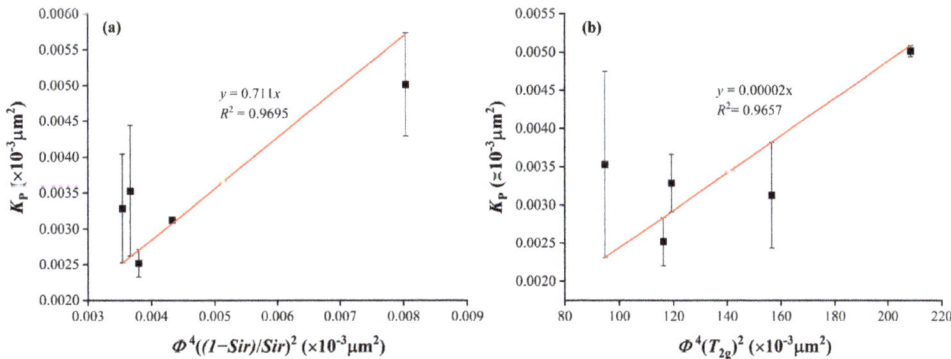

Figure 9. Determination of the Coates model constant C (**a**) and the SDR model constant C_1 (**b**).

Figure 10 shows the error between the Coates model and the SDR model and the measured permeability. The small error indicates that the two models can accurately predict the permeability of limestone samples. Permeability of cubic samples and cuttings samples is shown in Figure 11. For the four cubic samples, the prediction results of the Coates model and the SDR model are slightly different from the core permeability. For the two groups of cuttings, the prediction results of the two models are significantly different from the core results, with an error of one–two orders of magnitude; the smaller the particle size of the debris, the larger the error. This situation is highly similar to the change in PSD described above. And previous studies have suggested that the permeability is largely dependent on the grain size [39].

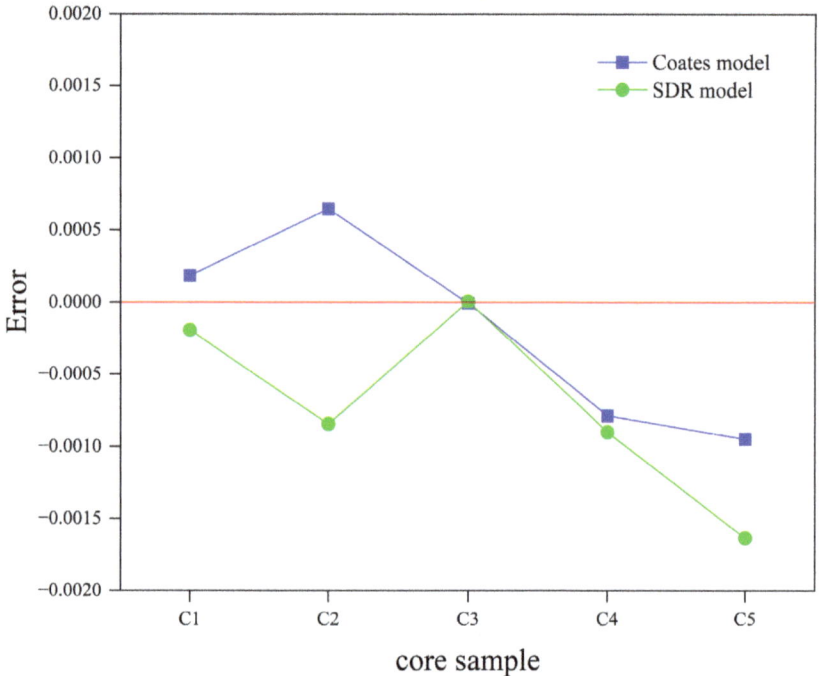

Figure 10. The error between the model's predicted value and the measured value.

At the same time, for these samples, the larger values of T_2 in the cuttings (especially powder-b) imply much larger permeabilities than the other samples, which is consistent with the larger pore size. For powder b, the permeability of the Coates model is significantly higher than that of the SDR model. This is because the Coates model considers that the permeability mainly depends on the macropores with good connectivity in the sample and considers the effect of movable porosity on the permeability, while the SDR model introduces T_{2g}, the contribution of pores of different sizes to permeability was averaged to reduce the effect of macropores on permeability. This reminds us that when using these two models for permeability prediction, the PSD characteristics and pores connectivity of the sample should be considered to improve the applicability of the model.

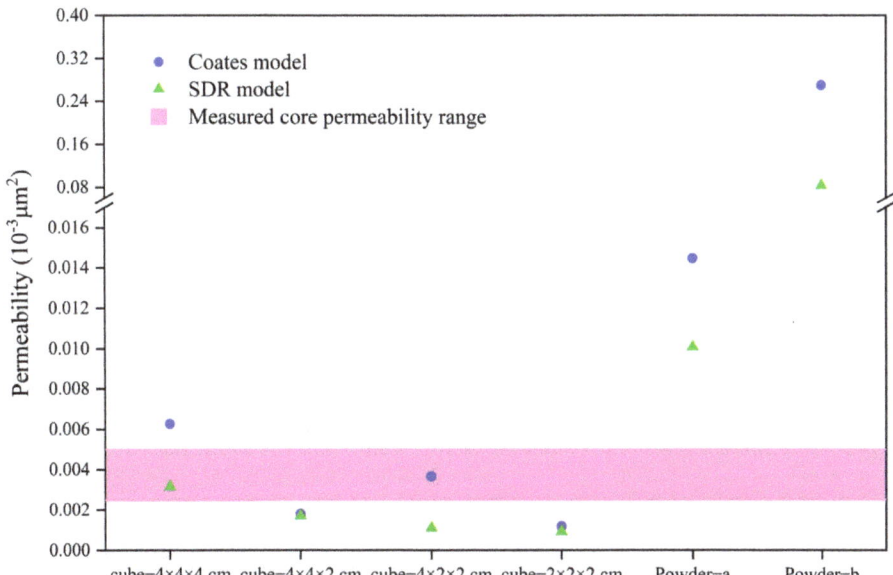

Figure 11. Permeability results of cubic samples and cuttings samples.

6. Applications

Cuttings analysis is a common practice in laboratory settings, and the selection of an appropriate cutting size to substitute intact cores is crucial in determining the petrophysical property of reservoir. The results of this study show that carbonate samples with >1 mm cutting can be used to obtain porosity; however, for PSD and permeability, the use of cuttings < 6.7 mm in diameter is not recommended.

Moreover, it is important to extrapolate the application of cuttings from the laboratory to the field. When the laboratory analysis is applied to the field, it must also consider the representative problems of the field cuttings, such as wellbore collapse, particle gravity differentiation, and drilling mud contamination. Due to gravitational particle divergence, the lag times calculation is used to identify which reservoir the cuttings are coming from. This method is often considered to be accurate in vertical Wells, with significant errors in horizontal and deviated Wells [12]. Some of the cuttings with particle sizes more than 5 mm may have resulted from the collapse of boreholes in unknown stratigraphic positions [10]. From a cuttings transport and dispersion point of view, cuttings with small particle size are easily contaminated by drilling mud [40]. Meanwhile, according to the results of this study, the cuttings permeability of 1–1.7 mm is much larger than standard core permeability. Thus, the use of cuttings with a particle size of less than 6.75 mm to represent the actual formation physical properties is not recommended.

7. Conclusions

The aim of this paper is to verify the feasibility of substituting intact cores with drilling cuttings in carbonate rocks and the effect of size on sample physical properties using low-field NMR techniques. The following conclusions are drawn from experimental results:

(1) In laboratory analyses using NMR techniques, cuttings with a size of 1–1.7 mm can accurately measure porosity;
(2) In laboratory analyses using NMR techniques, cuttings with a particle size of less than 6.75 mm do not accurately characterize the properties of extra-low porosity and extra-low permeability rocks, because of deviations in permeability and PSD;
(3) For millimeter-sized limestone cuttings, PSD and permeability are severely affected by mechanical pulverization effect and interparticle water.

Author Contributions: Conceptualization Y.C.; methodology, S.W.; validation, S.W.; investigation, S.W. and Z.W.; writing—original draft preparation, S.W.; and writing—review and editing, X.S. All authors have read and agreed to the published version of the manuscript.

Funding: This work was supported by the National Natural Science Foundation of China [42125205; 42202195] and the Fundamental Research Funds for the Central Universities [2652023001; 2652023066].

Data Availability Statement: Data is contained within the article.

Conflicts of Interest: The authors declare no conflicts of interest.

References

1. Roehl, P.O.; Choquette, P.W. *Carbonate Petroleum Reservoirs*; Springer Science & Business Media: Berlin/Heidelberg, Germany, 2012.
2. Santarelli, F.J.; Marsala, A.F.; Brignoli, M.; Rossi, E.; Bona, N. Formation Evaluation From Logging on Cuttings. *SPE Reserv. Eval. Eng.* **1998**, *1*, 238–244. [CrossRef]
3. Mirotchnik, K.; Kryuchkov, S.; Strack, K. A Novel Method To Determine Nmr Petrophysical Parameters From Drill Cuttings. In Proceedings of the SPWLA 45th Annual Logging Symposium, Noordwijk, The Netherlands, 6–9 June 2004.
4. Yu, Y.; Menouar, H. An Experimental Method to Measure the Porosity from Cuttings: Evaluation and Error Analysis. In Proceedings of the SPE Production and Operations Symposium, Oklahoma City, OK, USA, 1–5 March 2015.
5. Hübner, W. Studying the Pore Space of Cuttings by NMR and μCT. *J. Appl. Geophys.* **2014**, *104*, 97–105. [CrossRef]
6. Lenormand, R.; Fonta, O. Advances in Measuring Porosity and Permeability from Drill Cuttings. In Proceedings of the SPE/EAGE Reservoir Characterization and Simulation Conference, Abu Dhabi, United Arab Emirates, 28–30 October 2007.
7. Solano, N.A.; Clarkson, C.R.; Krause, F.F.; Lenormand, R.; Barclay, J.E.; Aguilera, R. Drill Cuttings and Characterization of Tight Gas Reservoirs—An Example from the Nikanassin Fm. in the Deep Basin of Alberta. In Proceedings of the SPE Canadian Unconventional Resources Conference, Calgary, AB, Canada, 30 October–1 November 2012.
8. Ortega, C.; Aguilera, R. A Complete Petrophysical-Evaluation Method for Tight Formations From Drill Cuttings Only in the Absence of Well Logs. *SPE J.* **2013**, *19*, 636–647. [CrossRef]
9. Fellah, K.; Utsuzawa, S.; Song, Y.Q.; Kausik, R. Porosity of Drill-Cuttings Using Multinuclear ^{19}F and ^{1}H NMR Measurements. *Energy Fuels* **2018**, *32*, 7467–7470. [CrossRef]
10. Chang, Y.H.; Yao, Y.B.; Liu, Y.; Zheng, S.J. Can cuttings replace cores for porosity and pore size distribution analyses of coal? *Int. J. Coal Geol.* **2020**, *227*, 15. [CrossRef]
11. Siddiqui, S.; Grader, A.S.; Touati, M.; Loermans, A.M.; Funk, J.J. Techniques for Extracting Reliable Density and Porosity Data From Cuttings. In Proceedings of the SPE Annual Technical Conference and Exhibition, Dallas, TX, USA, 9–12 October 2005.
12. Egermann, P.; Doerler, N.; Fleury, M.; Behot, J.; Deflandre, F.; Lenormand, R. Petrophysical Measurements From Drill Cuttings: An Added Value for the Reservoir Characterization Process. *SPE Reserv. Eval. Eng.* **2006**, *9*, 302–307. [CrossRef]
13. Munn, K.; Smith, D.M. A NMR Technique for the Analysis of Pore Structure-Numerical Inversion of Relaxation Measurements. *J. Colloid Interface Sci.* **1987**, *119*, 117–126. [CrossRef]
14. Al-Mahrooqi, S.H.; Grattoni, C.A.; Moss, A.K.; Jing, X.D. An Investigation of the Effect of Wettability on NMR Characteristics of Sandstone Rock and Fluid Systems. *J. Pet. Sci. Eng.* **2003**, *39*, 389–398. [CrossRef]
15. Hao, F.; Zhou, X.H.; Zhu, Y.M.; Zou, H.Y.; Yang, Y.Y. Charging of Oil Fields Surrounding the Shaleitian Uplift From Multiple Source Rock Intervals and Generative Kitchens, Bohai Bay Basin, China. *Mar. Pet. Geol.* **2010**, *27*, 1910–1926. [CrossRef]
16. Ye, T.; Wei, A.J.; Gao, K.S.; Sun, Z.; Li, F. New Sequence Division Method of Shallow Platform with Natural Gamma Spectrometry Data: Implication for Reservoir Distribution—A Case Study From Majiagou Formation of Bozhong 21–22 Structure, Bohai Bay Basin. *Carbonates Evaporites* **2020**, *35*, 14. [CrossRef]
17. SY/T 5162-2014; Analytical Method for Rock Sample by Scanning Electron Microscope. Petroleum Industry Press: Beijing, China, 2022.
18. Sun, X.; Yao, Y.; Liu, D.; Ma, R.; Qiu, Y. Effects of Fracturing Fluids Imbibition on CBM Recovery: In Terms of Methane Desorption and Diffusion. *SPE J.* **2024**, *29*, 505–517. [CrossRef]
19. Straley, C. In An Experimental Investigation of Methane in Rock Materials. In Proceedings of the SPWLA 38th Annual Logging Symposium, Houston, TX, USA, 15–18 June 1997.
20. Zheng, S.J.; Yao, Y.B.; Liu, D.M.; Cai, Y.D.; Liu, Y. Characterizations of Full-scale Pore Size Distribution, Porosity and Permeability of Coals: A Novel Methodology by Nuclear Magnetic Resonance and Fractal Analysis Theory. *Int. J. Coal Geol.* **2018**, *196*, 148–158. [CrossRef]
21. Wang, Z.F.; Yao, Y.B.; Ma, R.Y.; Zhang, X.A.; Zhang, G.B. Application of Multifractal Analysis Theory to Interpret T_2 Cutoffs of NMR Logging Data: A Case Study of Coarse Clastic Rock Reservoirs in Southwestern Bozhong Sag, China. *Fractal Fract.* **2023**, *7*, 20. [CrossRef]
22. Mai, A.; Kantzas, A. On the Characterization of Carbonate Reservoirs Using Low Field NMR Tools. In Proceedings of the SPE Gas Technology Symposium, Calgary, AB, Canada, 30 April–2 May 2002.

23. Coates, G.R.; Xiao, L.; Prammer, M.G. *NMR Logging: Principles and Interpretation*; Halliburton Energy Services: Huston, TX, USA, 1999.
24. Morriss, C.; Rossini, D.; Straley, C.; Tutunjian, P.; Vinegar, H. Core Analysis By Low-field Nmr. *Log Anal.* **1997**, *38*, 84–95.
25. Westphal, H.; Surholt, I.; Kiesl, C.; Thern, H.F.; Kruspe, T. NMR Measurements in Carbonate Rocks: Problems and an Approach to a Solution. *Pure Appl. Geophys.* **2005**, *162*, 549–570. [CrossRef]
26. Yao, Y.B.; Liu, D.M.; Che, Y.; Tang, D.Z.; Tang, S.H.; Huang, W.H. Petrophysical Characterization of Coals by Low-field Nuclear Magnetic Resonance (NMR). *Fuel* **2010**, *89*, 1371–1380. [CrossRef]
27. Lawal, L.O.; Adebayo, A.R.; Mahmoud, M.; Dia, B.; Sultan, A.S. A Novel NMR Surface Relaxivity Measurements on Rock Cuttings for Conventional and Unconventional Reservoirs. *Int. J. Coal Geol.* **2020**, *231*, 16. [CrossRef]
28. Wei, D.; Gao, Z.Q.; Zhang, C.; Fan, T.L.; Karubandika, G.M.; Meng, M.M. Pore Characteristics of the Carbonate Shoal from Fractal Perspective. *J. Pet. Sci. Eng.* **2019**, *174*, 1249–1260. [CrossRef]
29. Kenyon, W.E. Petrophysical Principles of Applications of NMR Logging. *Log Anal.* **1997**, *38*, 21–43.
30. Chen, T.; Wang, H.J.; Li, T.; Zheng, N. New insights into the formation of diagenetic illite from TEM studies. *Am. Miner.* **2013**, *98*, 879–887. [CrossRef]
31. Wilkin, R.T.; Barnes, H.L.; Brantley, S.L. The size distribution of framboidal pyrite in modern sediments: An indicator of redox conditions. *Geochim. Cosmochim. Acta* **1996**, *60*, 3897–3912. [CrossRef]
32. Palvanov, M.; Eren, M.; Kadir, S. EPMA analysis of a stalagmite from Küpeli Cave, southern Turkey: Implications on detrital sediments. *Carbonates Evaporites* **2024**, *39*, 11. [CrossRef]
33. Zhao, F.Y.; Jiang, S.H.; Li, S.Z.; Zhang, H.X.; Wang, G.; Lei, J.P.; Gao, S. Cenozoic Tectonic Migration in the Bohai Bay Basin, East China. *Geol. J.* **2016**, *51*, 188–202. [CrossRef]
34. Arson, C.; Pereira, J.M. Influence of Damage on Pore Size Distribution and Permeability of Rocks. *Int. J. Numer. Anal. Methods Geomech.* **2013**, *37*, 810–831. [CrossRef]
35. Han, M.L.; Wei, X.L.; Zhang, J.C.; Liu, Y.; Tang, X.; Li, P.; Liu, Z.Y. Influence of Structural Damage on Evaluation of Microscopic Pore Structure in Marine Continental Transitional Shale of the Southern North China Basin: A Method Based on the Low-temperature N_2 Adsorption Experiment. *Pet. Sci.* **2022**, *19*, 100–115. [CrossRef]
36. Smith, D.M.; Schentrup, S. Mercury Porosimetry of Fine Particles—Particle Interaction and Compression Effects. *Powder Technol.* **1987**, *49*, 241–247. [CrossRef]
37. Fleury, M.; Deflandre, F.; Godefroy, S. Validity of Permeability Prediction from NMR Measurements. *Comptes Rendus Acad. Sci. Ser. II Chem.* **2001**, *4*, 869–872. [CrossRef]
38. Amabeoku, M.O.; Funk, J.J.; Al-Dossary, S.M.; Al-Ali, H.A. Calibration of Permeability Derived from NMR Logs in Carbonate Reservoirs. In Proceedings of the SPE Middle East Oil Show, Manama, Bahrain, 17–20 March 2001.
39. Amaefule, J.O.; Altunbay, M.; Tiab, D.; Kersey, D.G.; Keelan, D.K. Enhanced Reservoir Description: Using Core and Log Data to Identify Hydraulic (Flow) Units and Predict Permeability in Uncored Intervals/Wells. In Proceedings of the SPE Annual Technical Conference and Exhibition, Houston, TX, USA, 3–6 October 1993.
40. Georgi, D.T.; Harville, D.G.; Robertson, H.A. Advances In Cuttings Collection And Analysis. In Proceedings of the SPWLA 34th Annual Logging Symposium, Calgary, AB, Canada, 28–30 September 1993.

Disclaimer/Publisher's Note: The statements, opinions and data contained in all publications are solely those of the individual author(s) and contributor(s) and not of MDPI and/or the editor(s). MDPI and/or the editor(s) disclaim responsibility for any injury to people or property resulting from any ideas, methods, instructions or products referred to in the content.

Article

Numerical Simulation Study on the Influence of Cracks in a Full-Size Core on the Resistivity Measurement Response

Hanwen Zheng [1,2], Zhansong Zhang [1,2,*], Jianhong Guo [1,2], Sinan Fang [1,2] and Can Wang [3]

1. College of Geophysics and Petroleum Resources, Yangtze University, Wuhan 430100, China; 2021710326@yangtzeu.edu.cn (H.Z.)
2. Key Laboratory of Exploration Technologies for Oil and Gas Resources, Ministry of Education, Yangtze University, Wuhan 430100, China
3. Hydrogeology and Engineering Geology Institute of Hubei Geological Bureau, Jingzhou 434007, China
* Correspondence: zhangzhs@yangtzeu.edu.cn

Abstract: The development of fractured oil fields poses a formidable challenge due to the intricate nature of fracture development and distribution. Fractures profoundly impact core resistivity, making it crucial to investigate the mechanism behind the resistivity response change in fracture cores. In this study, we employed the theory of a stable current field to perform a numerical simulation of the resistivity response of single-fracture and complex-fracture granite cores, using a full-size granite core with cracks as the model. We considered multiple parameters of the fracture itself and the formation to explore the resistivity response change mechanism of the fracture core. Our findings indicate that, in the case of a core with a single fracture, the angle, width, and length of the fracture (fracture occurrence) significantly affect core resistivity. When two fractures run parallel for a core with complex fractures, the change law of core resistivity is similar to that of a single fracture. However, if two fractures intersect, the relative position of the two fractures becomes a significant factor in addition to the width and length of the fracture. Interestingly, a 90° difference exists between the change law of core resistivity and the change law of the resistivity logging response. Furthermore, the core resistivity is affected by matrix resistivity and the resistivity of the mud filtrate, which emphasizes the need to calibrate the fracture dip angle calculated using dual laterolog resistivity with actual core data or special logging data in reservoirs with different geological backgrounds. In the face of multiple fractures, the dual laterolog method has multiple solutions. Our work provides a reference and theoretical basis for interpreting oil and gas in fractured reservoirs based on logging data and holds significant engineering guiding significance.

Keywords: fractured reservoir; core resistivity; numerical simulation; fracture occurrence; logging interpretation

Citation: Zheng, H.; Zhang, Z.; Guo, J.; Fang, S.; Wang, C. Numerical Simulation Study on the Influence of Cracks in a Full-Size Core on the Resistivity Measurement Response. *Energies* **2024**, *17*, 1386. https://doi.org/10.3390/en17061386

Academic Editor: Hossein Hamidi

Received: 24 February 2024
Revised: 10 March 2024
Accepted: 11 March 2024
Published: 13 March 2024

Copyright: © 2024 by the authors. Licensee MDPI, Basel, Switzerland. This article is an open access article distributed under the terms and conditions of the Creative Commons Attribution (CC BY) license (https://creativecommons.org/licenses/by/4.0/).

1. Introduction

As traditional oil and gas resources continue to dwindle, reservoir exploration and development technology is reaching new heights of sophistication [1,2]. The focus has shifted towards exploiting complex, unconventional reservoirs like volcanic oil and gas, shale oil and gas, and tight oil and gas on a global scale [3,4]. Among these reservoir types, fractures play a pivotal role as reservoir spaces and permeable pathways for oil and gas in igneous rock reservoirs, tight reservoirs, and shale reservoirs. The development of fractures stands out as a key characteristic influencing reservoir storage capacity [5]. In China's major oil and gas fields, low-permeability fractured reservoirs are prevalent and are poised to serve as a crucial resource base for bolstering reserves and production in the country's oil industry moving forward [6]. The increase in rock density in low-permeability fractured reservoirs leads to heightened rock strength and brittleness [7]. When subjected to tectonic stress, rocks exhibit varying degrees of fracturing, giving rise to fractured low-permeability reservoirs. The intricate development and distribution of fractures in these reservoirs pose challenges in terms of logging response and physical property evaluation, presenting significant technical hurdles in reservoir exploration and development endeavors [8]. Consequently, investigating how logging technology can be leveraged to economically and efficiently develop such reservoirs holds immense engineering importance.

For fractured reservoirs, core resistivity experiments play a critical role in evaluating hydrocarbon properties from logging data. However, due to laboratory constraints, these experiments often provide a limited perspective, making it challenging to fully characterize the resistivity response of cores with various fracture patterns [9]. As a result, an increasing number of scholars have turned to numerical simulation as a viable alternative. Presently, the most common approach involves characterization using imaging logging data. Consequently, scholars have undertaken successive numerical simulation studies on the computational model of electric imaging logging for fracture width, primarily employing the three-dimensional finite element method [10]. Factors examined include fracture inclination, formation resistivity, mud resistivity, distance between the instrument and the well wall, length of lateral extension of the fracture, and fracture spacing [11]. In 1990, Luthi et al. [12] pioneered the use of electric imaging instruments to study the relationship between electrical signals and fracture parameters. Their 3D finite element modeling simulated the current emitted from a single-button electrode in front of the fracture, solving Laplace differential equations for the electric field in and around the borehole on an adaptive 3D mesh. Their findings established a link between fracture width and anomalous current area, formation resistivity in the flushing zone, mud filtrate resistivity, and instrumentation-related coefficients. Subsequently, in 2001, Wang Dali [13] developed a simplified model of the electrode system structure for real micro-resistivity imaging logging and established a finite element numerical simulation method. His research revealed that mud resistivity significantly impacts the additional current relative to formation resistivity. However, the study's limitations were acknowledged, particularly regarding the finite element mesh node division of the orthotropic model, which was constrained by computer hardware conditions. In 2002, Ke et al. [14] advanced the field by developing a full-space numerical simulation program using the three-dimensional finite element method to investigate the relationship between logging response and fracture characteristics. They proposed a new formula for calculating fracture opening, showcasing the potential for innovation in this area of study. Continuing this trajectory, Aixin Chen's 2006 work involved numerical simulations of micro-resistivity imagers in non-uniform media [15], presenting further advancements in the 3D finite element method. This was followed by Yu Cao's [16] exploration of the response of micro-resistivity scanning imaging logging instruments to fractures, which delved into the instrument's resolution for different fracture scenarios. In 2015, Ponziani et al. [17] conducted a comprehensive investigation using a full borehole formation micro-resistivity imaging logging instrument through numerical simulation. Their findings demonstrated a linear relationship between fracture width and measured additional current, offering valuable insights into the measurement of fracture width and associated parameters. Ammar [18] utilized Comsol software (Comsol Multiphysics model, CMM, version 4.4, 2014) to construct a numerical simulation model for studying the resistivity of fracture-containing rock saturated with water. The study explored resistivity orthotropy and investigated the relationship between porosity and fracture parameters. Epov et al. [19] employed the finite element method to analyze the impact of tilted uniaxial anisotropy on electric logging, offering a new approach to studying the physical and electrical properties of carbonate reservoirs. He et al. [20] proposed a non-destructive method for testing the radial resistivity of cylindrical core samples, comparing different methods and using numerical simulations to highlight the variations in measurement results based on testing approaches, thereby enhancing the foundation for field interpretation. Tan et al. [21] utilized a three-dimensional finite element method to identify cracks in shale reservoirs and introduced a novel calculation method for crack porosity. Their conclusions, supported by comparisons with actual data, yielded significant results. Deng et al. [22] numerically simulated the lateral response of high-resolution azimuthal resistivity in fractured strata, linking azimuthal resistivity to fracture tilt and orientation and providing essential insights for logging interpretation in fractured reservoirs. With advancements in logging technologies, follow-drill logging has gained traction. Liu et al. [23] applied an adaptive hp-FEM algorithm to simulate the resistivity response of fractured reservoir cavities during follow-drill operations, offering advantages over traditional methods and establishing a theoretical basis for a quantitative evaluation in interpreting seam–hole type reservoirs. Kang et al. [24] used a three-dimensional finite element method to interpret the logging response of a drill-following resistivity imaging tool. By integrating a new borehole calculation model, they revealed relationships between the borehole diameter, current contrast, and formation-resistivity contrast, particularly influenced by fracture dip angles. Zhao et al. [25] conducted simulations using the finite element method on digital cores, highlighting the significant impact of fractures on partially saturated rock formations. They emphasized directional variations in resistivity within fractured samples and concluded that conventional methods may not accurately determine the saturation index in fractured reservoirs, necessitating the consideration of fracture characteristics. Wang et al. [26] established a stratigraphic fracture model for shale reservoirs using the finite element method and digital core analysis. Their study evaluated the effect of different fracture characteristics on bi-lateral logging responses, providing insights into the relationship between fracture parameters and resistivity for evaluating shale fractures. Kim et al. [27] proposed a new theoretical evaluation method for determining cylindrical core samples, including a theoretical framework for perimeter electrode setup and verification through finite element numerical simulations. Despite these notable advancements, the resistivity response measured using the instrument is influenced by numerous factors (for example, instrument type, surrounding rock pressure, mud filtrate, temperature, etc.), both in numerical simulations and actual field logging data [28]. For the core with cracks, the current research is mostly aimed at a single crack, and it is considered that the crack's length, width, and angle significantly affect the core resistivity. For the core of two complex fractures, there is little literature to conduct separate research and exploration, especially to discuss the influence of core resistivity when the angle of the two fractures changes. Such a comprehensive investigation would greatly enhance our understanding of fractured reservoirs and inform future exploration and development endeavors.

Therefore, considering the obvious influence of fractures on core resistivity, this study is guided by the theory of a stable current field and uses the full-diameter granite core with fractures as a model through the finite element method. Numerical simulations of the resistivity response of the single-fractured and complex-fractured granite core are carried out, respectively. Considering the multiple parameters of the fracture itself and the formation, these parameters include the matrix resistivity of the reservoir lithology, the resistivity of the mud filtrate, and the angle, length, and width of the fracture. Moreover, we explored the mechanism of the change in the resistivity response of the fracture core and summarized its change law. These mechanisms and conclusions can provide a reference and theoretical basis for the oil and gas interpretation of fractured reservoirs based on logging data. Especially for complex fractured reservoirs, the model can calculate fracture parameters more accurately and judge the physical changes in the formation corresponding to the change in resistivity, which has engineering guiding significance.

2. Methods and Principles

2.1. Fundamentals of Finite Element Theory

The finite element method (FEM) can be used to solve the approximate solution of the boundary value problem of partial differential equations. The basic idea of the finite element principle is to divide a continuous structure into a finite number of small elements and use a set of simple functions within each small element to approximately describe the behavior of the structure. The analytical results of the whole structure are finally obtained by decomposing the whole structure into a finite number of small elements and then analyzing each small element.

In the finite element boundary value problem, it is defined that there are control differential equations and boundary conditions in the solution region [29]:

$$L\Phi = f \qquad (1)$$

where L is the differential operator, Φ is the variable to be solved, and f is the source of excitation.

The Ritz and Galerkin methods are mainly used to solve boundary value problems [30–32]. The basic principle of the Ritz method is that the boundary value problem is expressed using a function. The minimum value of the function corresponds to the governing differential equation of the given boundary condition, and the solution of the equation is approximately obtained by solving the minimum value of the function. The Galerkin method is a weak-form solution, which transforms the partial differential equation into a weak form and then uses the weighted residual method to obtain an approximate solution. In this study, the Ritz method is used to solve the problem.

The finite element analysis process entails several key steps, beginning with establishing a geometric model and subdivision of the structure or region into a finite number of small elements, known as finite elements. Subsequently, the constraints of each node are determined based on the boundary conditions of the actual problem, encompassing aspects such as displacement and force constraints. This forms the foundation of the finite element model, where a suitable mathematical model and equation are selected to accurately describe the problem at hand, taking into account material parameters and loading conditions. Once the model is established, the equations representing the problem are systematically solved, yielding crucial information such as displacement, stress, and temperature at the nodes. This enables the derivation of analytical results for the problem under consideration. It is imperative to subject the calculation results to thorough verification and optimization, ensuring that the model and parameters are fine-tuned and resolved to uphold the accuracy and reliability of the obtained results. This iterative process serves to validate the analytical outcomes, ultimately enhancing their robustness and applicability.

2.2. The Basic Equation of a Stable Current Field

The stable current field refers to the electric field in which the strength and distribution of the current do not change with time. The current intensity is defined as the amount of electricity flowing through a certain section in a unit of time. The current intensity is also called current, expressed by I (Equation (2)):

$$I = \frac{dq}{dt} \qquad (2)$$

Current density: The current density is a vector used to describe the physical quantity of the current intensity and flow direction at a certain point in the circuit. Its size is the amount of charge that vertically passes through the unit area in unit time. The direction is the direction of movement of the positive charge, represented by J, and the international unit is ampere per square meter (A/m^2). The current density can be approximately proportional to the electric field, that is

$$J = \sigma E = E/R \qquad (3)$$

where σ is the conductivity reciprocal of the resistivity, E is the electric field, and R is the resistivity.

Therefore, the current intensity dI passing through any directed area element dS can be expressed as:

$$dI = J \times dS \qquad (4)$$

The current I through any section S is the integral of dI on the section S:

$$I = \int_S J \times dS \qquad (5)$$

The stable current field is similar to the electrostatic field, and its electric field strength E and potential u satisfy the following relationship:

$$E = -grad(u) \tag{6}$$

2.3. Simulation Principle and Model Construction of Core Resistivity Response

In conjunction with the density and occurrence characteristics of fractures within the formation, our study delved into numerical simulations of single and complex fractures (namely double fractures). A three-dimensional model of a full-diameter core replete with cracks was employed for the former. In contrast, the simulation of complex joints encompassed both parallel and cross-joint cases. It is worth noting that due to the substantial computational power required for simulating complex fractures, it is often difficult to meet such demands using conventional computer technology. As an alternative, a two-dimensional model may be constructed to substantially reduce the computational load without compromising the reliability of our findings.

Conventionally, core resistivity measurement involves the usage of a cylindrical core, whereby power supply and receiving electrodes are attached at both ends of the core. By measuring the total current intensity flowing through the core as well as the potential difference (voltage) at both ends, the resistance can be calculated and, subsequently, the resistivity of the core is derived using Equation (7):

$$\rho = K \times \frac{U}{I} \tag{7}$$

where ρ is the core resistivity; $K = \frac{S}{L}$ is the conversion coefficient; S is the core cross-sectional area, cm^2; L is the core length, cm; U is the voltage applied at both ends of the core; and I is the total current flowing through the core.

The measurement of core resistivity uses the electrode method. This method is the same as the conventional rock electrical experiment method. The core size is larger and larger, and it is generally a full-diameter core. This study simulates the full-diameter core resistivity measurement of the electrode method. According to the principle of core resistivity measurement and quasi-static electric field theory, the definite solution problem of the scalar potential field in the core area can be obtained:

$$\begin{cases} \frac{\partial}{\partial x}(\sigma \frac{\partial u}{\partial x}) + \frac{\partial}{\partial y}(\sigma \frac{\partial u}{\partial y}) + \frac{\partial}{\partial z}(\sigma \frac{\partial u}{\partial z}) = 0 \\ u\mid_{x=0} = \theta \\ u\mid_{x=L} = V_0 \end{cases} \tag{8}$$

where V_0 is the voltage of the power supply electrode, V; σ is the core conductivity, and its spatial distribution is $\sigma = \begin{cases} \sigma_1, Conductivity\ of\ fracture\ filling\ medium \\ \sigma_2, Rock\ resistivity \end{cases}$.

For the basic model, the model length of the given core is 10 cm, and the diameter of the core is 7.5 cm. Combined with the characteristics of granite, the matrix resistivity is set to 1000 Ωm, and the initial mud filtrate resistivity is set to 0.1 Ωm. The basic model uses the tetrahedral element, the solution space is meshed, and the corresponding potential distribution map is given. The schematic diagram of the single-fracture core model is shown in Figure 1.

The complex fracture model under investigation presents an intriguing scenario involving the consideration of two cracks. Moreover, we explore two distinct cases based on the relative position of these cracks, each offering unique insights into fracture behavior. The first case involves parallel cracks, as visually depicted in Figure 2, while the second case features intersecting cracks, as illustrated in Figure 3. To further delve into the simulation of intersecting cracks, this study explores two additional scenarios: one focuses on fixing one crack at a specific level, altering only the angle between the other crack and the central one, while the other scenario initiates with an initial dip angle for one crack, subsequently modifying the angle between the two cracks.

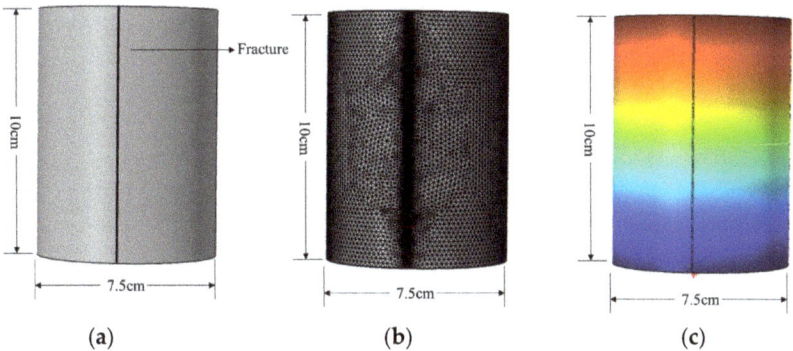

Figure 1. The single-fracture core model solves the regional grid subdivision results. (**a**) Three-dimensional core model; (**b**) grid subdivision diagram; and (**c**) potential distribution map.

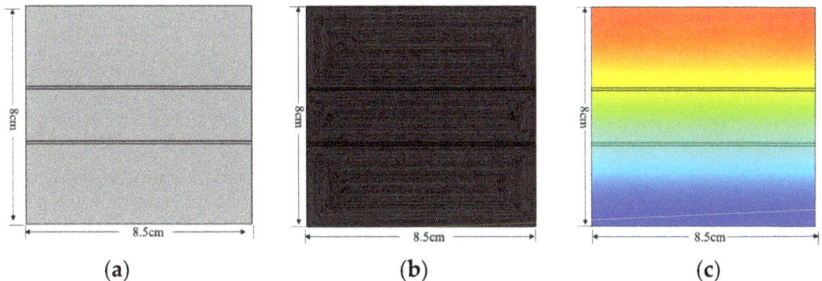

Figure 2. The parallel fractures core model solves the regional grid subdivision results. (**a**) Three-dimensional core model; (**b**) grid subdivision diagram; and (**c**) potential distribution map (Different colors represent different potential differences).

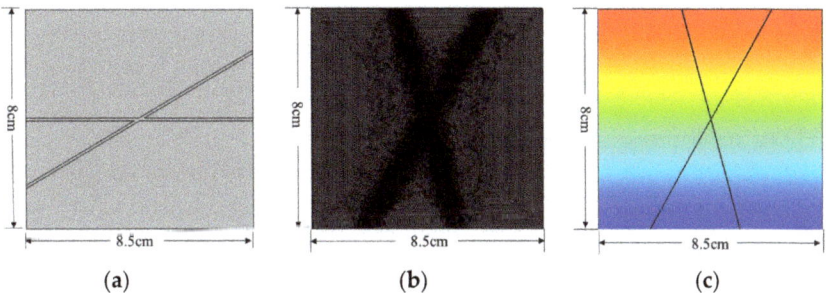

Figure 3. Model sketch of intersecting fractures. (**a**) Cross-seam example; (**b**) grid subdivision results; and (**c**) potential distribution diagram (Different colors represent different potential differences).

With careful attention to detail, numerical simulations were diligently performed for both the single-slit and complex-slit models. By meticulously analyzing the outcomes of these simulations, we aim to unravel the intricate dynamics of fractured formations and gain profound insights into their behavior.

The flow chart of this study is shown in Figure 4. Firstly, based on the finite element method of the stabilized current field, the single-fracture full-diameter core model and the double-fracture full-diameter core model were established, respectively.

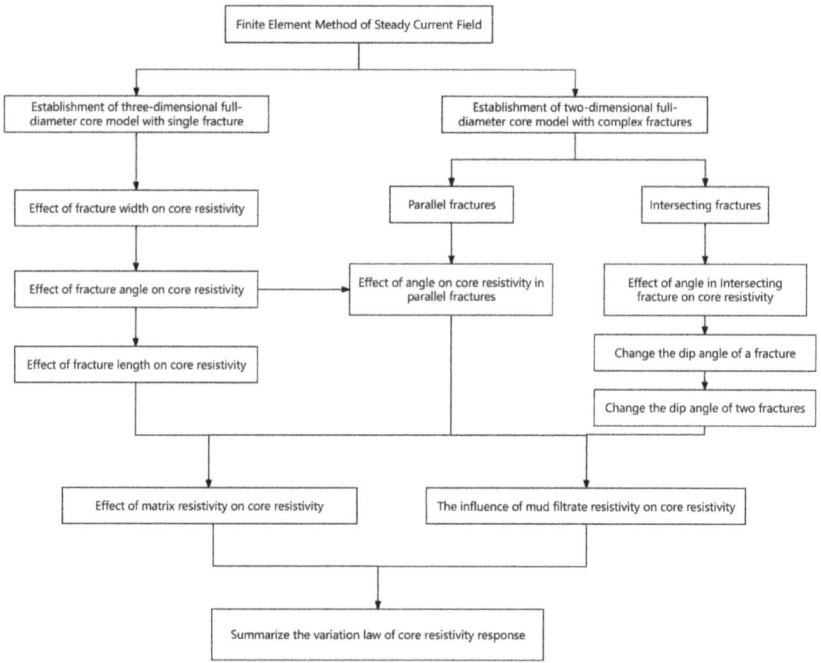

Figure 4. The research process diagram of this study.

In the second step, the effects of the fracture angle, width, and length on the core resistivity were investigated for the core with a single fracture.

In the third step, for the core with a double fracture, the effects of the core resistivity were investigated for the two cases: for the parallel fracture with different fracture angles and the intersecting fractures and the effect of two fractures with different angular relationships on core resistivity.

Finally, the effects of matrix resistivity and mud filtrate resistivity on core resistivity were investigated to summarize the laws of the above research contents and clarify the research's practical significance.

3. Results

3.1. The Resistivity Response Mechanism of a Single-Fracture Model

3.1.1. Core Resistivity Response Characteristics of Different Fracture Widths and Angles

For the influence of fracture width, the response of core resistivity is simulated when the fracture of different angles is close to the source and the fracture width is changed. The process is as follows: add 1A current source to the upper end of the core model, and the lower end is grounded (0 V). The voltage difference of the source surface is measured for each set of fracture parameters, and the core resistivity is calculated using Ohm's law. Considering different fracture dip angles, the corresponding core resistivity changes when the fracture width changes from 100 μm to 1000 μm are simulated. The results are shown in Figure 5. With the increase in the fracture dip angle, the resistivity of the core decreases. Compared with the low-angle fracture, the influence of changing the fracture width on the resistivity of the core is relatively large. When the high-angle fracture and the fracture width are within 400 μm, the response of the core resistivity changes the most.

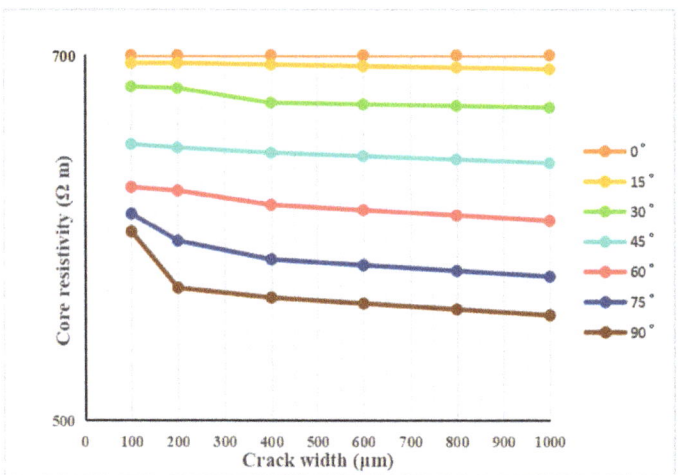

Figure 5. The relationship between the resistivity of the core and the width of the fracture under different fracture angles.

In exploring the impact of fracture angles on core resistivity response, a series of calculations were conducted by varying the fracture angle across different fracture widths. The dynamic shift in fracture angle is meticulously illustrated in Figure 6, offering a visual representation of the evolving characteristics. Accompanying this analysis, Figure 7 provides the potential distribution diagrams corresponding to various fracture angles, shedding light on the intricate relationship between geometry and resistivity response. Figure 8 showcases the calculated outcomes of core resistivity in response to changes in fracture angle, unveiling compelling insights into this phenomenon. Our analysis reveals a clear trend: as the fracture dip angle increases, core resistivity demonstrates a decreasing trend. Furthermore, as the fracture width expands, there is a corresponding decrease in core resistivity, albeit with a relatively modest range of change, aligning closely with the patterns observed in Figure 5. Delving deeper into the nuances of fracture angles, our findings unveil intriguing patterns. In fractures characterized by low angles (0–30°), variations in core resistivity remain minimal across different fracture widths, with decreases of less than 10%. Conversely, high-angle fractures (60–90°) exhibit a gradual stabilization in core resistivity reduction. Notably, the transition from low to high angles (30–60°) elicits the most substantial decrease in core resistivity, indicating heightened sensitivity of resistivity to changes in angle orientation. These findings underscore the intricate interplay between fracture characteristics and core resistivity dynamics, enriching our understanding of subsurface formations.

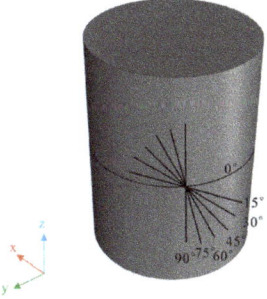

Figure 6. Schematic diagram of different fracture angle models.

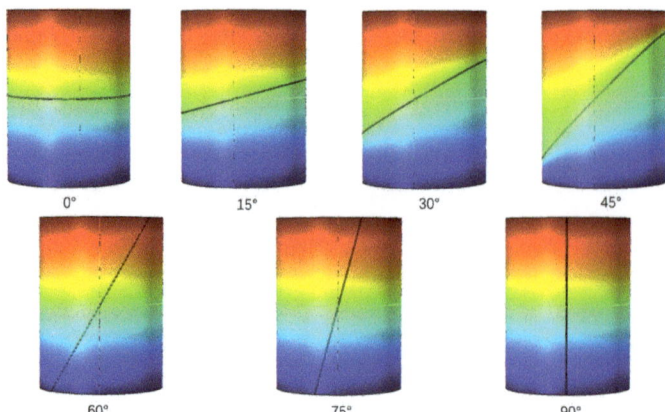

Figure 7. The potential distribution diagram of different fracture angle models (Different colors represent different potential differences).

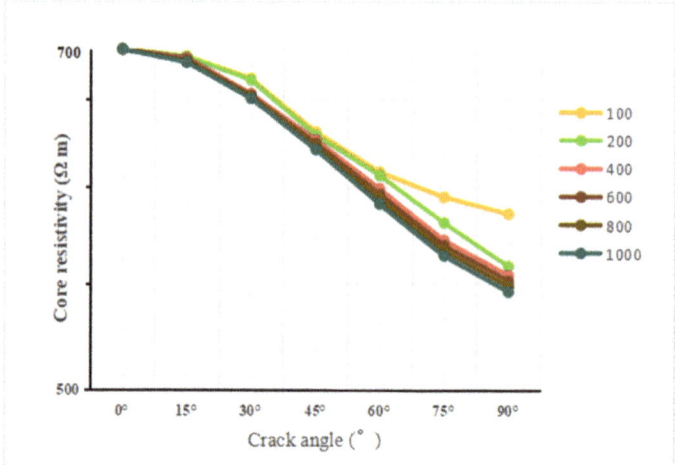

Figure 8. The relationship between the resistivity of the core and the fracture angle under different fracture widths.

3.1.2. Core Resistivity Response Characteristics of Fracture Length

In addition to considering the width and angle of the fracture, it is also necessary to consider the influence of the length of the fracture on the core resistivity. In the basic model, the models with fracture lengths of 4.5 cm, 5.5 cm, 6.5 cm, 7.5 cm, and 8.5 cm are constructed, respectively. The results for resistivity changes are shown in Figure 9. The analysis shows that, with an increase in fracture length, the core resistivity decreases, but when the fracture angle is 0°, that is, when the horizontal fracture is changed, the length of the fracture has the least influence on the core resistivity. When the fracture angle is vertical (the fracture angle is 90°), the length of the fracture has the greatest influence on the core resistivity. With the increase in the fracture angle, the influence of the fracture length on the core resistivity gradually increases.

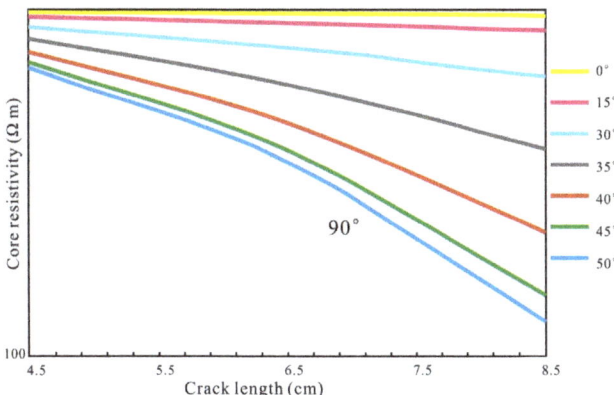

Figure 9. The relationship between the resistivity of the core and the length of the fracture at different fracture angles.

3.1.3. Core Resistivity Response Characteristics of Mud Resistivity

The influence of different mud types on core resistivity was also analyzed for mud types with different salinity. For fractures with different angles, the core resistivity response characteristics of mud resistivity are simulated by changing the mud resistivity at the fracture position, as shown in Figure 10. The results show that the resistivity of low-angle fractured cores is not sensitive to the change in mud resistivity. The increased fracture angle makes the core resistivity more sensitive to the mud resistivity. The resistivity of high-angle fractured cores increases significantly with the increase in mud resistivity. In the non-logarithmic scale diagram, it can be found that, the higher the fracture angle, the greater the degree of influence. The change in mud resistivity is not a simple linear change for the same angle of fracture.

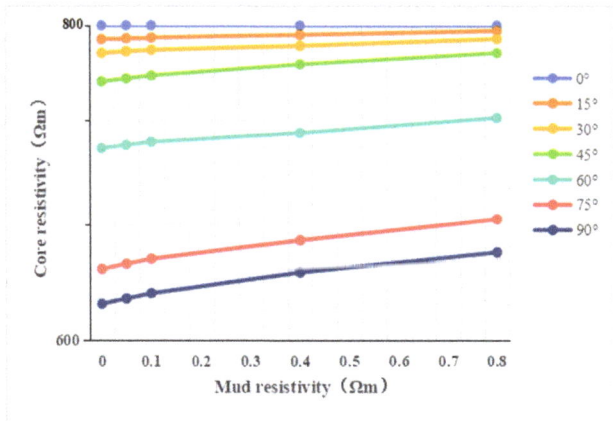

Figure 10. The relationship between the resistivity of the core and the resistivity of the mud at different fracture angles.

3.2. Resistivity Response Mechanism of Complex Fracture Model

3.2.1. Core Resistivity Response Characteristics of Parallel Fractures

The initial simulation in our investigation addresses the complex fracture scenario involving two parallel fractures. Within this simulation, we systematically vary the dip angle of the fractures from 15° to 90°, with a step size of 15°. Notably, the dip angle is defined as the angle between the horizontal line and the upper crack. Readers can refer to Figures 6 and 7, which show a schematic diagram of the seven dip angles. It is imperative to note that when the angle exceeds 90°, the crack assumes a symmetrical shape relative to the angle preceding 90°, thus warranting exclusive consideration of angles within 90°. Furthermore, the simulation encompasses a range of fracture widths, specifically 100 μm, 200 μm, 400 μm, 600 μm, 800 μm, and 1000 μm. The results are presented in Figure 11. Our analysis unveils a consistent decrease in core resistivity with increasing fracture angle. Notably, within the 30–60° interval of fracture dip angle, the core resistivity exhibits the most significant variation. Beyond a 60° angle, the resistivity demonstrates a tendency toward stabilization. Additionally, as fracture width increases, there is a discernible decrease in core resistivity, albeit with a minimal magnitude of change.

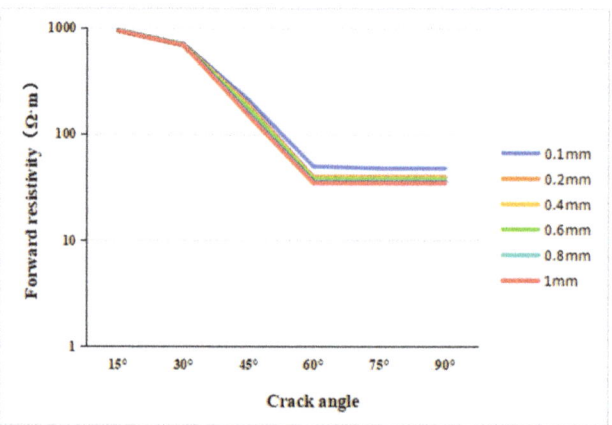

Figure 11. The relationship between the resistivity of the core and the fracture angle under different fracture widths in the parallel fracture model.

3.2.2. Core Resistivity Response Characteristics of Intersecting Fractures

For the simulation of cross cracks, the first is to consider different crack widths, fix one of the crack levels, and realize the simulation of mesh cracks by changing the angle between another crack and the middle crack. The calculation formula of resistivity is as follows:

$$R = U/J \times S \tag{9}$$

However, considering that the complex seam is simulated in a two-dimensional case, that is, because the area S in Equation (9) does not exist, the coefficient S that does not exist needs to be calibrated. The calibration method adopted by S is to remove the crack, known potential, and resistance (the resistance is the established matrix resistivity); a certain current value can be obtained, and then the current value and the current density are divided, and the coefficient between the two is calculated, which is the coefficient to be calibrated. Combined with the matrix resistivity of 1000 Ωm in the simulation of this study, the calibrations of 0.001, 0.1, 10, and 1000 Ωm were carried out, and the calibration coefficient in a range was obtained (Table 1). It should be noted that the selection of this value is based on the resistivity of the mud filtrate and the resistivity of the matrix. This calibration method is a geometric simplification method. According to the cross-slit experiments at different angles, it is found that the change in the angle in the core is actually a reflection of the change in the volume of the fluid. Therefore, the coefficient can be used to simplify the volume change, and the purpose is to extract the resistivity calculation formula of the complex seam.

It can be determined that the calculation coefficient of the matrix resistivity of 1000 Ωm is 0.08, which can be substituted into Equation (9) to calculate the resistivity, and thus the relationship diagram, as shown in Figure 12, is obtained: as the angle increases, the core resistivity decreases. The attenuation amplitude of the core resistivity within the angle of less than 45° is significantly higher than that of the angle of more than 45°. The attenuation of the resistivity is the largest when the angle is between 30° and 45°. When the angle is greater than 75°, the attenuation of the core resistivity is the smallest. The core resistivity is the smallest for the same fracture width when the angle is 90°. As the fracture width increases, the core resistivity decreases relatively.

Table 1. Coefficient calibration statistical table.

U	R	I	J	Coefficient
100	0.001	100,000	1,250,000	0.08
100	0.1	1000	12,500	0.08
100	10	10	125	0.08
100	1000	0.1	1.25	0.08

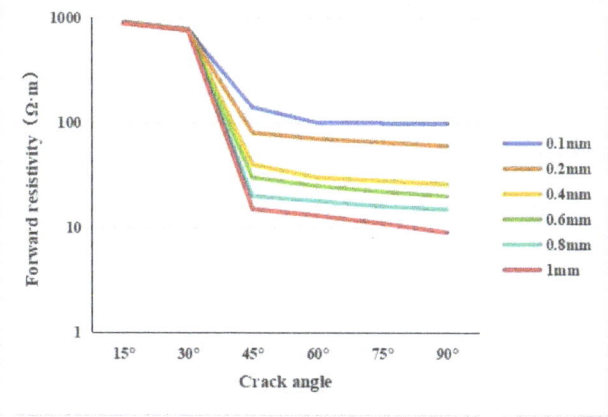

Figure 12. The plotted relationship between the resistivity of the core and the angle of the fracture under different fracture widths in the mesh fracture model with a fixed fracture level.

This study also simulated a given initial dip angle of a crack and then changed the angle of the two cracks. Because the simulated matrix resistivity does not change, the calibrated coefficient in Table 1 can still be used. In this simulation, the initial inclination angle is 30° and 60°, the corresponding angle is 15–165°, with 15° as the change step, and the crack width is 100 μm, 200 μm, 400 μm, 600 μm, 800 μm, and 1000 μm. The matrix resistivity is still 1000 Ωm, and the mud filtrate resistivity is 0.1 Ωm. The simulation results are shown in Figure 13 when the initial inclination angles are 30° and 60°. For the initial dip angle of 30° (Figure 13a), when the angle is between 15° and 105°, the core resistivity changes with 60° as the symmetry center. When the angle is greater than 75°, the core resistivity continues to increase with the increase in the angle, and between 105° and 120°, the core resistivity increases sharply. For the initial dip angle of 60° (Figure 13b), the core resistivity shows obvious fluctuation changes, and the core resistivity changes in the angle range of 75–90° and 150–165° are the most obvious compared with other angles. The change in core resistivity in the range of 75–165° is symmetrical. Similarly, with the increase in fracture width, the change range of core resistivity decreases. It should be noted that when the initial dip angle is 60°, the resistivity curve of the core resistivity with the angle is symmetrical when the angle is 75–160°. Then, if one of the two fractures is vertical, the resistivity curve changes with the angle of 90° as the symmetry center. As the angle increases, the core resistivity increases first and then decreases, showing a symmetrical change.

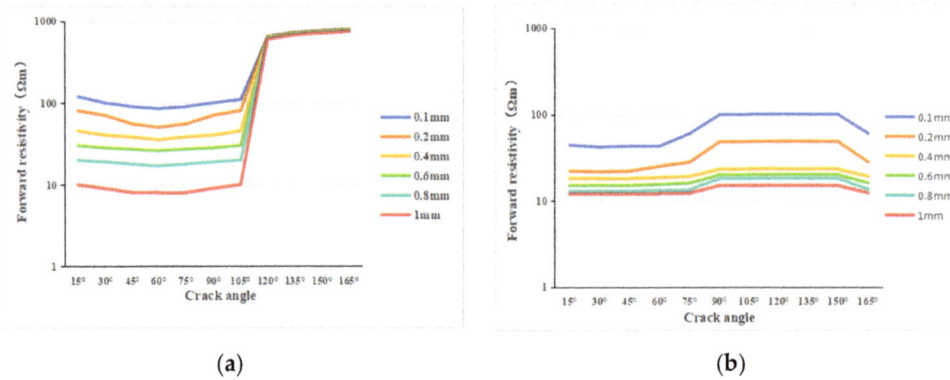

(a) (b)

Figure 13. The relationship between the resistivity of the core and the angle of the fracture under different fracture widths in the mesh fracture model under different initial dip angles. (**a**) The initial inclination angle is 30°; (**b**) the initial inclination angle is 60°.

In the realm of core resistivity analysis, the presence of a crack introduces an angle parameter ranging from 0 to 90°. Notably, when considering a uniform mud filtrate composition, core resistivity exhibits a gradual decline with increasing crack angle. This observed trend holds true for both single fractures and parallel fractures, highlighting a consistent behavior across varying fracture configurations. However, the dynamics shift when examining cross fractures, where core resistivity modulation is contingent upon the geometric relationship between the two fractures. Specifically, a notable decrease in core resistivity is observed in scenarios featuring high-angle fractures on both planes. Furthermore, when the sum of the dip angles of the two fractures totals less than 180°, a distinct pattern emerges: the core resistivity diminishes as the combined dip angles increase. This intricate interplay underscores the nuanced influence of fracture orientation on core resistivity responses, shedding light on the complex interactions within subsurface formations and their electrical properties.

4. Discussion

4.1. The Difference between Core Resistivity and Logging Resistivity

There is a difference between the response of core resistivity and the response of actual logging resistivity. The measurement angle of core resistivity is 90° different from the logging angle. As shown in Figure 14, the logging data correspond to the radial direction (Figure 14a), while the core resistivity is vertical (Figure 14b); that is, there is a difference of 90° between the simulation results and the response of logging data. For the response of logging data, taking a single fracture as an example, the resistivity response increases with the increase in fracture angle, and the corresponding core resistivity is a low-angle fracture, which is also consistent [33]. For the solution of fracture angle, many scholars use dual laterolog data. The research results show that for a single fracture, this method can distinguish the angle categories of fractures without considering mud invasion [21,26]. However, there are multiple solutions for multiple cracks, especially when the angle between the two cracks is large. In the actual logging data, the fractured reservoir is obviously affected by the invasion. Therefore, when using the dual laterolog data to calculate the fracture angle, it is also necessary to analyze whether the formation is invaded. Considering the gap between drilling time and logging time, the larger the porosity is, the more developed the fracture is, the faster the invasion is, the deeper the depth is, and the more obviously the resistivity logging formation is invaded.

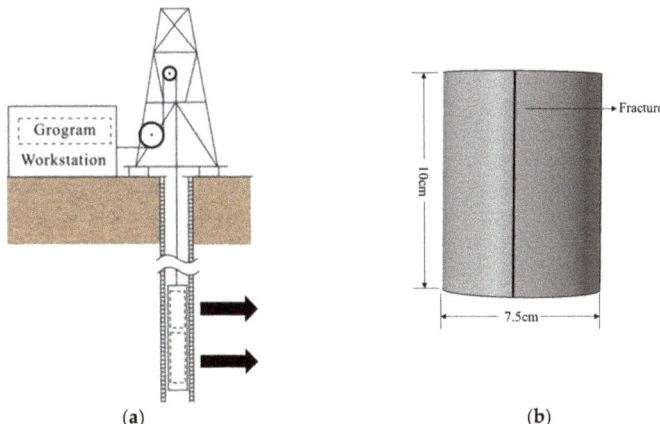

(a) (b)

Figure 14. Logging data and core resistivity measurement diagram. (**a**) Logging data measurement diagram (The black arrow represents the direction of the resistivity logging measurement.); (**b**) core resistivity measurement diagram.

4.2. The Core Resistivity Response of The Cross Fracture Is Affected by the Background Value of the Matrix and the Resistivity of the Mud

For complex reservoirs, there are also differences in rock mineral content. For different lithologies, resistivity logging response will also be affected. Therefore, it is considered to be studied by changing the matrix resistivity in the simulation. Based on the mesh fracture model with an initial dip angle of 60°, the matrix resistivity of 1000 Ωm is changed to the matrix resistivity of 450 Ωm, and then the simulation is carried out to explore the response law of the fracture angle and the fracture width to the core resistivity under the change in the matrix resistivity. The results are shown in Figure 15. Through comparative analysis, the simulation results of the matrix resistivity of 450 Ωm are the same as those of the matrix resistivity of 1000 Ωm, but there is a numerical difference in the measured core resistivity in the two cases. The difference in core resistivity in different angle fractures is different. For the same angle, the larger the fracture width, the smaller the influence of matrix resistivity change on core resistivity. On the contrary, the smaller the fracture width, the more sensitive to the change in matrix resistivity. Therefore, when using dual laterolog resistivity to solve the fracture angle, in addition to considering the influence of invasion, it is also necessary to consider the lithology of the reservoir. The conductivity difference of different lithologies will also affect the reliability of the calculation results.

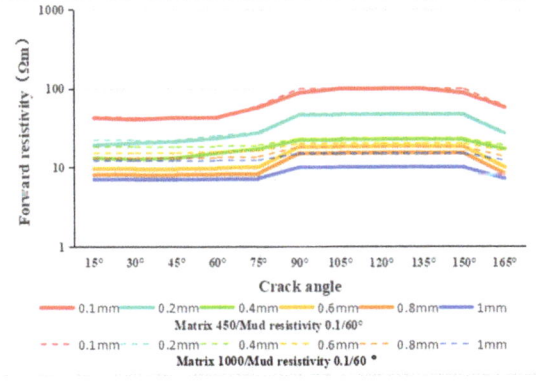

Figure 15. Comparison of mesh fracture simulation results under different matrix resistivity conditions.

Similarly, the resistivity of the mud will also have an impact on the measurement of core resistivity. For the mesh fracture model with an initial dip angle of 60°, the resistivity of the set 0.1 Ωm mud is changed to 0.5 Ωm, and the mesh fracture model is explored. In the case of changes in mud resistivity, the response of fracture angle and fracture width changes the core resistivity. Considering that the matrix resistivity of this simulation is 1000 Ωm, the resistivity calculation coefficient of 0.08 calibrated by the previous matrix 1000 Ωm can be used to calculate the resistivity of this simulation. The results obtained are compared with the simulation results of the mesh fracture with the initial dip angle of 60° and the mud resistivity of 0.1 Ωm (Figure 16). The overall law and change trend is the same as that of the mud resistivity of 0.1 Ωm, without any change, and the core resistivity value increases in multiples; when the mud resistivity increases from 0.1 Ωm to 0.5 Ωm, the corresponding core resistivity increases with the increase in fracture width, but it is not five times, which is different due to the change in fracture angle. The influence of mud resistivity change on core resistivity has a greater relationship with fracture width and a smaller relationship with fracture angle.

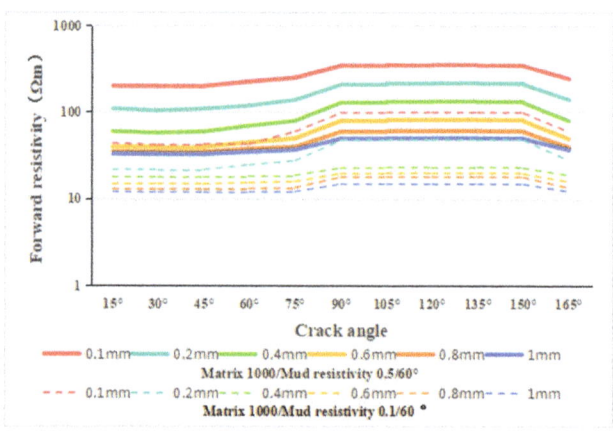

Figure 16. The comparison diagram shows the simulation results of the mesh seam under different mud filtrate resistivity conditions.

4.3. The Contribution of This Study and the Limitations of the Research

This work studies and summarizes the influence of a single fracture and two fractures on core resistivity when the fracture angle, width, and length change. Compared with previous studies, most research has been conducted from the perspective of logging, and analysis of the core is relatively rare.

Compared with previous work, this research studies the factors that can affect the resistivity of fractured cores, especially in the case of two fractures. The results also show that when there is only one fracture, the dual laterolog resistivity can be used to calculate the fracture angle, but when there are two non-parallel fractures, the traditional calculation method is not applicable. This study reveals this phenomenon through the results of numerical simulation. The same resistivity will correspond to different fracture angles, and it is difficult to determine the specific occurrence of the two fractures. At the same time, for different lithologies, the fracture angle is demonstrated by changing the resistivity of the matrix, which also shows the complexity of the fractured reservoir and needs to be simulated in combination with the actual work area background.

The work of this study also provides a theoretical basis for the interpretation of the oilfield site. For the fractured reservoir, the change in resistivity will be affected by the fracture. The results show that the angle of the fracture has the most obvious influence on the core resistivity, and the fracture angle also has an effect on the reservoir permeability. The results of this study provide more possibilities for the logging interpretation of fractured reservoirs and a further accurate understanding of resistivity logging. It is also helpful for saturation evaluation. After the fracture information is clarified, the correction model of the saturation calculation method can be determined according to the actual situation. The influence of fractures on the cementation index or the coefficient in the conductivity efficiency model can be further studied to improve the accuracy of reservoir evaluation and provide a theoretical basis and model support for the formulation of subsequent development plans.

The work of Reference [20] is aimed at the measurement of core resistivity, but it is aimed at the influence of the measurement method on resistivity response. In the study, the resistivity measurement of different fractured cores is carried out using the same method, aiming to reveal the influence of fractures on core resistivity. However, for the form of numerical simulation and the content of the study, the intersection point of the cross fracture is located in the center of the fracture, which is different from the corresponding situation of the actual logging data measurement. Therefore, the trend is usually consistent in the application of the research conclusion, and additional correction is needed. In the face of the actual data of the fracture development zone, there will be differences between the quantitative calculation results and the actual data. The research is mainly based on theoretical work and explores the influence of fracture parameters on core resistivity. In practical work, resistivity logging is affected by many factors, which is also supplemented in the introduction. At present, the fractured reservoir is extremely complex, and it is difficult to link the two. With the development and maturity of digital core technology, the characterization of fractures will gradually be refined [4,34].

5. Recommendations

For the actual field operation, the research results of this paper prove that it is feasible to use the dual laterolog resistivity to calculate the fracture angle for a single fracture, but for two fractures, there will be multiple solutions, which also shows that the fieldwork needs to consider the number of fractures when calculating the fracture angle, rather than calculating the dip angle only by resistivity logging. The results of this study can also provide suggestions for on-site work:

1. For fractured reservoirs, especially igneous reservoirs, imaging logging is necessary because when the fractures are crossed and irregular, using the resistivity method to determine the fracture occurrence is unreliable, and there are multiple solutions. At this time, imaging logging can be used to solve the fracture parameters finely, which provides more reliable data support for the subsequent calculation of permeability and saturation.
2. The discussion part of Reference [6] reveals the influence of the invasion phenomenon on resistivity measurement, and the fractured reservoir is more obviously affected by the invasion. In order to ensure the reliability of resistivity logging response value, the interval between drilling time and logging time should be shortened as much as possible.
3. The calculation of permeability can be corrected by referring to the angle of fracture, and the permeability model can be corrected according to the actual working area and the research results of this study, which can improve the reliability of reservoir evaluation.

6. Conclusions

This study used the finite element method to conduct numerical simulations of core resistivity responses in single-crack and complex-crack configurations. Using full-diameter cores containing cracks as models, we comprehensively considered various parameters related to the cracks and subsurface formations. Our results demonstrate that the inclination angle, width, and length significantly affect core resistivity for single cracks. An increase in crack angle under typical background values of resistivity and mud filtration leads to a decrease in core resistivity, whereas increasing crack width and length has the opposite effect. Moreover, there is a 90° difference between core resistivity and resistivity logging response, whereby an increase in fracture angle corresponds to a decrease in core resistivity but an increase in resistivity logging response. When two cracks are parallel for complex fractures, the pattern of core resistivity change mirrors that of a single crack. However, in cross fractures, the position of the two cracks plays a more prominent role than crack width and length in determining core resistivity. Calculating the crack angle using bilateral resistivity may lead to polytropic results, making it challenging to accurately determine the crack angle. Additionally, the resistivity response to varying lithology and mud filtrate is nonlinearly correlated with fracture production, necessitating the integration of fracture parameterization with actual logging responses to account for different geological backgrounds. The research results of this study show the influence of complex fractures on resistivity logging results and prove that, for a single fracture, the dual laterolog method can be used to calculate the fracture angle. At the same time, it shows that the traditional method has multiple solutions when calculating the angle of cross fractures, which provides a new idea for field interpretation in the face of abnormal resistivity changes. Our research provides a reference and theoretical basis for interpreting hydrocarbon content in fractured reservoirs based on logging data.

Author Contributions: Conceptualization, H.Z. and Z.Z.; methodology, H.Z., Z.Z. and S.F.; software, S.F.; validation, J.G. and H.Z.; investigation, H.Z. and S.F.; resources, Z.Z. and S.F.; data curation, H.Z. and C.W.; writing—original draft preparation, H.Z.; writing—review and editing, H.Z. and Z.Z.; supervision, Z.Z.; project administration, Z.Z.; funding acquisition, Z.Z. and J.G. All authors have read and agreed to the published version of the manuscript.

Funding: This work was financially sponsored by the Open Fund of Key Laboratory of Exploration Technologies for Oil and Gas Resources, Ministry of Education (No. K2023–02).

Data Availability Statement: Data are contained within the article.

Acknowledgments: The authors would also like to thank the anonymous reviewers for their valuable comments and suggestions. The data of the figures and tables in this paper can be obtained by contacting the corresponding author.

Conflicts of Interest: The authors declare no conflicts of interest.

References

1. Lee, M.W.; Collett, T.S. Gas hydrate saturations estimated from fractured reservoir at Site NGHP-01-10, Krishna-Godavari Basin, India. *J. Geophys. Res. Solid Earth* **2009**, *114*. [CrossRef]
2. Lai, J.; Wang, G. Fractal analysis of tight gas sandstones using high-pressure mercury intrusion techniques. *J. Nat. Gas Sci. Eng.* **2015**, *24*, 185–196. [CrossRef]
3. Li, P.; Zheng, M.; Bi, H.; Wu, S.; Wang, X. Pore throat structure and fractal characteristics of tight oil sandstone: A case study in the Ordos Basin, China. *J. Pet. Sci. Eng.* **2017**, *149*, 665–674. [CrossRef]
4. Cai, J.; Wei, W.; Hu, X.; Wood, D.A. Electrical conductivity models in saturated porous media: A review. *Earth-Sci. Rev.* **2017**, *171*, 419–433. [CrossRef]
5. Wang, G.; Carr, T.R. Organic-rich Marcellus Shale lithofacies modeling and distribution pattern analysis in the Appalachian Basin. *AAPG Bull.* **2013**, *97*, 2173–2205. [CrossRef]
6. Zhao, Q.; Guo, J.; Zhang, Z. A method for judging the effectiveness of complex tight gas reservoirs based on geophysical logging data and using the L block of the Ordos Basin as a case study. *Processes* **2023**, *11*, 2195. [CrossRef]
7. Li, M.; Zhang, C. An Improved Method to Accurately Estimate TOC of Shale Reservoirs and Coal-Measures. *Energies* **2023**, *16*, 2905. [CrossRef]
8. Chu, H.; Liao, X.; Chen, Z.; Zhao, X.; Liu, W.; Zou, J. Pressure transient analysis in fractured reservoirs with poorly connected fractures. *J. Nat. Gas Sci. Eng.* **2019**, *67*, 30–42. [CrossRef]
9. Yang, Y.; Yang, J.; Yang, G.; Tao, S.; Ni, C.; Zhang, B.; He, X.; Lin, J.; Huang, D.; Liu, M.; et al. New research progress of Jurassic tight oil in central Sichuan Basin, SW China. *Pet. Explor. Dev. Online* **2016**, *43*, 954–964. [CrossRef]
10. Khoshbakht, F.; Azizzadeh, M.; Memarian, H.; Nourozi, G.H.; Moallemi, S.A. Comparison of electrical image log with core in a fractured carbonate reservoir. *J. Pet. Sci. Eng.* **2012**, *86*, 289–296. [CrossRef]
11. Aghli, G.; Moussavi-Harami, R.; Tokhmechi, B. Integration of sonic and resistivity conventional logs for identification of fracture parameters in the carbonate reservoirs (A case study, Carbonate Asmari Formation, Zagros Basin, SW Iran). *J. Pet. Sci. Eng.* **2020**, *186*, 106728. [CrossRef]
12. Luthi, S.M.; Souhaite, P. Fracture apertures from electrical borehole scans. *Geophysics* **1990**, *55*, 821–833. [CrossRef]

13. Wang, D. Study on Interpretation Method of Micro-Resistivity Imaging Logging in Fractured Reservoir. Ph.D. Thesis, China University of Petroleum, Beijing, China, 2001.
14. Ke, S.; Sun, G. On quantitative evaluation of fractures by STAR Imager II. *Well Logging Technol.* **2002**, *26*, 101–103.
15. Chen, A. Numerical simulation and analysis of microresistivity imaging tool. *Nat. Gas Ind.* **2006**, *26*, 83–85.
16. Cao, Y. The Numerical Stimulation and Experimental Verification of Electrical Imaging Logging in Fractured Formations. Master's Thesis, Yangtze University, Jingzhou, China, 2014.
17. Ponziani, M.; Slob, E.; Luthi, S.; Bloemenkamp, R.; Nir, I.L. Experimental validation of fracture aperture determination from borehole electric microresistivity measurements. *Geophysics* **2015**, *80*, 175–181. [CrossRef]
18. Ammar, A.I. Development of numerical model for simulating resistivity and hydroelectric properties of fractured rock aquifers. *J. Appl. Geophys.* **2021**, *189*, 104319. [CrossRef]
19. Epov, M.I.; Moskaev, I.A.; Nechaev, O.V.; Glinskikh, V.N. Effect of Tilted Uniaxial Electrical Anisotropy Parameters on Signals of Electric and Electromagnetic Logging Soundings according to Results of Numerical Simulation. *Russ. Geol. Geophys.* **2023**, *64*, 735–742. [CrossRef]
20. He, J.; Liu, T.; Wen, L.; He, T.; Li, M.; Li, J.; Wang, L.; Yao, X. Numerical simulation analysis of difference from a radial resistivity testing method for cylindrical cores and a conventional testing method. *Mathematics* **2022**, *10*, 2885. [CrossRef]
21. Tan, M.; Wang, P.; Li, J.; Liu, Q.; Yang, Q. Numerical simulation and fracture evaluation method of dual laterolog in organic shale. *J. Appl. Geophys.* **2014**, *100*, 1–13. [CrossRef]
22. Deng, S.; Li, L.; Li, Z.; He, X.; Fan, Y. Numerical simulation of high-resolution azimuthal resistivity laterolog response in fractured reservoirs. *Pet. Sci.* **2015**, *12*, 252–263. [CrossRef]
23. Liu, D.; Ma, Z.; Xing, X.; Li, H.; Guo, Z. Numerical simulation of LWD resistivity response of carbonate formation using self-adaptive hp-FEM. *Appl. Geophys.* **2013**, *10*, 97–108. [CrossRef]
24. Kang, Z.; Li, X.; Ni, W.; Li, F.; Hao, X. Using logging while drilling resistivity imaging data to quantitatively evaluate fracture aperture based on numerical simulation. *J. Geophys. Eng.* **2021**, *18*, 317–327. [CrossRef]
25. Zhao, J.; Sun, J.; Liu, X.; Chen, H.; Cui, L. Numerical simulation of the electrical properties of fractured rock based on digital rock technology. *J. Geophys. Eng.* **2013**, *10*, 055009. [CrossRef]
26. Wang, N.; Li, K.; Sun, J.; Wang, D.; He, X.; Xiang, Z.; Liu, H.; Wang, P. Research on dual lateral log simulation of shale bedding fractures under different influencing conditions. *Front. Energy Res.* **2023**, *11*, 1249985. [CrossRef]
27. Kim, J.; Hong, C.; Kim, J.; Chong, S. Theoretical and Numerical Study on Electrical Resistivity Measurement of Cylindrical Rock Core Samples Using Perimeter Electrodes. *Energies* **2021**, *14*, 4382. [CrossRef]
28. Aghli, G.; Moussavi-Harami, R.; Mohammadian, R. Reservoir heterogeneity and fracture parameter determination using electrical image logs and petrophysical data (a case study, carbonate Asmari Formation, Zagros Basin, SW Iran). *Pet. Sci.* **2020**, *17*, 51–69. [CrossRef]
29. Geuzaine, C.; Remacle, J.F. Gmsh: A 3-D finite element mesh generator with built-in pre-and post-processing facilities. *Int. J. Numer. Methods Eng.* **2009**, *79*, 1309–1331. [CrossRef]
30. Zhao, X.; Lee, Y.Y.; Liew, K.M. Free vibration analysis of functionally graded plates using the element-free kp-Ritz method. *J. Sound Vib.* **2009**, *319*, 918–939. [CrossRef]
31. Beirão da Veiga, L.; Brezzi, F.; Marini, L.D.; Russo, A. Virtual element method for general second-order elliptic problems on polygonal meshes. *Math. Models Methods Appl. Sci.* **2016**, *26*, 729–750. [CrossRef]
32. Hao, R.; Lu, Z.; Ding, H.; Chen, L. A nonlinear vibration isolator supported on a flexible plate: Analysis and experiment. *Nonlinear Dyn.* **2022**, *108*, 941–958. [CrossRef]
33. Guo, J.; Zhang, Z.; Xiao, H.; Zhang, C.; Zhu, L.; Wang, C. Quantitative interpretation of coal industrial components using a gray system and geophysical logging data: A case study from the Qinshui Basin, China. *Front. Earth Sci.* **2023**, *10*, 1031218. [CrossRef]
34. Zhu, L.; Zhang, C.; Zhang, C.; Zhou, X.; Zhang, Z.; Nie, X.; Liu, W.; Zhu, B. Challenges and prospects of digital core-reconstruction research. *Geofluids* **2019**, *2019*, 7814180. [CrossRef]

Disclaimer/Publisher's Note: The statements, opinions and data contained in all publications are solely those of the individual author(s) and contributor(s) and not of MDPI and/or the editor(s). MDPI and/or the editor(s) disclaim responsibility for any injury to people or property resulting from any ideas, methods, instructions or products referred to in the content.

Article

Intelligent Identification Method for the Diagenetic Facies of Tight Oil Reservoirs Based on Hybrid Intelligence—A Case Study of Fuyu Reservoir in Sanzhao Sag of Songliao Basin

Tao Liu [1], Zongbao Liu [2], Kejia Zhang [1,*], Chunsheng Li [1], Yan Zhang [1], Zihao Mu [1], Fang Liu [1], Xiaowen Liu [2], Mengning Mu [1] and Shiqi Zhang [2]

1 School of Computer & Information Technology, Northeast Petroleum University, Daqing 163318, China; ltao0403@163.com (T.L.)
2 School of Earth Sciences, Northeast Petroleum University, Daqing 163318, China
* Correspondence: zkj@nepu.edu.cn

Abstract: The diagenetic facies of tight oil reservoirs reflect the diagenetic characteristics and micropore structure of reservoirs, determining the formation and distribution of sweet spot zones. By establishing the correlation between diagenetic facies and logging curves, we can effectively identify the vertical variation of diagenetic facies types and predict the spatial variation of reservoir quality. However, it is still challenging work to establish the correlation between logging and diagenetic facies, and there are some problems such as low accuracy, high time consumption and high cost. To this end, we propose a lithofacies identification method for tight oil reservoirs based on hybrid intelligence using the Fuyu oil layer of the Sanzhao depression in Songliao Basin as the target area. Firstly, the geological characteristics of the selected area were analyzed, the definition and classification scheme of diagenetic facies and the dominant diagenetic facies were discussed, and the logging response characteristics of various diagenetic facies were summarized. Secondly, based on the standardization of logging curves, the logging image data set of various diagenetic facies was built, and the imbalanced data set processing was performed. Thirdly, by integrating CNN (Convolutional Neural Networks) and ViT (Visual Transformer), the C-ViTM hybrid intelligent model was constructed to identify the diagenetic facies of tight oil reservoirs. Finally, the effectiveness of the method is demonstrated through experiments with different thicknesses, accuracy and single-well identification. The experimental results show that the C-ViTM method has the best identification effect at the sample thickness of 0.5 m, with Precision of above 86%, Recall of above 90% and $F1$ score of above 89%. The calculation result of the Jaccard index in the identification of a single well was 0.79, and the diagenetic facies of tight reservoirs can be identified efficiently and accurately. At the same time, it also provides a new idea for the identification of the diagenetic facies of old oilfields with only logging image data sets.

Keywords: tight oil reservoirs; diagenetic phases; log recognition; hybrid intelligence; reservoir prediction

1. Introduction

With the continuous exploration and development of unconventional oil and gas resources, tight oil reservoirs have attracted increasing attention [1,2]. China's tight oil reservoirs are mainly distributed in Songliao Basin, Ordos Basin, Bohai Bay Basin and Junggar Basin [3,4]. Tight oil reservoirs are deeply buried and have the characteristics of coexistence of source rock and reservoir, low porosity and low permeability, complex pore structure and strong heterogeneity after complex diagenetic processes [5–9]. The analysis of the influence of diagenetic change on reservoir quality can effectively guide the exploration and development of tight oil reservoirs, so the study of diagenetic facies is of great significance for the quality evaluation of tight oil reservoirs and "sweet spots" prediction [10,11].

At present, there is no standard definition for diagenetic facies. The meaning of diagenetic facies has been expressed by scholars from different perspectives, but it is generally believed that diagenetic facies are the integration of diagenetic type, diagenetic degree and diagenetic mineral combination [12,13]. The high cost and discontinuity of coring make it relatively difficult to study the diagenetic facies space [14,15]. For this purpose, many scholars have utilized the multidimensional nature and continuity of logging data to establish the correlation model between diagenetic facies and logging data, identify diagenetic facies and screen favorable reservoirs [16]. Common methods for identifying diagenetic facies logging include rendezvous diagram method, spider diagram method (also known as radar chart method), diagenetic facies strength calculation method and mathematical method. The rendezvous diagram method establishes the identification criteria of each type of diagenetic facies by selecting different types and quantities of logging curves. Ran [17] and Shi [18] et al. used the logging rendezvous diagram method to quantitatively characterize the diagenetic facies of tight oil reservoirs in Ordos Basin and established the logging quantitative identification standard of diagenetic facies. The spider diagram method maps the logging responses of various types of diagenetic facies to the axis to identify diagenetic facies types. Lai [19] et al. proposed the method for identifying diagenetic facies with spider diagram logging and distinguished different types of diagenetic facies with graphs. The identification plates of the rendezvous diagram method and spider diagram method are simple to make, but require a lot of time for experienced domain experts to complete, and the identification effect is affected by the number of logging curves. The diagenetic facies strength calculation method achieves quantitative characterization of diagenetic facies by calculating diagenetic strength parameters such as compactional and cementational porosity loss. Ozkan [20] et al. calculated the intergranular volume, compactional porosity loss and cementational porosity loss based on established point data, and realized diagenetic facies type judgment and reservoir quality evaluation in combination with logging data. This method has a high identification accuracy, but requires a large amount of experimental data support and high cost. Based on the logging response characteristics, the mathematical method realizes the identification of typical diagenetic facies types by identifying the morphological characteristics of logging curves. It mainly includes principal component analysis and neural network prediction methods. Cui [21] and Zhu [22] et al. established a diagenetic facies prediction model based on principal component analysis, but this method only applies to the case where the degree of overlap is less than 15%. Qi [5] et al. proposed a logging identification method for the diagenetic facies of tight oil reservoirs based on CNN. This method completes the identification of diagenetic facies type by extracting local features from logging curve images. However, due to the insufficient learning of the overall features of logging curve images by CNN and the long training time, it still needs to be further improved. In order to improve the deficiency of CNN algorithm in learning the overall features of images, Dosovitskiy [23] et al. proposed the Vision Transformer (ViT) method and applied the image patch sequence converter to complete the task of image classification and identification.

In summary, the existing methods have problems such as dependence on domain experts, high time cost and the need to improve identification accuracy. To this end, we referred to the hybrid intelligent algorithm design idea of "complementary advantages" [23–25] and proposed the intelligent identification method for the diagenetic facies of tight oil reservoirs based on the integration of CNN and Vision Transformer, C-ViTM. The overall process of the method is shown in Figure 1. The hybrid intelligent algorithm idea of this method is mainly reflected in the following: the local features of logging curves in samples are extracted with CNN, and the ViT is applied to learn the global features of each logging curve, realizing the rapid and accurate intelligent identification of diagenetic facies.

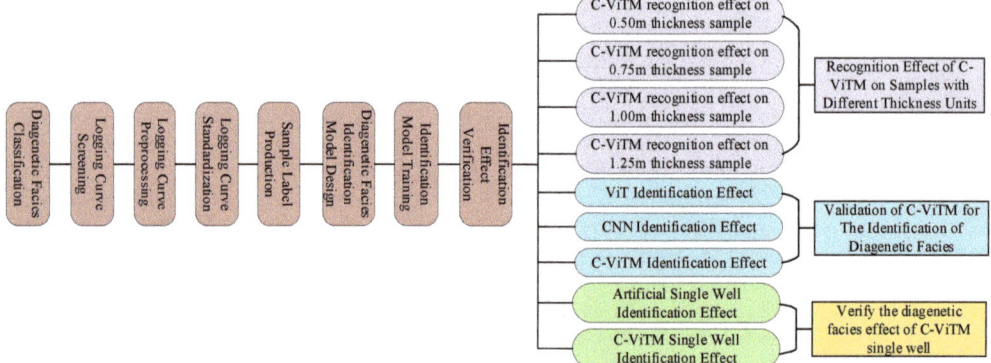

Figure 1. C-ViTM method process.

The remaining parts of this study are organized as follows. The second section elaborates the geological setting. The diagenetic facies types and logging response characteristics are analyzed in the third section. The fourth section provides a detailed introduction to the C-ViTM method. The experimental scheme is described in the fifth section. The experimental results are discussed in the sixth section. The seventh section is about the summary of the research work and the prospects for future work.

2. Geological Setting

Fuyu reservoir (K1q4) at the top of Quantou Formation (K1q) is a typical tight oil reservoir in Sanzhao Sag of north Songliao Basin, China, and it belongs to the lacustrine delta sedimentary system [26], as shown in Figure 2. The depth of reservoir rock is 1043–2302 m, and the diagenesis mainly includes compaction, cementation and dissolution, and the lithology is mainly siltstone, mudstone and fine sandstone [27,28]. Due to strong compaction, Fuyu reservoir has a low porosity and permeability, with an average porosity of 10.8% and air permeability of 0.64 mD [29]. There is a good linear relationship between porosity and permeability, indicating that Fuyu reservoir is a typical porous tight sandstone reservoir [30].

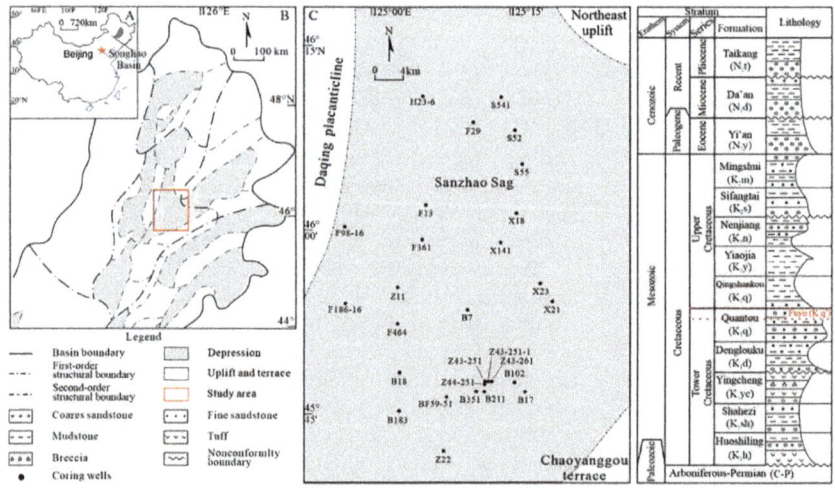

Figure 2. Structural locations of Fuyu reservoir of Sanzhao sag in Songliao Basin: (**A**) Location of Songliao Basin; (**B**) Study area Fuyu oil layer location; (**C**) Location of research wells.

Fuyu reservoir in Sanzhao Sag has the characteristics of fine particle size and high shale content [31]. Generally, reservoir compaction strength is positively correlated with the burial depth of sediments [28]. By observing cast thin sections, it is found that due to compaction, the brittle particles are broken, the clastic particles are mainly in linear contact, and the long particles are arranged semi-directionally, as shown in Figure 3a. The types of cement in the reservoir include carbonate cement, authigenic clay minerals, quartz secondary enlargement, authigenic feldspar, etc., all of which produce destructive effects on the physical properties of the reservoir, as shown in Figure 3b,d,e. Through the dissolution of feldspar, lithic fragment, calcite cement, argillaceous matrix and other components, new pores are created or the original pores are enlarged, and the physical properties of the reservoir are improved, as shown in Figure 3c,e,f.

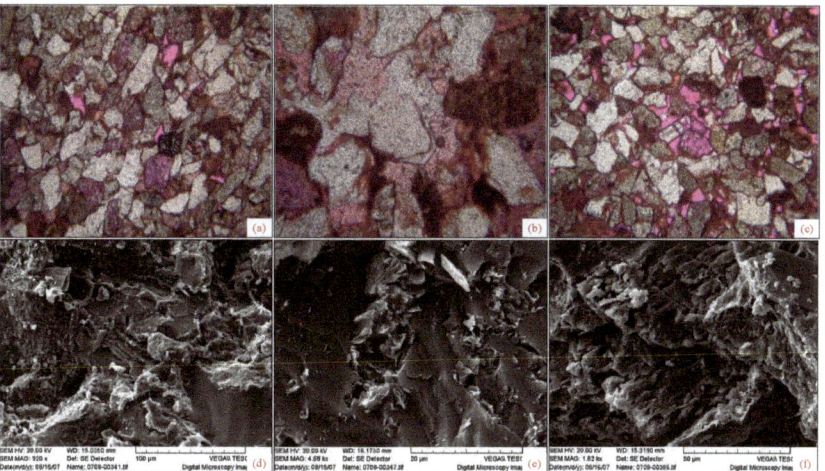

Figure 3. Photo of Fuyu reservoir in Sanzhao Sag: (**a**) Fang 186-32 Well, 1848.09 m, cast thin section, unipolar light, ×40, feldspar fine sandstone. Due to compaction, the clastic particles are in linear contact, and the long particles are arranged semi-directionally; (**b**) Fang 188-138 Well, 1789.4 m, cast thin sheet, unipolar light, ×100, feldspathic lithic fine sandstone, quartz authigenic enlargement, calcite stained. It is presumed that calcite cementation is later than quartz secondary enlargement; (**c**) Fang 186-132 Well, 1847.0 m, cast thin sheet, unipolar light, ×40, lithic feldspar fine sandstone. Interparticle and intragranular dissolution pores are visible, feldspar particles undergo dissolution, and calcite is stained; (**d**) Fang 186-32 Well, 1831.59 m, ×920, intergranular calcite cement; (**e**) Fang 186-32 Well, 1835.6 m, ×4680. Soda feldspar particles undergo dissolution to form intragranular dissolution pores, which are filled with lamellar illite; (**f**) Fang 188-138 Well, 1793.0 m, ×1820. Feldspar particles undergo dissolution to form secondary pores, with filamentous illite visible on the surface of particles.

3. Diagenetic Facies Classification and Logging Response Analysis

This section elaborates the definition and classification scheme of diagenetic facies. Based on the screening of logging curves, the logging response characteristics of different types of diagenetic facies were analyzed.

3.1. Definition of Diagenetic Facies

As a response to the combined results of different diagenetic histories, diagenetic facies reflect the comprehensive features of diagenetic types, diagenetic degrees, diagenetic mineral composition types and their evolution, and determine the formation and distribution of sweet spots in tight reservoirs, and it is an important basis for the evaluation of tight reservoirs [21,32–34]. Zou et al. [13] defined diagenetic facies as a comprehensive description of sedimentary diagenesis and diagenetic mineral evolution under diagenesis and tectonism. Zhang et al. [35] defined diagenetic facies as the product of diagenesis and evolution of sediments under fluid and tectonic action, which is a comprehensive characterization of minerals such as rock particles, cements, fabrics and fracture-caves. Duan et al. [36] defined diagenetic facies as the product of sediments reflecting petrological and geochemical characteristics after diagenetic and tectonic processes in specific sedimentary and physicochemical environments, including comprehensive characteristics such as rock particles, cements, fabrics, pores and fractures. We define diagenetic facies as the stratigraphic unit of sediments that reflects the diagenetic degree, diagenetic type and diagenetic mineral composition through fluid, tectonic and diagenetic processes in sedimentary and physicochemical environments, which can be identified based on thin section analysis and logging response, including compaction, lithology, cementation and dissolution [2,20,21,37–39].

3.2. Classification of Diagenetic Facies

The diagenetic facies types of Fuyu reservoir in Sanzhao Sag were classified based on compaction, lithology, cementation and dissolution. The compaction effect increased with the burial depth, and the porosity decreased from 25.5% to 17.5% when the burial depth reached 1800 m. Cements such as carbonate minerals and quartz secondary enlargement fill the reservoir, which inhibits the compaction and pressure dissolution to some extent. The porosity increased from less than 15% to 20.7% when the burial depth exceeded 1800 m, because the dissolution of feldspar, rock debris and carbonate cement results in the formation of interparticle dissolution pores and intragranular dissolution pores to form secondary pore development zones, which improves the reservoir performance [31]. According to the buried depth of 1800 m, Fuyu reservoir compaction was divided into weak compaction (buried depth ≤ 1800 m) and medium to strong compaction (buried depth > 1800 m). According to the logging response, the lithology is mainly divided into fine sandstone, siltstone and mudstone [3]. Cementation and dissolution are divided into weak cementation dissolution and weak dissolution cementation according to the action strength. In addition, diagenetic facies whose lithology type is mudstone have poor reservoir performance, so they are classified into one class. According to the above standards, the diagenetic facies types of Fuyu reservoir in Sanzhao Sag are divided as follows: weakly compacted weakly cemented dissolved siltstone phase (Wip), weakly compacted weakly cemented dissolved fine sandstone phase (Wap), medium to strong compaction of weakly cemented dissolved fine sandstone phase (Map), medium to strong compaction of weakly cemented dissolved siltstone phase (Mip), medium to strong compaction of weakly dissolved colluvial fine sandstone phase (Msap) and mudstone phase (Mp).

Wip: buried depth ≤ 1800 m, mainly siltstone, with loose samples and good particle sorting. The pores are relatively developed, mainly interparticle dissolution pores, with a maximum of 60 μm, and have connectivity. There are a few calcite cements, which are distributed in a scattered way. The rock exhibits the oil impregnation phenomenon and good reservoir properties.

Wap: buried depth ≤ 1800 m, mainly fine sandstone, with loose samples and good particle sorting. The pores are relatively developed, mainly interparticle dissolution pores, with a maximum of 60 μm, and have connectivity. There are a few calcite-filled interparticle metasomatic particles. The rock exhibits the oil impregnation phenomenon and good reservoir properties.

Map: buried depth > 1800 m, mainly fine sandstone, with loose samples and good particle sorting. The pores are relatively developed, mainly interparticle dissolution pores, with a maximum of 80 μm, and have connectivity. There are a few calcite-filled interparticle metasomatic particles. The rock exhibits the oil impregnation phenomenon and good reservoir properties.

Mip: buried depth > 1800 m, mainly siltstone, with loose samples and good particle sorting. The dissolution pores are relatively developed, mainly interparticle dissolution pores, granular dissolution pores and intragranular dissolution pores. The interparticle dissolution pores can reach up to 50 μm, and have certain connectivity. There are calcite-filled interparticle metasomatic particles. The rock exhibits good reservoir properties.

Msap: buried depth > 1800 m, mainly fine sandstone, with loose samples and general particle sorting. The pores are relatively developed, mainly interparticle dissolution pores, with a maximum of 60 μm, and have certain connectivity. There are calcite-filled interparticle metasomatic particles. The rock exhibits good reservoir properties.

Mp: mainly mudstone, with relatively dense or dense samples and general particle sorting. The pores are undeveloped and unevenly distributed, and there are interparticle dissolution pores and muddy micropores, which have poor connectivity. The interstitial materials are argillaceous, calcite and quartz secondary enlargement. The rock exhibits poor reservoir performance.

According to the reservoir properties of different diagenetic facies types, Class I reservoir, Class II reservoir and Class III reservoir were classified. Class I reservoir corresponds to Wip, Wap and Map. Class II reservoir corresponds to Mip and Msap. Class III reservoir corresponds to Mp.

3.3. Diagenetic Facies Logging Response Characteristics

In order to establish the correlation between diagenetic facies and logging, and analyze the logging response characteristics of various diagenetic facies, the importance of each logging curve was scored by using the decision tree scoring method. Gamma ray (GR), spontaneous potential (SP), borehole compensated acoustilog (AC), deep lateral resistivity (RLLD), shallow lateral resistivity (RLLS) and caliper (CAL) were selected as the response curves [40].

Wip: GR shows low-amplitude micro-tooth shape, RLLD and RLLS present small-amplitude frame shape, and AC has low-amplitude micro-tooth shape characteristics. Wap: GR has a stable shape, and RLLD and RLLS exhibit a high-amplitude bell shape. Map: GR shows low-amplitude smoothness, RLLD and RLLS are box or bell shaped, and AC is in a symmetrical tooth shape. Mip: GR shows low-amplitude smoothness, RLLD and RLLS values are lower, and AC exhibits a low-amplitude finger type. Msap: GR shows low-amplitude smoothness, RLLD and RLLS are box or toothed curves, and AC is finger shaped. Mp: GR value is higher, RLLD and RLLS values are lower, and AC is bell shaped. See Figure 4 for details.

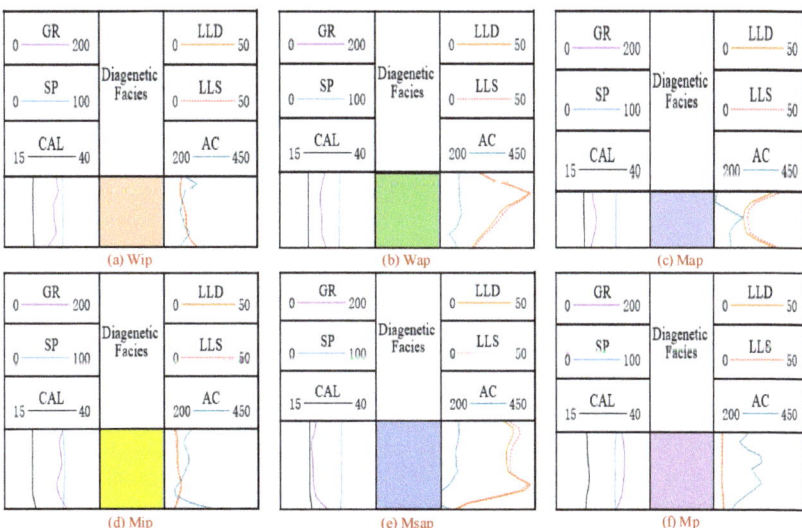

Figure 4. Logging response characteristics of different diagenetic facies.

4. C-ViTM Method

In this section, the research process of the C-ViTM method is introduced in detail, including five steps of logging data analysis, data set establishment, imbalanced data set processing, diagenetic facies identification model design and model training.

4.1. Logging Data Analysis

Logging data and core analysis data provide accurate lithology and physical property information, and they are the main basis for diagenesis and diagenetic facies logging identification [3,40]. The cumulative logging depth of 30 wells in Fuyu reservoir is 3064.5 m, as shown in Table 1. The well locations in the study area are shown in Figure 1.

Table 1. Log data statistics of Fuyu reservoir in Sanzhao depression.

Order	Well	Coring Depth (m)		Length (m)	Order	Well	Coring Depth (m)		Length (m)
		Top	Bottom				Top	Bottom	
1	B7	1872.700	2081.100	208.40	16	F464	1835.039	1939.989	104.95
2	B17	1858.025	2043.475	185.45	17	H23-6	1800.020	1831.170	31.15
3	B18	1836.001	2139.951	303.95	18	S52	1719.000	1872.000	153.00
4	B102	1914.500	1963.400	48.90	19	S541	1818.025	1946.975	128.95
5	B183	1895.012	1906.962	11.95	20	S55	1754.000	1793.950	39.95
6	B211	1775.000	1792.550	17.55	21	X18	1943.025	2021.975	18.95
7	B351	1982.981	2021.181	38.20	22	X21	2182.000	2228.000	46.00
8	BF59-51	1720.690	1854.940	134.25	23	X23	2068.000	2138.000	70.00
9	F13	1787.300	1998.800	211.50	24	X141	2030.000	2095.981	69.95
10	F27	1755.125	1843.125	88.00	25	Z11	1810.000	2004.700	194.70
11	F29	1843.000	1875.000	32.00	26	Z22	1695.800	1798.900	103.10
12	F98-16	1760.000	1870.950	110.95	27	Z43-251	1800.030	1861.830	61.80
13	F186-16	1920.040	2024.990	104.95	28	Z43-251-1	1800.009	1876.959	76.95
14	F188-138	1767.100	1858.050	90.95	29	Z43-261	1800.049	1885.999	85.95
15	F361	1765.325	1890.475	125.15	30	Z44-251	1773.950	1880.900	106.95

The quantity and thickness of various diagenetic facies were preliminarily counted according to the logging curve data of the above 30 wells, and the statistical results are shown in Figure 5. According to the statistical results, there is an imbalance in the data of various diagenetic facies, among which the number of Mip is the highest (486), and the number of Wap is the lowest (57), with a quantity difference of 429. The maximum thickness of Mp is 753.45 m, and the minimum thickness of Wap is 68.75 m, with a thickness difference of 684.70 m. According to the thickness statistics, the thickness of various diagenetic facies is mainly in the range of 0 m–1.50 m. The external factor leading to this result may be that the logging data are not standardized, so the standardization of logging curves is necessary in the data establishment process to verify the accuracy of logging data.

4.2. Establishment of the Diagenetic Facies Image Data Set

To ensure the accuracy of the logging data, the GeoSoftwareSuite9.1 software was applied to standardize the logging data. The standardized results of Well f188-138 are shown in Figure 6. After the completion of standardization, the Plot method was used to convert logging curve data into logging curve images, and the Resize method was applied to convert logging curve images into a 224-pixel × 224-pixel logging curve image data set. In addition, based on the statistical results of diagenetic facies thickness (Figure 5a), we sampled the logging curve images according to the longitudinal uniform thickness units of 0.50 m, 0.75 m, 1.00 m and 1.25 m to obtain the best diagenetic facies identification effect. The obtained image sample data set is shown in Table 2 (1.50 m samples are not displayed due to insufficient quantity). According to Table 2, there is an imbalance in the number of samples of different diagenetic facies types. This imbalance will result in higher identification accuracy of diagenetic facies type (Mp) with a large number of samples, and lower identification accuracy of diagenetic facies type (Wap) with a small number of samples. For this reason, we conducted imbalance processing on the sample data set.

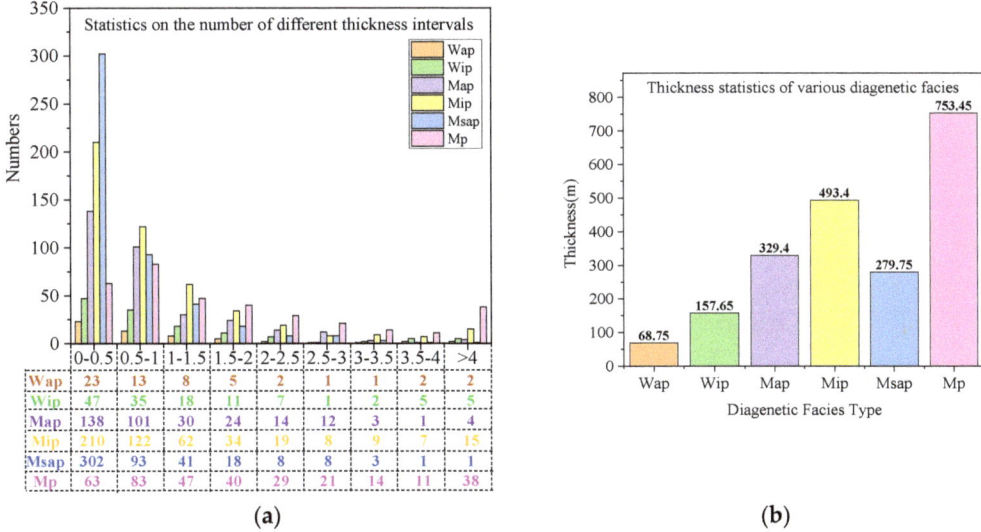

(a)   (b)

Figure 5. Statistical results of quantity and thickness of various diagenetic facies in Fuyu reservoir: (**a**) Statistics on the number of different thickness intervals; (**b**) Thickness statistics of various diagenetic facies.

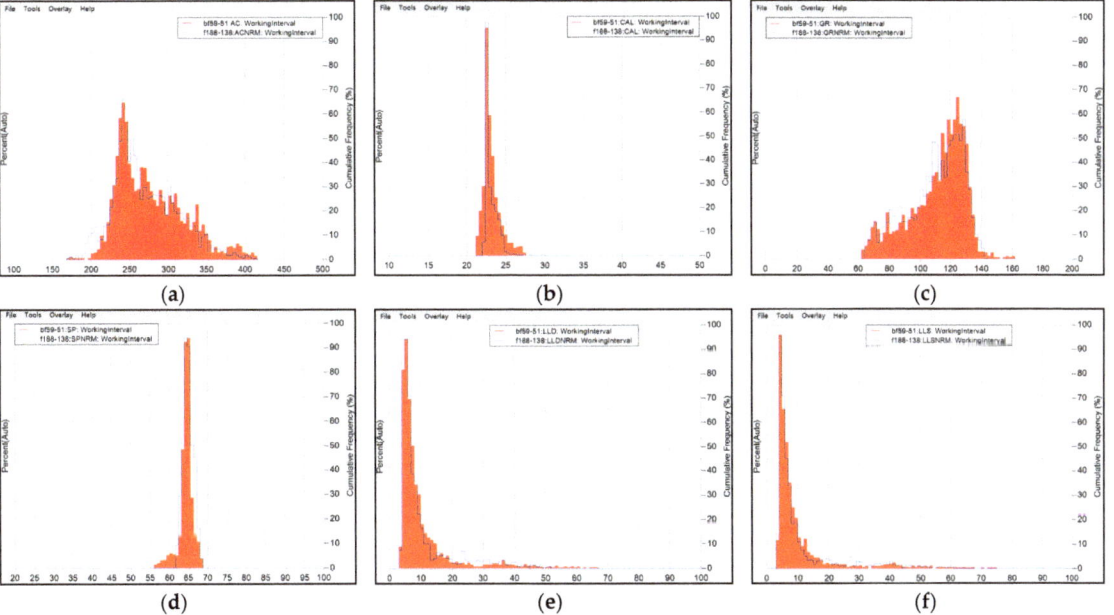

Figure 6. Standardization results of each logging curve of Well f188-138 in Fuyu reservoir: (**a**) AC standardization results; (**b**) CAL standardization results; (**c**) GR standardization results; (**d**) SP standardization results; (**e**) RLLD standardization results; (**f**) RLLS standardization results.

Table 2. Statistics of the number of logging curve image samples of different thickness units.

Diagenetic Facies	Longitudinal Thickness Unit Interval			
	0.50 (m)	0.75 (m)	1.00 (m)	1.25 (m)
Wip	267	151	97	72
Wap	110	64	45	31
Map	442	242	150	105
Mip	768	449	290	194
Msap	364	183	114	70
Mp	1342	831	587	439

4.3. Processing of Imbalanced Data Sets

Over-sampling and under-sampling were combined to process various diagenetic facies logging curve image samples, so as to reduce the influence of sample quantity imbalance on the identification effect of diagenetic facies.

For Wap, Wip (0.50 m) and Msap (0.50 m) with a small number of samples, sliding overlap-tile sampling was used for up-sampling. That is, based on the original diagenetic facies logging curve image samples, a sliding window with a certain step size was set on the image samples, and the sample interval was taken as the sampling window for overlap-tile along the depth downwards to increase the number of samples [41]. We used 50% of the sample interval length as the sliding step size for overlap-tile sampling, as shown in Figure 7. The total number of samples after sampling can be calculated by Equation (1).

$$N = \frac{2 \times (B-T)}{SI} \quad (1)$$

where N represents the total number of samples, B represents the bottom depth, T represents the top depth, and SI represents the sample interval.

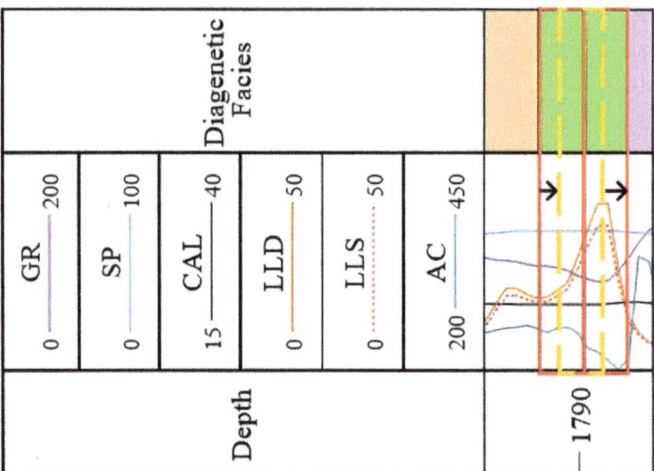

Figure 7. Wap sliding overlap-tile sampling process.

The random under-sampling method was used for the sampling of diagenetic facies (Mp) with a large number of samples, so that the sample size was the same as the average of the total number of other 5 types of diagenetic facies samples, to ensure the overall sample balance. The number of samples of various diagenetic facies obtained after screening is shown in Table 3. We divided the training set and the testing set at a ratio of 8:2.

Table 3. Statistics of the number of samples of various diagenetic facies after imbalance processing.

Diagenetic Facies	Longitudinal Thickness Unit Interval			
	0.50 (m)	0.75 (m)	1.00 (m)	1.25 (m)
Wip	600	150	95	70
Wap	525	150	110	80
Map	440	240	150	105
Mip	765	445	290	190
Msap	515	180	110	70
Mp	570	235	150	105

4.4. Design of Identification Model for the Diagenetic Facies of Tight Oil Reservoirs

To improve the identification efficiency and accuracy of the diagenetic facies of tight oil reservoirs, we transformed the problem of diagenetic facies type identification into the problem of logging curve image identification based on the characteristics of different diagenetic facies logging curves. By referring to the ideas of Mo Zhao et al. [25], the identification method for the diagenetic facies of tight oil reservoirs (C-ViTM) based on hybrid intelligence was constructed by integrating the CNN network model and ViT model structure, as shown in Figure 8. The C-ViTM model uses the ResNet101 network in CNN to extract features of logging curves from image samples and applies ViT to learn the overall features of each logging curve in image samples. The C-ViTM model gives full play to the local feature extraction ability of ResNet101 and the global information control ability of ViT. The specific process is described as follows:

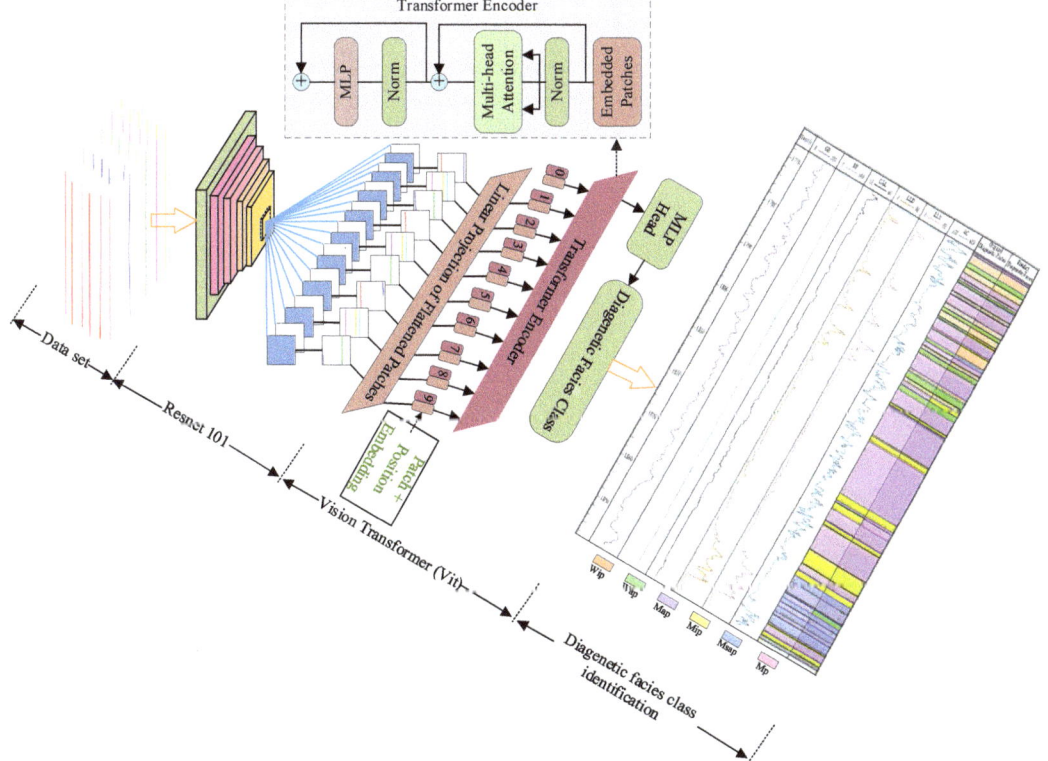

Figure 8. C-ViTM model structure.

The local feature extraction of logging curve images using the ResNet101 network includes 1 image input layer, 1 convolutional layer, 1 BatchNorm, 1 Relu activation function, 1 maximum pooling layer, 3 Conv Block layers and 27 Density Block layers. The ResNet101 network model structure is shown in Figure 9. ViT learns the global information of logging curve images, including 1 Patch Embedding layer, 1 Class Embedding layer, 1 Position Embedding layer, 2 Layer Norms, 1 Multi-Head Attention, 2 Dropout layers and 1 MLP Head. See Figure 8 for details.

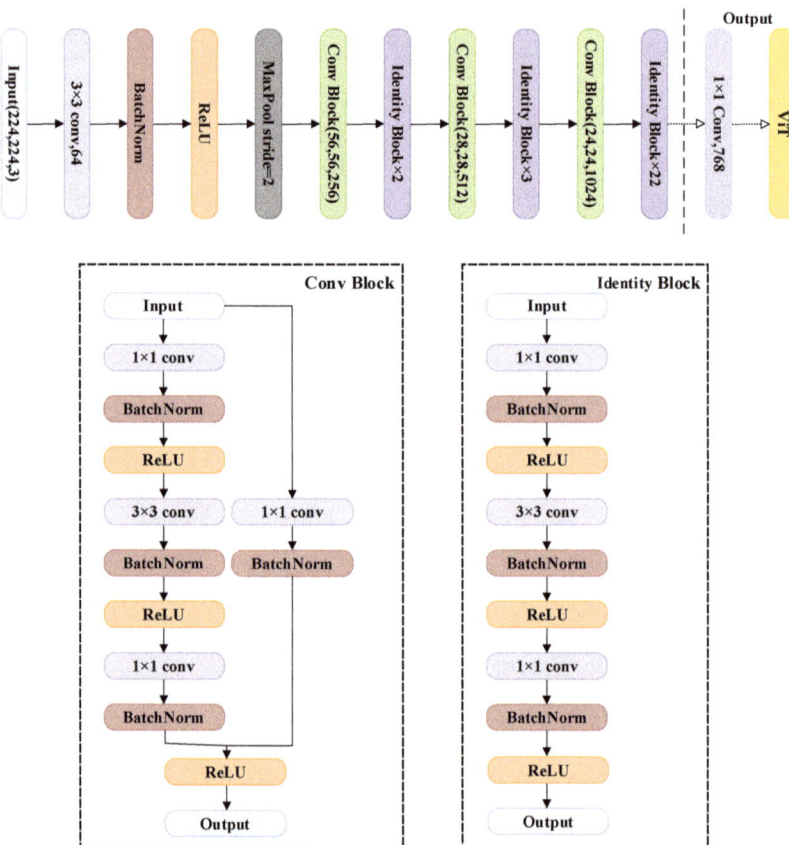

Figure 9. ResNet101 network model structure.

Patch Embedding was used to divide the three-dimensional matrix [224, 224, 3] of the image after feature extraction using the ResNet101 network into a two-dimensional vector matrix [196, 768] according to the patch size of 16 × 16, to provide a basis for meeting the input requirements of ViT. The specific process is shown below:

$$\underset{Images}{[224, 224, 3]} \rightarrow \underset{\substack{patch\ num \\ patches}}{196 \times [16, 16, 3]} \rightarrow \underset{\substack{patch\ num\ token \\ tokens}}{196 \times [768]} \quad (2)$$

where patches represent the total number of patches. Token represents a one-dimensional vector mapped from a single patch. Tokens represent the total number of tokens.

Class Embedding represents the sequence for the classification of diagenetic facies. The principle is to concatenate a vector of length 168 with tokens [196, 768] generated from logging curve images, to generate the two-dimensional vector [197, 768] that meets the input requirements of ViT. The specific process is shown below:

$$[196, 768] \rightarrow Concat(token, [class]token) \rightarrow Concat([196, 768], [1, 768]) \rightarrow [197, 768] \quad (3)$$

Position Embedding superimposes the position information encoding for curve features in each patch to record the position of features. Layer Norm normalizes vector data containing logging curve information to accelerate the learning speed. The role of Multi-Head Attention is to improve the learning ability. Dropout mitigates the phenomenon of model over-fitting. MLP Head outputs the diagenetic facies identification results.

4.5. Training of the Identification Model for the Diagenetic Facies of Tight Oil Reservoirs

The training process of the C-ViTM method includes two stages. In the first stage, the transfer learning technology [42] was applied for pre-training of the C-ViTM model to improve the learning efficiency of the model. In the second stage, the adaptive gradient descent method (Adam) [43] was used to update the model training parameter θ_t and complete iterative training. The update process of training parameter θ_t is described below:

Step 1: Calculate the gradient information g_t of small batch samples.

$$g_t = \nabla_\theta J(\theta, X_t, y_t) \quad (4)$$

where $J(\theta, X_t, y_t)$ represents the objective function and X_t and y_t represent the features and labels of a small batch of samples, respectively.

Step 2: Calculate the first matrix estimator S_t and the second matrix vector R_t.

$$S_t = \beta_1 S_{t-1} + (1 - \beta_1) g_t \quad (5)$$

$$R_t = \beta_2 R_{t-1} + (1 - \beta_2) g_t^2 \quad (6)$$

where S_t represents the first matrix estimation vector and R_t represents the second matrix estimation vector. β_1 and β_2 represent the decay rate, with values of 0.9 and 0.999, respectively.

Step 3: Calculate the gradient information g_t' after bias correction.

$$\hat{S}_t = \frac{S_t}{1 - \beta_1^t} \quad (7)$$

$$\hat{R}_t = \frac{R_t}{1 - \beta_2^t} \quad (8)$$

$$g_t' = \frac{\alpha \hat{S}_t}{\sqrt{\hat{R}_t} + \epsilon} \quad (9)$$

where $\hat{S}_t$ and $\hat{R}_t$ represent the first matrix estimation vector and the second matrix estimation vector after bias correction, respectively. α indicates the learning rate. ϵ represents the smoothing item, with a value of 10^{-8}, preventing division by 0.

Step 4: Update the model parameter θ_t.

$$\theta_t = \theta_{t-1} - g_t' \quad (10)$$

The iterative process loss is shown in Figure 10. It can be seen from the figure that the 70th epoch model tends to be stable.

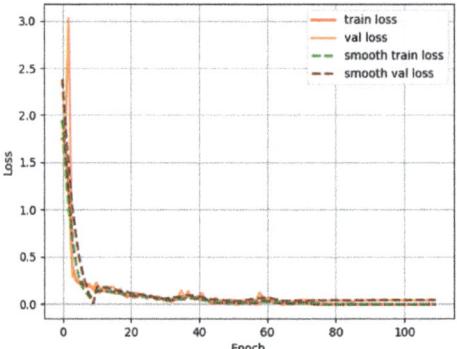

Figure 10. Iterative process loss.

5. Experimental Scheme

The experiment was divided into three parts: experiment of identification effect for different thickness units, accuracy comparison experiment and single-well identification effect. The specific parameters of the experimental devices are shown below: Intel Xeon Silver 4210R CPU, 64 G memory, RTX 6000/8000 GPU; Ubuntu 20.04.3 operating system; PyTorch 1.7.1 experimental framework.

5.1. Experiment of the Identification Effect for Different Thickness Units

The experiment of identification effect for different thickness units took mean Average Accuracy (mAA) [44] and mean Average Precision (mAP) [45] as evaluation indicators.

mAA denotes the average of the identification accuracy of multiple diagenetic facies types. Accuracy represents the proportion of the number correctly identified in the total, which is expressed by Equation (11), where TP is the number of correctly identified diagenetic facies, and N is the total number.

$$\text{Accuracy} = \frac{TP}{N} \tag{11}$$

mAP represents the average of the average precision (AP) of multiple diagenetic facies types. AP refers to the area below the Precision—Recall curve. Precision represents the proportion of the correct number in the identified diagenetic facies types, which can be represented by Equation (12). Recall represents the proportion of the correct number in the actual composition, which can be expressed by Equation (13). The value of mAP is in the range of [0, 1]. A larger value indicates a better identification effect.

$$\text{Precision} = TP/(TP+FP) \tag{12}$$

$$\text{Recall} = TP/(TP+FN) \tag{13}$$

where TP represents a true positive test; FP represents a false positive test; FN represents a false negative test.

In the experiment process, data sets of the same number with intervals of 0.50 m, 0.75 m, 1.00 m and 1.25 m were selected to calculate the Accuracy and AP values of the C-ViTM algorithm for the identification results of various diagenetic facies, and the mAA and mAP values were calculated. The number of various diagenetic facies is shown in Table 3.

5.2. Accuracy Comparison Experiment

In the accuracy comparison experiment, the Precision, Recall and F1 score [46] of the identification results of diagenetic facies types were calculated as evaluation indicators by establishing the confusion matrix. The meanings and calculation methods of Precision and Recall are the same as those in Section 5.1. The $F1$ score was calculated based on Precision and Recall, and can be expressed by Equation (14). The value of $F1$ score is in the range of [0, 1]. A larger value indicates a better identification effect.

$$F1 = \frac{2 \times (Precision \times Recall)}{(Precision + Recall)} \tag{14}$$

In the experiment process, data sets of the same number were selected, CNN algorithm, ViT algorithm and C-ViTM algorithm were used to identify diagenetic facies, and Precision, Recall and $F1$ score were calculated to verify the effectiveness of the C-ViTM algorithm.

5.3. Single-Well Identification Effect Experiment

The Jaccard index [47] was used as the evaluation indicator in the single-well identification effect experiment. The Jaccard index completes the judgment by calculating the intersection over union between the thickness of various diagenetic facies in a single well judged by geologists (A) and the thickness of various diagenetic facies identified by C-ViTM (B). The Jaccard index can be expressed by Equation (15).

$$J(A,B) = \frac{|A \cap B|}{|A \cup B|}, 0 \leq J(A,B) \leq 1 \tag{15}$$

6. Results and Discussion

6.1. Experimental Results of the Identification Effect for Different Thickness Units

The calculation results of mAA and mAP in the identification effect of diagenetic facies with different thickness units are shown in Figure 11. Figure 11 shows that the identification effect of various diagenetic facies is the best at the thickness unit of 0.5 m. The Accuracy and mAA values of the identification result of various diagenetic facies are all above 0.9, and the calculation results of AP value and mAP value are above 0.76. This is because the mAA and mAP values of different diagenetic facies are related to the thickness of various diagenetic facies and the thickness of sample intervals. In addition, Mip and Mp have higher scores at the thickness units of 1.00 m and 1.25 m, possibly due to the larger thickness of their diagenetic facies samples. This phenomenon is also verified by the statistical results in Figure 4.

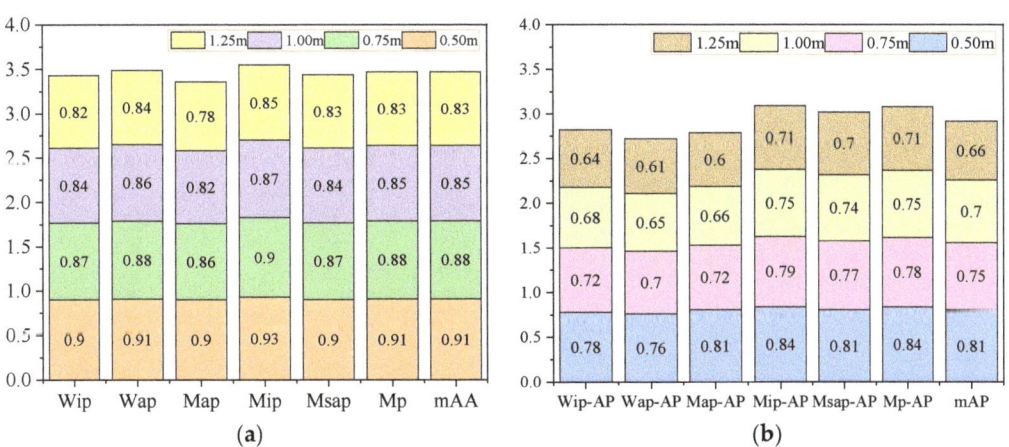

Figure 11. Calculation results of mAA and mAP in the identification effect of various diagenetic facies at different thickness units: (**a**) Calculation results of Accuracy and mAA values in the identification effect of various diagenetic facies; (**b**) Calculation results of AP value and mAP value in the identification effect of various diagenetic facies.

Since the 0.5 m thickness unit realized the best identification result, the sample of 0.5 m thickness unit was used for the accuracy experiment.

6.2. Accuracy Comparison Experiment Results

To verify the accuracy of the C-ViTM method in identifying diagenetic facies, the confusion matrix of the C-ViTM, CNN and ViT methods was constructed, and the influence of geological characteristics on the method was analyzed, as shown in Figure 12. The calculation results of Precision, Recall and $F1$ score are shown in Figure 13.

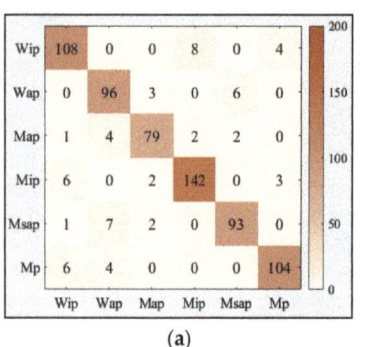
(a)

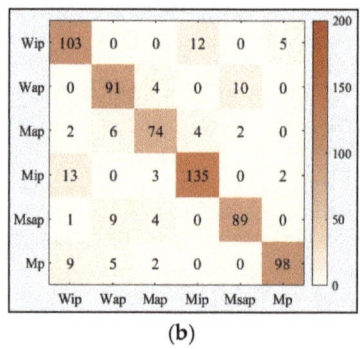
(b)

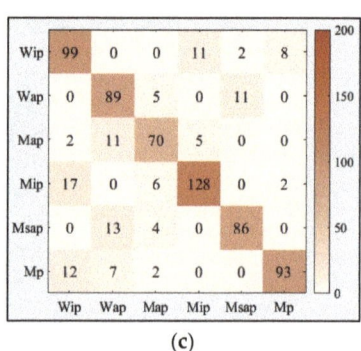
(c)

Figure 12. Identification results of diagenetic facies: (**a**) C-ViTM identification results; (**b**) CNN identification results; (**c**) ViT identification results.

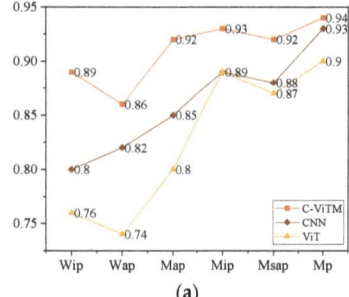
(a)

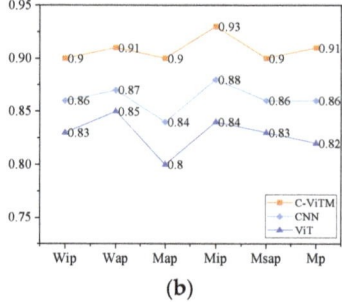
(b)

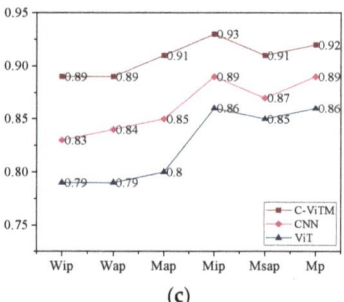
(c)

Figure 13. Calculation results of Precision, Recall and $F1$ score: (**a**) Precision calculation results; (**b**) Recall calculation results; (**c**) $F1$ score calculation results.

According to Figure 12, the C-ViTM method has the best effect on the identification of diagenetic facies, and the CNN method has a higher identification accuracy than ViT, proving that the proposed method can identify diagenetic facies of tight reservoirs better than a single CNN model and ViT model. Due to the limited number of selected target data sets, the CNN method shows a better identification effect compared to the ViT method. This phenomenon has been verified in natural image identification [23]. In addition, it can be seen from Figure 12 that Wip is mainly misjudged as Mip and Mp; Wap is mainly misjudged as Msap and Map; Map is mainly misjudged as Wap, Mip and Wip; Mip is mainly misjudged as Wip, Mp and Map; Msap is mainly misjudged as Wap and Map; Mp is mainly misjudged as Wip and Wap. The cause of misjudgment is related to the internal structural characteristics during the formation of diagenetic facies, which makes the logging curves have similar characteristics [46]. At the same time, it is found that although the diagenetic facies of the same type are subjected to similar compaction, cementation and dissolution, the different action intensity and diagenetic grade result in different particle size sorting and pore connectivity of their components, which affect the reservoir performance.

Figure 13 shows that the Precision, Recall and F1 score of C-ViTM method are higher than those of the CNN method and ViT method, and those of the CNN method are higher than those of the ViT method. It is known from Figure 13a that the three methods have the lowest efficiency for Wap, mainly because Wap and Msap logging curves have similar characteristics, so that some Msaps are misjudged as Wap, reducing the Precision. The same phenomenon also exists in Wip and Mip. The C-ViTM method has a Precision of over 86% for various diagenetic facies, indicating its high Precision. It is learned from Figure 13b that the three methods have the lowest efficiency for Map, mainly because its sample size is relatively small, and some Maps are misjudged. The C-ViTM method has a Recall of over 90% for various diagenetic facies, indicating its good Recall. In Figure 13c, the C-ViTM method has the highest FI scores, which are all above 89%, indicating that it has a good identification effect on various diagenetic facies and can meet the accuracy requirements for identifying the diagenetic facies of tight reservoirs. In addition, the labeling of sample labels also affects the identification efficiency of the three methods.

6.3. Experimental Results of the Single-Well Identification Effect

To further validate the effectiveness of the C-ViTM method, a well from Fuyu reservoir in Sanzhao Sag, Songliao Basin (not involved in the training) was randomly selected for verification, and the identification results are shown in Figure 14. The Original Diagnostic Facies in the figure are the results of manual identification, and the division of diagenetic facies is mainly based on core data and logging response characteristics. Predicted Diagenetic Facies are the identification results of the C-ViTM method (the color is slightly deepened for distinction). The calculation results of the Jaccard index for various diagenetic facies are shown in Table 4. Table 4 shows that the Jaccard index of various diagenetic facies in a single well is above 0.74, and the average Jaccard index is 0.79, indicating that the C-ViTM method has a good single-well identification effect and can be applied to the identification of diagenetic facies of tight reservoirs, with good application effect. Since the selected single well is the well in the target area and has the same depositional environment, it has a good identification effect. For the identification of diagenetic facies of other tight reservoirs with similar geological characteristics, further research is needed according to the logging data.

Table 4. Calculation results of Jaccard index for various diagenetic facies.

	Diagenetic Facies					
	Wip	Wap	Mip	Map	Msap	Mp
Jaccard	0.78	0.74	0.75	0.74	0.81	0.91

6.4. Application Prospect and Limitation Analysis of the C-ViTM Method in Diagenetic Facies Identification

As a diagenetic facies identification method based on hybrid intelligence, the C-ViTM method has higher identification accuracy, and can replace the manual identification of the diagenetic facies of tight reservoirs to a certain extent to determine the location of high-quality reservoirs. However, due to the influence of sedimentation and diagenesis on the characteristics of tight reservoirs, various diagenetic facies have similar characteristics to some extent, resulting in misjudgment. At the same time, the identification effect of the C-VITM method is easily affected by the quantity of various diagenetic facies samples, the geometric characteristics of logging curves and the design of model structure, so it needs to be further optimized in the identification of the diagenetic facies of other types of reservoirs such as carbonate reservoir and volcanic reservoir.

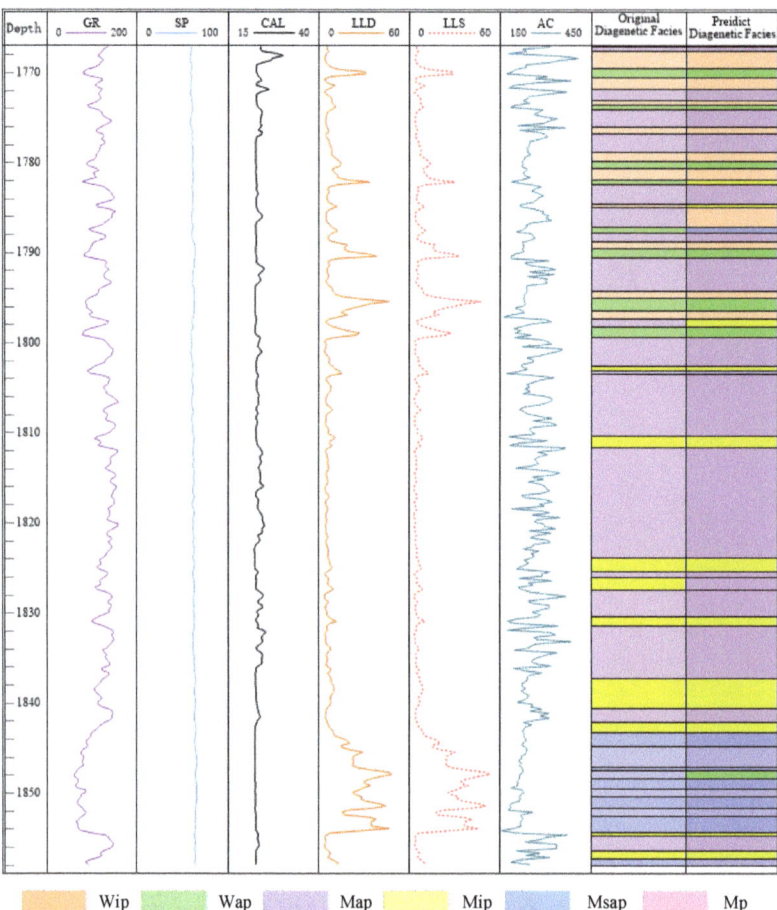

Figure 14. Identification results of single-well diagenetic facies.

7. Conclusions

Based on the intersection of geological big data and artificial intelligence, a new C-ViTM method for the identification of the diagenetic facies of tight oil reservoirs was proposed in this study, which solves the problems such as difficulty in the identification of the diagenetic facies of tight oil reservoirs, time cost and human cost.

(1) Based on core data and logging response characteristics, the diagenetic facies of tight reservoirs of Fuyu reservoir in Sanzhao Sag were classified into seven types: Wip, Wap, Mip, Map, Msap, Mp, etc. The relationship between diagenetic facies and reservoir performance was established. Wip, Wap and Mip were classified as Class I reservoirs; Map and Msap were classified as Class II reservoirs; Mp was classified as a Class III reservoir. The reservoir performance was completed while realizing diagenetic facies identification.

(2) By comparing the identification results of diagenetic facies at different thickness intervals of 0.50 m, 0.75 m, 1.00 m and 1.25 m, it was found that the best identification effect can be realized at the sample thickness of 0.50 m, indicating that the identification results are related to the thickness of various diagenetic facies and the thickness of sample intervals.

(3) Compared with the single methods of CNN and ViT, C-ViTM has a better identification effect, with Precision of above 86%, Recall of above 90% and F1 score of above 89%.

The C-ViTM method is suitable for the identification of the diagenetic facies of tight reservoirs, but the identification effect is easily affected by the number of samples and the similarity of the internal structural features of diagenetic facies (the similarity of logging curve features), such as Wip and Mip.

(4) The average Jaccard index calculated by using the C-ViTM method in diagenetic facies identification of a single well is 0.79, indicating that the C-ViTM method has a good identification effect and wide application prospects.

In future work, we expect to optimize the C-ViTM method so that it can be better applied to the identification and evaluation of diagenetic facies in other reservoirs such as shale oil reservoirs, and evaluate the potential application of this method.

Author Contributions: T.L.: Methodology, Software, Investigation, Data Curation, Writing—Original Draft, Visualization, Project administration. Z.L.: Conceptualization, Methodology, Resources, Writing—Review & Editing, Funding acquisition. K.Z.: Formal analysis, Writing—Review & Editing, Funding acquisition. C.L.: Methodology, Writing—Review & Editing. Y.Z.: Formal analysis, Visualization. Z.M.: Data Curation, Visualization. F.L.: Resources, Supervision. X.L.: Investigation, Project administration, Funding acquisition. M.M.: Software, Validation, Visualization. S.Z.: Data Curation, Investigation. All authors have read and agreed to the published version of the manuscript.

Funding: This research is supported by the National Natural Science Foundation of China (42172161), the Youth Fund of the National Natural Science Foundation of China (42102173), the CNPC Innovation Foundation (2020D-5007-0102), the Heilongjiang Provincial Natural Science Foundation of China (YQ2020D001), the Heilongjiang Provincial Natural Science Foundation of China (LH2020F003) and the Heilongjiang Provincial Department of Education Project of China (UNPYSCT-2020144).

Data Availability Statement: The original contributions presented in the study are included in the article, further inquiries can be directed to the corresponding authors.

Conflicts of Interest: The authors declare that they have no known competing financial interests or personal relationships that could have appeared to influence the work reported in this paper.

References

1. Huang, L.; Yan, J.; Liu, M.; Zhang, Z.; Ye, S.; Zhang, F.; Zhong, G.; Wang, M.; Wang, J.; Geng, B. Diagenetic facies logging identification and application of deep tight sandstone gas reservoir: A case study of the third member of Xujiahe formation in Dayi area of western Sichuan depression. *J. China Univ. Min. Technol.* **2022**, *51*, 107–123.
2. Lai, J.; Wang, G.; Wang, S.; Cao, J.; Li, M.; Pang, X.; Zhou, Z.; Fan, X.; Dai, Q.; Yang, L.; et al. Review of diagenetic facies in tight sandstones: Diagenesis, diagenetic minerals, and prediction via well logs. *Earth-Sci. Rev.* **2018**, *185*, 234–258. [CrossRef]
3. Liu, F.; Wang, X.; Liu, Z.; Tian, F.; Zhao, Y.; Pan, G.; Peng, C.; Liu, T.; Zhao, L.; Zhang, K.; et al. Identification of tight sandstone reservoir lithofacies based on CNN image recognition technology: A case study of Fuyu reservoir of Sanzhao Sag in Songliao Basin. *Geoenergy Sci. Eng.* **2023**, *222*, 211459. [CrossRef]
4. Guo, Q.; Wang, S.; Chen, X. Assessment on tight oil resources in major basins in China. *J. Asian Earth Sci.* **2019**, *178*, 52–63. [CrossRef]
5. Qi, M.; Han, C.; Ma, C.; Liu, G.; He, X.; Li, G.; Yang, Y.; Sun, R.; Cheng, X. Identification of Diagenetic Facies Logging of Tight Oil Reservoirs Based on Deep Learning—A Case Study in the Permian Lucaogou Formation of the Jimsar Sag, Junggar Basin. *Minerals* **2022**, *12*, 913. [CrossRef]
6. Liu, T.; Li, C.; Liu, Z.; Zhang, K.; Liu, F.; Li, D.; Zhang, Y.; Liu, Z.; Liu, L.; Huang, J. Research on Image Identification Method of Rock Thin Slices in Tight Oil Reservoirs Based on Mask R-CNN. *Energies* **2022**, *15*, 5818. [CrossRef]
7. Lai, J.; Wang, G.W.; Ran, Y.; Zhou, Z.L.; Cui, Y.F. Impact of diagenesis on the petrophysical properties of tight oil reservoirs: The case of Upper Triassic Yanchang Formation Chang 7 oil layers in Ordos Basin. *J. Pet. Sci. Eng.* **2016**, *145*, 54–65. [CrossRef]
8. Lai, J.; Wang, G.; Fan, Z.; Zhou, Z.; Chen, J.; Wang, S. Fractal analysis of tight shaly sandstones using nuclear magnetic resonance measurements. *AAPG Bull.* **2018**, *102*, 175–193. [CrossRef]
9. Ma, P.; Lin, C.; Zhang, S.; Dong, C.; Zhao, Y.; Dong, D.; Shehzad, K.; Awais, M.; Guo, D.; Mu, X. Diagenetic history and reservoir quality of tight sandstones: A case study from Shiqianfeng sandstones in upper Permian of Dongpu Depression, Bohai Bay Basin, eastern China. *Mar. Pet. Geol.* **2018**, *89*, 280–299. [CrossRef]
10. Su, Y.; Zha, M.; Ding, X.; Qu, J.; Gao, C.; Jin, J.; Iglauer, S. Petrographic, palynologic and geochemical characteristics of source rocks of the Permian Lucaogou formation in Jimsar Sag, Junggar Basin, NW China: Origin of organic matter input and depositional environments. *J. Pet. Sci. Eng.* **2019**, *183*, 106364. [CrossRef]
11. Lin, M.; Xi, K.; Cao, Y.; Liu, Q.; Zhang, Z.; Li, K. Petrographic features and diagenetic alteration in the shale strata of the Permian Lucaogou Formation, Jimusar sag, Junggar Basin. *J. Pet. Sci. Eng.* **2021**, *203*, 108684. [CrossRef]

12. Lai, J.; Wang, G.; Chai, Y.; Ran, Y. Prediction of diagenetic facies using well logs: Evidences from Upper Triassic Yanchang Formation Chang 8 sandstones in Jiyuan region, Ordos basin, China. *Oil Gas Sci. Technol.–Rev. D'ifp Energ. Nouv.* **2016**, *71*, 34. [CrossRef]
13. Zou, C.-N.; Tao, S.-Z.; Zhou, H.; Zhang, X.-X.; He, D.-B.; Zhou, C.-M.; Wang, L.; Wang, X.-S.; Li, F.-H.; Zhu, R.-K.; et al. Genesis, classification and evaluation method of diagenetic facies. *Pet. Explor. Dev.* **2008**, *35*, 526–540. [CrossRef]
14. Lai, J.; Fan, X.; Pang, X.; Zhang, X.; Xiao, C.; Zhao, X.; Han, C.; Wang, G.; Qin, Z. Correlating diagenetic facies with well logs (conventional and image) in sandstones: The Eocene–Oligocene Suweiyi Formation in Dina 2 Gasfield, Kuqa depression of China. *J. Pet. Sci. Eng.* **2019**, *174*, 617–636. [CrossRef]
15. Li, Y.; Chang, X.; Yin, W.; Sun, T.; Song, T. Quantitative impact of diagenesis on reservoir quality of the Triassic Chang 6 tight oil sandstones, Zhenjing area, Ordos Basin, China. *Mar. Pet. Geol.* **2017**, *86*, 1014–1028. [CrossRef]
16. Ma, B.; Yang, S.; Zhang, H.; Kong, Q.; Song, C.; Wang, Y.; Bai, Q.; Wang, X. Diagenetic facies quantitative evaluation of low-permeability sandstone: A case study on Chang 82 reservoirs in the Zhenbei area, Ordos basin. *Energy Explor. Exploit.* **2018**, *36*, 414–432. [CrossRef]
17. Ran, Y.; Wang, G.W.; Lai, J.; Zhou, Z.L.; Cui, Y.F.; Dai, Q.Q.; Chen, J.; Wang, S.C. Quantitative characterization of diagenetic facies of tight sandstone oil reservoir by using logging crossplot: A case study on Chang 7 tight sandstone oil reservoir in Huachi area, Ordos Basin. *Acta Sedimentol. Sin.* **2016**, *34*, 694–706.
18. Shi, Y.; Xiao, L.; Mao, Z.; Guo, H. An identification method for diagenetic facies with well logs and its geological significance in low-permeability sandstones: A case study on Chang 8 reservoirs in the Jiyuan region, Ordos Basin. *Acta Pet. Sin.* **2011**, *32*, 820–828.
19. Lai, J.; Wang, G.W.; Wang, S.N.; Xin, Y.; Wu, Q.K.; Zheng, Y.Q.; Li, J.; Cang, D. Overview and research progress in logging recognition method of clastic reservoir diagenetic facies. *J. Cent. South Univ. (Sci. Technol.)* **2013**, *44*, 4942–4953.
20. Ozkan, A.; Cumella, S.P.; Milliken, K.L.; Laubach, S.E. Prediction of lithofacies and reservoir quality using well logs, late cretaceous Williams fork formation, Mamm creek field, Piceance basin, Colorado. *AAPG Bull.* **2011**, *95*, 1699–1723. [CrossRef]
21. Cui, Y.; Wang, G.; Jones, S.J.; Zhou, Z.; Ran, Y.; Lai, J.; Li, R.; Deng, L. Prediction of diagenetic facies using well logs–A case study from the upper Triassic Yanchang Formation, Ordos Basin, China. *Mar. Pet. Geol.* **2017**, *81*, 50–65. [CrossRef]
22. Zhu, P.; Lin, C.Y.; Wu, P.; Fan, R.F.; Wang, D.R.; Liu, X.L. Logging quantitative identification of diagenetic facies by using principal component analysis: A case of Es_3x^1 in Zhuang62-66 Area, Wu Hao-zhuang Oil field. *Prog. Geophys.* **2015**, *30*, 2360–2365.
23. Dosovitskiy, A.; Beyer, L.; Kolesnikov, A.; Weissenborn, D.; Zhai, X.; Unterthiner, T.; Dehghani, M.; Minderer, M.; Heigold, G.; Gelly, S.; et al. An image is worth 16x16 words: Transformers for image recognition at scale. *arXiv* **2020**, arXiv:2010.11929.
24. Carion, N.; Massa, F.; Synnaeve, G.; Usunier, N.; Kirillov, A.; Zagoruyko, S. End-to-end object detection with transformers. In Proceedings of the Computer Vision–ECCV 2020: 16th European Conference, Glasgow, UK, 23–28 August 2020; Proceedings, Part I 16; Springer International Publishing: Cham, Switzerland, 2020; pp. 213–229.
25. Zhao, M.; Cao, G.; Huang, X.; Yang, L. Hybrid Transformer-CNN for Real Image Denoising. *IEEE Signal Process. Lett.* **2022**, *29*, 1252–1256. [CrossRef]
26. Meng, Q.; Zhao, B.; Chen, S.; Lin, T.; Zhou, Y.; Qiao, W. Sedimentary enrichment mode and effect analysis of exploration and development: A case study of Fuyu reservoir tight oil in northern Songliao Basin. *Acta Sedimentol. Sin.* **2021**, *39*, 112–125.
27. Deng, Q.; Hu, M.; Hu, Z.; Wu, Y. Sedimentary characteristics of shallow-water delta distributary channel sand bodies: A case from II-I formation of Fuyu oil layer in the Sanzhao depression, Songliao Basin. *Oil Gas Geol.* **2015**, *36*, 118–127.
28. Wang, B.Q.; Xu, W.F.; Liu, Z.L.; Kong, F.Z.; Chang, Z.Y.; Wang, C.R. Diagenesis of reservoirs in Fuyu and Yangdachengzi of Sanzhao region. *Oil Gas Geol.* **2001**, *22*, 82–87.
29. Tang, Z.; Zhao, J.; Wang, T. Evaluation and key technology application of "sweet area" of tight oil in south Songliao Basin. *Nat. Gas Geosci.* **2019**, *30*, 1114–1124.
30. Cao, Y.; Xi, K.; Zhu, R.; Zhang, S.; Zhang, X.; Zheng, X. Microscopic pore throat characteristics of tight sandstone reservoirs in Fuyu layer of the fourth member of Quantou Formation in southern Songliao Basin. *J. China Univ. Pet.* **2015**, *39*, 7–17.
31. Liu, Y.; Zhu, X.; Zhang, S.; Zhao, D. Diagenesis and pore evolution of reservoir of the Member 4 of Lower Cretaceous Quantou Formation in Sanzhao Sag, northern Songliao Basin. *J. Palaeogeogr. (Chin. Ed.)* **2010**, *12*, 480–488.
32. Fu, G.-M.; Qin, X.-L.; Miao, Q.; Zhang, T.-J.; Yang, J.-P. Division of diagenesis reservoir facies and its control—Case study of Chang-3 reservoir in Yangchang formation of Fuxian exploration area in northern Shaanxi. *Min. Sci. Technol. (China)* **2009**, *19*, 537–543. [CrossRef]
33. Du, H.Q.; Zhu, R.K.; He, Y.B. The diagenesis of the 2nd Member reservoirs of Xujiahe Formation and its influence on reservoirs of Hechuan area. *Acta Petrol. Et Mineral.* **2012**, *31*, 403–411.
34. Olivarius, M.; Weibel, R.; Hjuler, M.L.; Kristensen, L.; Mathiesen, A.; Nielsen, L.H.; Kjøller, C. Diagenetic effects on porosity–permeability relationships in red beds of the Lower Triassic Bunter Sandstone Formation in the North German Basin. *Sediment. Geol.* **2015**, *321*, 139–153. [CrossRef]
35. Zhang, X.X.; Zou, C.N.; Tao, S.Z.; Xu, C.; Song, J.; Li, G. Diagenetic facies types and semiquantitative evaluation of low porosity and permeability sandstones of the Fourth Member Xujiahe Formation Guangan Area, Sichuan basin. *Acta Sedimentol. Sin.* **2010**, *28*, 50–57.
36. Duan, X.; Song, R.; Li, G.; Li, N. Research of integrated diagenetic facies characteristics of T3 x2 reservoir in Sichuan Basin. *J. Southwest Pet. Univ. Sci. Technol. Ed.* **2011**, *33*, 7–14.

37. Rushing, J.A.; Newsham, K.E.; Blasingame, T.A. Rock typing: Keys to understanding productivity in tight-gas sands. In Proceedings of the Society of Petroleum Engineers Unconventional Reservoirs Conference, Keystone, CO, USA, 10–12 February 2008; SPE Paper 114164; p. 31. [CrossRef]
38. Wang, J.; Cao, Y.; Liu, K.; Liu, J.; Kashif, M. Identification of sedimentary-diagenetic facies and reservoir porosity and permeability prediction: An example from the Eocene beach-bar sandstone in the Dongying Depression, China. *Mar. Pet. Geol.* **2017**, *82*, 69–84. [CrossRef]
39. Liu, H.; Zhao, Y.; Luo, Y.; Chen, Z.; He, S. Diagenetic facies controls on pore structure and rock electrical parameters in tight gas sandstone. *J. Geophys. Eng.* **2015**, *12*, 587–600. [CrossRef]
40. Zhang, J.; Ambrose, W.; Xie, W. Applying convolutional neural networks to identify lithofacies of large-n cores from the Permian Basin and Gulf of Mexico: The importance of the quantity and quality of training data. *Mar. Pet. Geol.* **2021**, *133*, 105307. [CrossRef]
41. Gu, Y.; Zhang, D.; Bao, Z.; Feng, Z.; Li, J. Permeability prediction using Gradient Boosting Decision Tree (GBDT): A case study of tight sandstone reservoirs of member of Chang 4+5 in western Jiyuan Oilfeld. *Prog. Geophys.* **2021**, *36*, 0585–0594.
42. Tan, C.; Sun, F.; Kong, T.; Zhang, W.; Yang, C.; Liu, C. A survey on deep transfer learning. In Proceedings of the International Conference on Artificial Neural Networks, Rhodes, Greece, 4–7 October 2018; Springer: Cham, Switzerland, 2018; pp. 270–279.
43. Kingma, D.P.; Ba, J. Adam: A method for stochastic optimization. *arXiv* **2014**, arXiv:1412.6980.
44. Antariksa, G.; Muammar, R.; Lee, J. Performance evaluation of machine learning-based classification with rock-physics analysis of geological lithofacies in Tarakan Basin, Indonesia. *J. Pet. Sci. Eng.* **2022**, *208*, 109250. [CrossRef]
45. Pereira, N. PereiraASLNet: ASL letter recognition with YOLOX taking Mean Average Precision and Inference Time considerations. In Proceedings of the 2022 2nd International Conference on Artificial Intelligence and Signal Processing (AISP), Vijayawada, India, 12–14 February 2022; pp. 1–6. [CrossRef]
46. Maria Navin, J.R.; Pankaja, R. Performance analysis of text classification algorithms using confusion matrix. *Int. J. Eng. Tech. Res. IJETR* **2016**, *6*, 75–78.
47. Jaccard, P. Nouvelles recherches sur la distribution florale. *Bull. Soc. Vaud. Sci. Nat.* **1908**, *44*, 223–270.

Disclaimer/Publisher's Note: The statements, opinions and data contained in all publications are solely those of the individual author(s) and contributor(s) and not of MDPI and/or the editor(s). MDPI and/or the editor(s) disclaim responsibility for any injury to people or property resulting from any ideas, methods, instructions or products referred to in the content.

Article

Study on Sedimentary Environment and Organic Matter Enrichment Model of Carboniferous–Permian Marine–Continental Transitional Shale in Northern Margin of North China Basin

Hanyu Zhang [1,2], Yang Wang [1,2,*], Haoran Chen [1,2], Yanming Zhu [1,2], Jinghui Yang [1,2], Yunsheng Zhang [1,2], Kailong Dou [1,2] and Zhixuan Wang [1,2]

[1] Key Laboratory of Coalbed Methane Resources and Reservoir Formation Process, Ministry of Education, China University of Mining and Technology, Xuzhou 221008, China; ts21010061a31@cumt.edu.cn (H.Z.); ts22010039a31@cumt.edu.cn (H.C.); ym63z@cumt.edu.cn (Y.Z.); ts21010152p31@cumt.edu.cn (J.Y.); ts21010161p31@cumt.edu.cn (Y.Z.); ts22010043a31@cumt.edu.cn (K.D.); ts22010062a31@cumt.edu.cn (Z.W.)
[2] School of Resources and Geoscience, China University of Mining and Technology, Xuzhou 221116, China
* Correspondence: wangy89@cumt.edu.cn; Tel.: +86-15896422124

Abstract: The shales of the Taiyuan Formation and Shanxi Formation in the North China Basin have good prospects for shale gas exploration and development. In this study, Well KP1 at the northern margin of the North China Basin was used as the research object for rock mineral, organic geochemical, and elemental geochemical analyses. The results show that brittle minerals in the shales of the Taiyuan Formation and Shanxi Formation are relatively rare (<40%) and that the clay mineral content is high (>50%). The average TOC content is 3.68%. The organic matter is mainly mixed and sapropelic. The source rocks of the Taiyuan Formation and Shanxi Formation are mainly felsic, and the tectonic background lies in the continental island arc area. The primary variables that influenced the enrichment of organic materials during the sedimentary stage of the Taiyuan Formation were paleosalinity and paleoproductivity. Paleosalinity acted as the primary regulator of organic matter enrichment during the sedimentary stage of the Shanxi Formation.

Keywords: North China Basin; Taiyuan and Shanxi Formations; marine–continental transitional shale; depositional environment; organic matter enrichment

1. Introduction

In recent years, China has made major breakthroughs in the exploration and development of marine and continental shale, and it has discovered several huge shale gas fields [1–3]. Compared with the development research of marine shale, the exploration and development of marine–continental transitional shale is still in the early stages [4–6]. The sedimentary environment of marine–continental transitional shale is relatively complex, generally deposited in the transitional environment from the surface ocean to the river delta [7–10]. Marine–continental transitional shale is widely distributed in China, including in the North China Basin, Qinshui Basin, Ordos Basin, and Bohai Bay Basin [11–13]. These resources are rich in shale gas reserves, accounting for about one-quarter of the total shale gas resources, and they represent potentially important areas for unconventional oil and gas exploration [4,5]. Among them, multiple sets of marine–continental transitional shale distributed in the Carboniferous–Permian system in the northern North China Basin are characterized by diverse kerogen types, large changes in organic matter abundance, high maturity, and medium clay abundance. They have become the current research hotspots for the exploration and development of shale gas [14–16].

The abundance of shale organic matter affects the generation and accumulation of shale oil and gas, but the enrichment and preservation of organic matter are complex

Citation: Zhang, H.; Wang, Y.; Chen, H.; Zhu, Y.; Yang, J.; Zhang, Y.; Dou, K.; Wang, Z. Study on Sedimentary Environment and Organic Matter Enrichment Model of Carboniferous–Permian Marine–Continental Transitional Shale in Northern Margin of North China Basin. *Energies* 2024, 17, 1780. https://doi.org/10.3390/en17071780

Academic Editors: Reza Rezaee and Nikolaos Koukouzas

Received: 24 January 2024
Revised: 23 March 2024
Accepted: 4 April 2024
Published: 8 April 2024

Copyright: © 2024 by the authors. Licensee MDPI, Basel, Switzerland. This article is an open access article distributed under the terms and conditions of the Creative Commons Attribution (CC BY) license (https://creativecommons.org/licenses/by/4.0/).

physical and chemical processes [17–21]. Previous studies have shown that organic matter enrichment and preservation are jointly controlled by the depositional environment (terrigenous input, sedimentation rate, paleoclimate, paleoredox, paleosalinity, and primary productivity) and the source of organic matter [19–24]. The controlling factors of the depositional environment and organic matter enrichment of marine–terrestrial transitional shale are more complex [21,25]. Due to the intense tectonic movement of the North China Basin in the Late Paleozoic, the North China Basin transitioned from a surface sea to a river delta environment [14,26]. This led to changes in the depositional environment and provenance. Changes in the sedimentary environment and provenance affect the growth of paleontology in sedimentary water bodies and the enrichment and preservation of organic matter. Furthermore, this leads to differences in the abundance of organic matter in shale at different depositional stages [27–31]. Therefore, it is necessary to reveal the main factors controlling the enrichment of organic matter in the marine–continental transitional shale in the northern part of the North China Basin and to determine the optimal conditions for the enrichment of organic matter in marine–continental transitional shale.

The primary purpose of this study is threefold: (1) to reveal and compare the abundance of organic matter and the characteristics of reservoir physical properties in Late Paleozoic marine–continental transitional shale (specifically that of the Taiyuan and Shanxi Formations) based on rock mineral analysis and organic geochemistry; (2) to identify the source characteristics, structural background, and paleoenvironmental factors of the marine–continental transitional shale within the Taiyuan and Shanxi Formations of the Late Paleozoic through an elemental geochemical analysis combined with numerical analysis methods, and to determine the main controlling factors of organic matter enrichment in these shale formations; and (3) to establish an organic matter enrichment model for marine–terrestrial transitional shale and determine the optimal conditions for shale organic matter enrichment. Our study holds great significance for evaluating the potential of marine–continental transitional shale and provides a theoretical basis for optimizing reservoir selection for marine–continental transitional shale gas exploration in the North China Basin.

2. Geological Setting

The North China Basin is adjacent to the Taihang Mountain uplift to the west, the Bohai Sea to the east, and the Yanshan Fold uplift to the north (Figure 1a). Because of the staggered ups and downs of faults, several secondary tectonic units have developed [32]. The sedimentary evolution process of the North China Basin can be roughly divided into four stages: the stages of early Paleozoic uplift and late Paleozoic sea–land alternation; the early Mesozoic filling of the intracontinental lake basin depression; and the late Mesozoic and Cenozoic rifting stages [33,34].

The North China Basin was in the land–sea alternation stage during the Carboniferous–Permian period, which was crucial for the basin's ecosystem to change from an intracontinental lake to an epicontinental sea. Well KP1 (latitude N39° 45′ 46.8″, longitude E118° 21′ 25.2″) in the study area is located in Fengnan District, Tangshan City, Hebei Province, and its structural position belongs to the east wing of the Kaiping syncline. The Taiyuan Formation has a thickness of 71.33 m and is primarily made up of sandstone, thin limestone, coal seams, and black shale, according to KP1 drilling data in the research region. The Shanxi Formation has a thickness of 69.5 m and is primarily made up of sandstone, shale, and thin coal seams (Figure 1b).

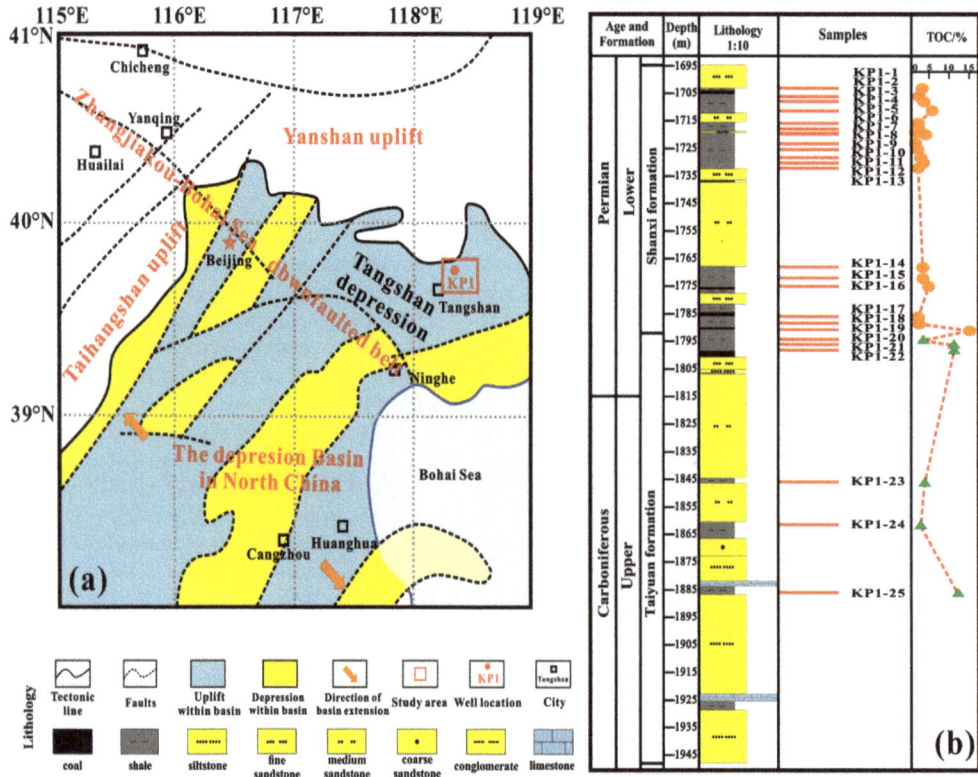

Figure 1. Structural maps (**a**) and borehole lithology histograms (**b**).

3. Samples and Analysis

3.1. Samples and Experimental Method

A total of 25 representative samples were collected and analyzed in this study: 19 samples from the Shanxi Formation and 6 samples from the Taiyuan Formation. The samples were taken from the shale core of Well KP1, with sampling depths ranging between 1695 and 1886 m. The main lithology of the samples comprises shale and carbonaceous shale (Figure 1b).

The analysis of shale samples includes total organic carbon content (TOC) detection, major- and trace-element content detection, XRD whole-rock mineral quantitative analysis detection, clay mineral X-ray diffraction quantitative analysis detection, rock pyrolysis analysis detection, and kerogen maceral identification. The above analysis and testing are carried out in the laboratory of the Shandong Coal Geological Planning and Exploration Institute.

The content of major elements in the 25 samples was determined according to the Chinese standard GB/T21114-2019 by using the refractory X-ray fluorescence spectrum chemical analysis melting glass method [35]. The XRD whole-rock mineral quantitative analysis test and clay mineral X-ray diffraction quantitative analysis test were carried out using a Rigaku SmartLab9 X-ray diffractometer (Rigaku Company, Tokyo, Japan). According to the national standard SY/T5163-2018, 17 samples were analyzed and tested to further determine the brittleness of the shale [36]. By using an OGE-VI rock pyrolysis instrument, 25 samples were analyzed according to the Chinese standard experimental method GB/T18602-2012, and the hydrocarbon generation potential of the shale organic matter was further analyzed [37]. The process of identifying and categorizing kerogen macerals by the use of transmission light fluorescence is known as identification. According to the Chinese standard experimental method SY/T5125-2014, 25 samples of shale were analyzed, and the analysis results were further used for kerogen-type identification [38].

3.2. Data Processing

The majority of trace elements in sediments are made up of terrigenous debris and authigenic components. Only the authigenic components in sediments can reflect the evolutionary characteristics of the sedimentary environment in geological history. However, rock composition is complex and diverse, and there is some deviation in judging its enrichment only by the content of trace elements and standard shale. To eliminate terrigenous detrital components from having an impact on authigenic components, the trace-element concentration is usually standardized by using the Al element with relatively stable properties during diagenesis [39]. In this study, the upper crust content (UCC) was used for a standardized calculation. The enrichment factor (EF) of elements was calculated as follows:

$$X\text{-EF} = (X/Al)_{sample}/(X/Al)_{UCC} \quad (1)$$

The EF of element X is represented by X-EF, the measured concentrations of elements X and Al in the examined samples are represented by $(X/Al)_{sample}$, and the X/Al ratio in the UCC is represented by $(X/Al)_{UCC}$. X-EF greater than 1 means that some element is richer than the UCC, while X-EF less than 1 means that an element is more deficient than the UCC.

To assess paleoproductivity more accurately, the Ba_{bio} content in the sediments was also calculated [40] utilizing the equation below:

$$Ba_{bio} = Ba_{sample} - Al_{sample} \times (Ba/Al)_{PAAS} \quad (2)$$

Here, Ba_{sample} is the measured content of element Ba in the samples under study, $(Ba/Al)_{PAAS}$ are the ratios of Ba/Al in post-Archean Australian shale (PAAS), and Ba_{bio} represents the Ba content produced by biological processes; the estimated content in terrigenous debris is generally subtracted from the total amount in the sample.

There are multiple indicators that can characterize the paleoredox conditions of sedimentary water bodies in the shale deposition stage, such as V/Cr, V/Sc, and Ce/La. In this study, the data sets composed of different element indexes representing paleoredox conditions were processed using the normalization method. The average value obtained after the normalization of the different element indexes was used as the representative parameter of the paleoredox conditions. The calculation formula is as follows:

$$\text{Paleo-} = \{[(V/Cr)_{sample} - (V/Cr)_{min}]/[(V/Cr)_{max} - (V/Cr)_{min}] + [(V/Sc)_{sample} - (V/Sc)_{min}]/[(V/Sc)_{max} - (V/Sc)_{min}] + [(Ce/La)_{sample} - (Ce/La)_{min}]/[(Ce/La)_{max} - (Ce/La)_{min}]\}/3 \quad (3)$$

In the formula, Paleo- is the representative parameter of the paleoredox conditions; $(V/Cr)_{sample}$, $(V/Sc)_{sample}$, and $(Ce/La)_{sample}$ are the ratios of V/Cr, V/Sc, and Ce/La, respectively; $(V/Cr)_{min}$, $(V/Sc)_{min}$, and $(Ce/La)_{min}$ are the minimum values of the V/Cr, V/Sc, and Ce/La ratios, respectively; and $(V/Cr)_{max}$, $(V/Sc)_{max}$, and $(Ce/La)_{max}$ are the maximum values of V/Cr, V/Sc, and Ce/La, respectively.

Based on the above calculation data, by using Origin (2021) software and combining the discriminant templates and analysis methods developed by previous researchers, the mudstone provenance and tectonic background characteristics, sedimentary environment, and water retention environment of the Taiyuan Formation and Shanxi Formation are discussed. By using SPSS Pro (Education version) software, the main controlling factors of shale organic matter enrichment in the Taiyuan Formation and Shanxi Formation are discussed and analyzed.

4. Results

4.1. Lithological Characteristics

Through a detailed observation of the core samples from Well KP1, the source rock sections were found to have strong heterogeneity. The core lithology of the Shanxi Formation's interbedded sandstone and mudstone in Well KP1 is mostly composed of dark-grey sandy mudstone with a thin overlay of dark-grey fine sandstone (Figure 2a). The vertical variation in the core lithology of the Taiyuan Formation shows that grey-white coarse sandstone and medium sandstone were transformed into dark-grey and black-grey silty mudstone (Figure 2b). The core lithology of the Taiyuan Formation is interbedded with dark-grey siltstone, fine sandstone, and grey-black sandy mudstone, mainly sandstone (Figure 2c).

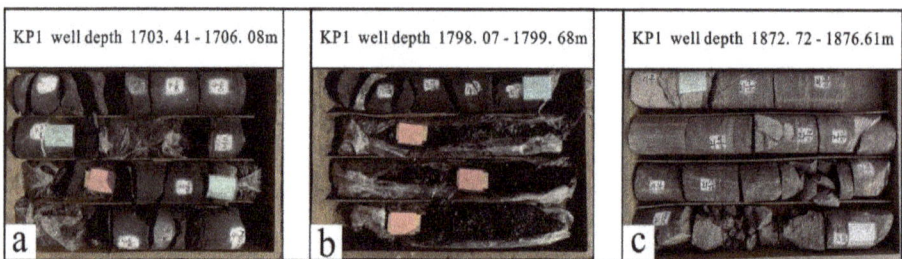

Figure 2. Core lithology maps of the Taiyuan Formation and Shanxi Formation of Well KP1 in the study area. (**a**) The dark-grey fine sandstone of the Shanxi Formation is interbedded with a thin layer of dark-grey sandy mudstone. (**b**) Grey-black sandy mudstone of the Taiyuan Formation. (**c**) The interbedded deep grey siltstone, fine sandstone, and grey-black sandy mudstone of the Taiyuan Formation.

The brittle mineral content of shale in the Shanxi Formation was between 22.5% and 48.4% (average = 40.02%), mainly quartz minerals (Figure 3b). The Taiyuan Formation shale had a comparatively high brittle mineral concentration, ranging from 34.1% to 58.5% (average = 47.7%) (Figure 3b). The brittle mineral content of shale in the Taiyuan and Shanxi Formations fulfilled the minimum fracturing feasibility standard (brittle mineral = 40%). However, the contents of clay minerals in the shales of the Taiyuan and Shanxi Formations were high, with averages of 46.1% and 40.6%, respectively [41] (Figure 3b).

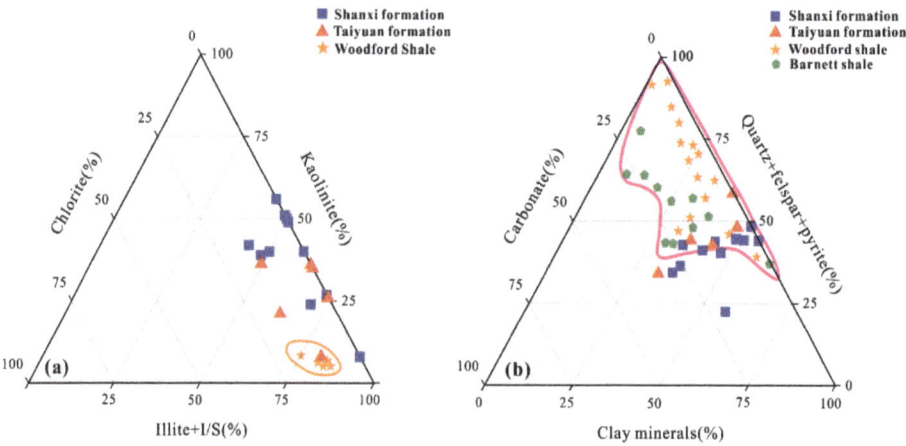

Figure 3. Ternary diagrams of clay mineral content and mineral content in mud shale of Well KP1. (**a**) Ternary diagram of clay mineral content. (**b**) Ternary diagram of mineral content.

The clay minerals in the Taiyuan and Shanxi Formations were mainly illite–montmorillonite mixed layer minerals (average values of 57.27% and 65.83%, respectively) and kaolinite (average values of 38.72% and 27%, respectively). This slightly differs from the clay mineral composition of the Woodford shale (Figure 3a). Therefore, compared with the North American marine gas shale (Barnett and Woodford shales), the brittle mineral content of the Taiyuan and Shanxi Formations shale is relatively low, barely meeting the hydraulic fracturing standard, and the clay mineral content is relatively high, making fracturing more difficult (Figure 3a).

4.2. Organic Matter Abundance

According to the experimental results, the TOC content of the shale of the Shanxi Formation was between 0.20% and 15.39% (average = 2.60%). Furthermore, 47.37% of the shale samples met the standards for excellent source rocks, and 89.47% of the shale samples met the standards for medium source rocks (Figure 4; Table 1). The TOC content of the shale of the Taiyuan Formation was between 1.71% and 12.15%, with an average of 6.84%. Moreover, 85.71% of the shale samples met the standards for excellent source rocks, and all samples met the standards for good source rocks (Figure 4; Table 1).

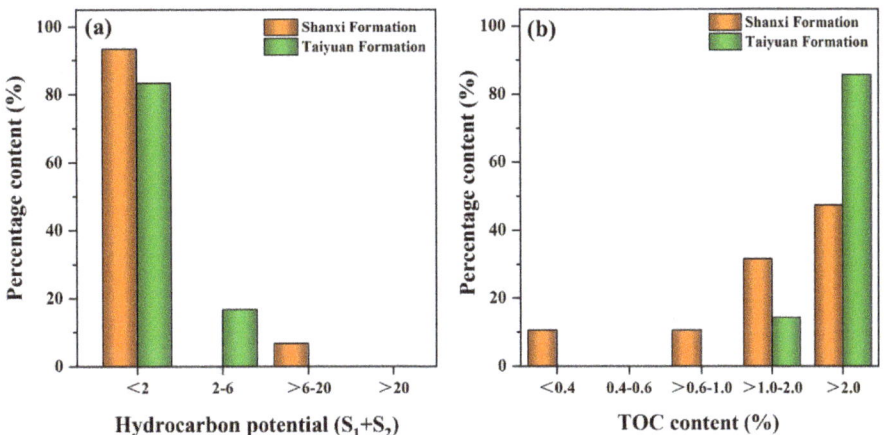

Figure 4. The frequency distributions of organic matter abundance of source rocks in Well KP1. (**a**) Hydrocarbon potential ($S_1 + S_2$) diagram. (**b**) TOC content (%) diagram.

Table 1. Evaluation criteria for organic matter abundance of continental source rocks.

Index	Non-Source Rocks	Poor Source Rocks	Medium Source Rocks	Good Hydrocarbon Source Rocks
TOC (%)	<0.4	0.4–0.6	>0.6–1.0	>1.0–2.0
$S_1 + S_2$ (mg/g)	-	<2	2–6	>6–20

According to the curve fitting equation of shale samples, $\log10(S_1 + S_2) = 1.49\log10(TOC) - 1.18$ ($R^2 = 0.86$), indicating that TOC content has a good correlation with hydrocarbon generation potential. According to Figure 5, most of the Well KP1 shale samples fell into the organic-rich and non-organic-rich shale areas, and only one sample fell into the non-effective shale area [42].

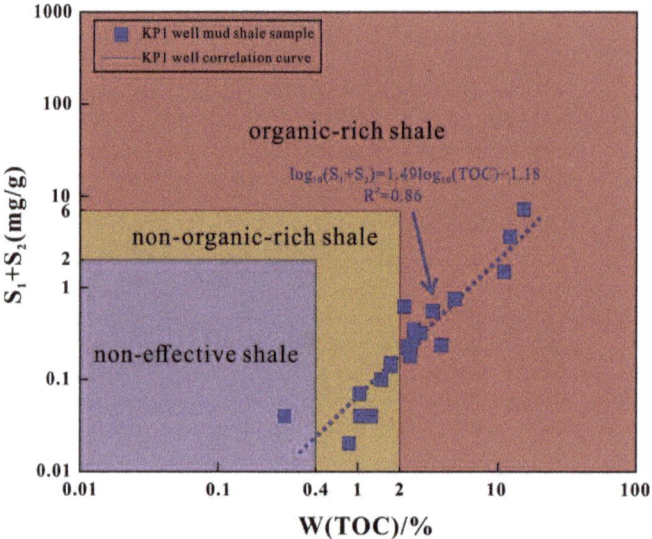

Figure 5. Plot of the hydrogen index ($S_1 + S_2$) versus the TOC, outlining the source potential of different lithotypes and the relationship between the TOC and $S_1 + S_2$ from the KP1 shale.

4.3. Organic Matter Type

The maceral characteristics of kerogen can effectively provide the biogenic composition of organic matter and determine the kerogen type. According to the discriminant diagram, the shale samples in Well KP1 were mainly of the mixed type and a small amount of the sapropel type (Figure 6).

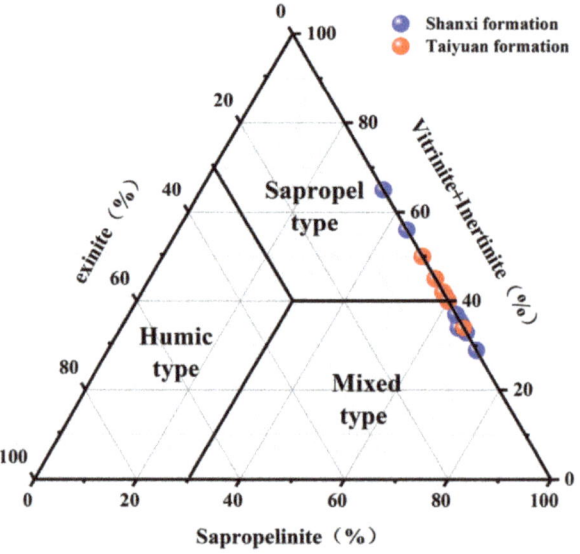

Figure 6. Triangle diagram of kerogen maceral composition of source rock in Well KP1.

This study showed that the shale of the Taiyuan and Shanxi Formations has abundant sapropelic components, indicating that organic matter mainly originates from lower aquatic algae. The analysis showed that the source rocks of the Taiyuan and Shanxi Formations have rich oil and gas generation potential.

4.4. Element Geochemical Characteristics

According to the test results for the major elements, the shale samples of the Taiyuan and Shanxi Formations had the highest Al, Si, and Fe contents. The Si content was the highest. The Si contents in the Taiyuan and Shanxi Formations were between 44.52% and 66.57% (average = 56.22%) and between 52.81% and 64.92% (average = 56.42%), respectively. In addition to these three elements, there were some elements with relatively high contents, such as Mg, K, Ca, and Ti, with average contents of 2.72%, 4.30%, 2.59%, and 1.04%, respectively (Table 2).

The trace-element content results show that the shale samples are mainly enriched in V, Sr, Zr, and Ba, with contents generally greater than 100 ppm. Their average contents were 138.45 ppm (48.2–175 ppm, $n = 25$), 201.88 ppm (98–434 ppm, $n = 25$), 199.68 ppm (131–430 ppm, $n = 25$), and 473.2 ppm (276–880 ppm, $n = 25$), respectively. The total rare earth element abundance ($\sum$REE) of the Well KP1 shale samples ranged from 179.16 to 390.52 µg/g (average = 249.73 µg/g) (Table 3).

Table 2. Organic matter content (TOC %) and whole-rock major-element data of Taiyuan and Shanxi Formations shale in KP1 well.

Stratum	Sample No.	Depth (m)	TOC (%)	Na [a] (wt%)	Mg [b] (wt%)	Al [c] (wt%)	Si [d] (wt%)	Mn [e] (wt%)	K [f] (wt%)	Ca [g] (wt%)	Ti [h] (wt%)	P [i] (wt%)	Fe [j] (wt%)
	KP1-1	1703.56	2.15	0.56	1.36	25.92	59.12	0.11	4.02	0.40	1.25	0.09	7.17
	KP1-2	1706.38	1.22	0.50	1.88	26.70	59.36	0.04	3.80	0.43	1.17	0.15	5.97
	KP1-3	1708.5	2.52	0.52	2.51	22.07	57.84	0.20	4.10	1.18	1.15	0.13	10.31
	KP1-4	1711.73	4.89	0.64	6.07	18.31	50.64	0.23	4.34	9.91	0.84	0.14	8.88
	KP1-5	1716.06	0.87	0.62	1.47	21.60	66.57	0.02	4.27	0.23	0.81	0.09	4.32
	KP1-6	1717.97	1.47	0.55	1.69	22.41	65.04	0.02	4.87	0.21	0.83	0.06	4.31
	KP1-7	1719.68	0.2	0.93	3.75	24.80	53.56	0.06	5.74	4.76	1.00	0.13	5.26
	KP1-8	1720.37	3.43	0.48	4.21	18.37	60.53	0.14	2.83	5.63	1.11	0.16	6.56
	KP1-9	1723.72	0.3	0.36	1.60	17.68	45.81	0.55	2.87	0.93	0.89	0.15	29.17
Shanxi Formation	KP1-10	1725.77	0.86	0.43	1.70	22.86	60.13	0.26	3.94	0.57	1.32	0.13	8.66
	KP1-11	1728.37	1.72	0.40	4.46	19.09	51.60	0.29	3.45	6.01	1.17	0.22	13.32
	KP1-12	1730.26	2.52	0.52	4.99	20.61	55.57	0.13	3.96	7.58	1.26	0.13	5.24
	KP1-13	1732.41	1.03	0.54	1.72	22.88	58.20	0.19	4.40	1.05	1.12	0.15	9.75
	KP1-14	1768.15	2.27	0.44	1.71	24.32	55.21	0.37	4.17	0.86	1.14	0.18	11.59
	KP1-15	1772.01	2.31	0.42	1.60	25.92	50.68	0.46	4.06	0.88	1.21	0.20	14.57
	KP1-16	1774.97	3.95	0.57	0.73	29.92	59.73	0.02	2.57	0.14	1.31	0.05	4.97
	KP1-17	1785.92	1.04	0.52	2.70	21.81	57.48	0.15	5.14	2.53	0.95	0.30	8.42
	KP1-18	1788.44	1.24	0.49	3.52	19.77	55.23	0.36	5.15	2.96	0.88	0.15	11.48
	KP1-19	1790.80	15.39	0.67	1.38	20.98	44.52	0.06	2.92	1.57	1.00	0.06	26.82
	KP1-20	1794.50	2.36	0.55	6.91	17.08	52.81	0.18	3.75	10.58	0.84	0.18	7.12
	KP1-21	1796.43	11.05	0.63	2.78	19.08	56.22	0.19	4.71	2.58	0.77	0.12	12.92
Taiyuan Formation	KP1-22	1798.35	11	0.79	1.89	20.50	52.94	0.11	4.52	1.47	0.88	0.08	16.83
	KP1-23	1846.15	2.82	0.42	2.29	20.67	53.04	0.19	5.13	0.59	0.98	0.17	16.51
	KP1-24	1861.47	1.71	0.44	2.31	19.68	61.69	0.08	7.13	0.29	1.00	0.06	7.32
	KP1-25	1885.79	12.15	0.47	1.41	22.64	64.92	0.02	4.67	0.57	1.31	0.10	3.90

[a] Na: Na$_2$O; [b] Mg: MgO; [c] Al: Al$_2$O$_3$; [d] Si: SiO$_2$; [e] Mn: MnO; [f] K: K$_2$O; [g] Ca: CaO; [h] Ti: TiO$_2$; [i] P: P$_2$O$_5$; [j] Fe: TFe$_2$O$_3$.

Table 3. The whole-rock trace-element analysis data of Taiyuan and Shanxi Formations shale in KP1 well.

Stratum	Sample No.	Depth (m)	$Sc(10^{-6})$	$V(10^{-6})$	$Co(10^{-6})$	$Ni(10^{-6})$	$Cu(10^{-6})$	$Zn(10^{-6})$	$Rb(10^{-6})$	$Sr(10^{-6})$	$Y(10^{-6})$	$Zr(10^{-6})$	$Nb(10^{-6})$	$Mo(10^{-6})$	$Cs(10^{-6})$	$Ba(10^{-6})$
Shanxi Formation	KP1-1	1703.56	11.6	134	47	72.3	49.3	88.6	73.4	142	30.3	288	15.8	6.17	8.36	668
	KP1-2	1706.38	11.3	145	11.7	23.7	66	109	68.6	128	24.1	242	15	0.52	6.14	433
	KP1-3	1708.50	20.5	155	22.5	36	65.7	132	73.8	136	32.6	220	15.2	1.08	7.42	485
	KP1-4	1711.73	18	166	13.3	21	26.2	103	87.7	434	28.6	155	12.3	1.04	6.35	645
	KP1-5	1716.06	12.4	102	10.5	19.9	55	82.8	80.4	98	19	161	14.4	0.7	6.69	505
	KP1-6	1717.97	13.5	136	16.8	28.2	58.9	68.4	92.2	118	26.1	151	15.6	1.58	10.3	507
	KP1-7	1719.68	18.1	48.2	6.66	9.06	14.9	114	81.6	257	43.9	430	18.4	2.02	5.57	880
	KP1-8	1720.37	18.4	113	11.5	18.7	45.3	141	74.4	193	37	246	13.5	2.72	5.47	416
	KP1-9	1723.72	25.8	163	32.6	30.9	52.9	87.8	60.3	138	34.4	150	12.5	0.56	7.38	383
	KP1-10	1725.77	18.3	171	20.8	37	75.6	107	64.9	136	32.9	215	16.7	0.38	8.46	366
	KP1-11	1728.37	24.9	144	20.7	35.1	66.5	121	83.4	175	37.1	185	14.4	0.7	6.69	414
	KP1-12	1730.26	13.4	118	31.1	53.8	64.5	139	81.8	165	26.5	172	17.2	1.73	7.07	364
	KP1-13	1732.41	16.3	175	23.6	35	66.4	91.8	75.2	146	30.8	189	14.9	0.57	11.2	386
	KP1-14	1768.15	17.6	139	22.4	33	57.8	108	97	230	30.7	173	14.8	1.25	7.83	394
	KP1-15	1772.01	7.26	164	21.8	27	64.1	96	47.5	192	22.2	202	16.3	1.14	6.75	313
	KP1-16	1774.97	9.89	159	25	43.2	18.9	224	46	193	16.3	183	13.8	1.62	5.38	363
	KP1-17	1785.92	22	164	24	43.4	45.9	98.5	118	282	30.2	174	14.4	0.96	10	514
	KP1-18	1788.44	21.3	157	20.2	28.9	41.4	54	93.3	260	32	145	14	1.26	8.97	511
	KP1-19	1790.80	12.9	93.4	16.9	41.2	34.2	77.4	46.3	247	20.3	134	9.43	3.18	4.57	276
Taiyuan Formation	KP1-20	1794.50	15.6	116	18.7	27.9	39.5	153	77.5	304	23.5	162	11.4	0.44	5.79	514
	KP1-21	1796.43	16.9	144	18.7	58.5	46.4	128	81.9	294	32.4	131	10.7	7.29	7.17	602
	KP1-22	1798.35	15.1	148	18.6	57.2	43.1	92.8	71.3	260	26.3	131	10.3	2.27	5.15	474
	KP1-23	1846.15	21.4	162	18.5	29.4	50.2	120	93.8	223	38.7	202	15.2	0.79	11.4	535
	KP1-24	1861.47	16.6	132	19.6	25	48.4	114	114	168	32.4	239	17.6	0.68	12.8	568
	KP1-25	1885.79	15	114	8.25	22.3	58.5	24.6	75.1	128	30.1	312	16.3	0.78	10.6	314

5. Discussion

5.1. Provenance and Tectonic Setting

The location of source rock deposition in petroliferous basins influences the preservation state of organic matter in the source rocks [43,44]. The shale material composition is easily altered by sedimentary cycles [45]. Therefore, determining the sedimentary cycle rate of shale samples is a prerequisite for an accurate reflection of their provenance characteristics and tectonic background [43]. Previous studies have shown that the trace elements Zr, Th, and Sc can effectively represent the degree of sediment recycling and changes in rock mineral composition [45,46]. Based on this, a Th/Sc–Zr/Sc diagram was established to determine the degree of sediment recycling [46]. The results show that the average Th/Sc and Zr/Sc ratios of the Well KP1 samples in the northern North China Basin were 0.78 and 13.13, respectively (Figure 7), which are close to the elemental composition characteristics of the PAAS samples (Th/Sc = 0.91; Zr/Sc = 13.12) [47]. This indicated that the shale samples were less affected by weathering, denudation, transportation, and sedimentary recycling. Geochemical shale data can be used to accurately identify the source rock and its tectonic background characteristics.

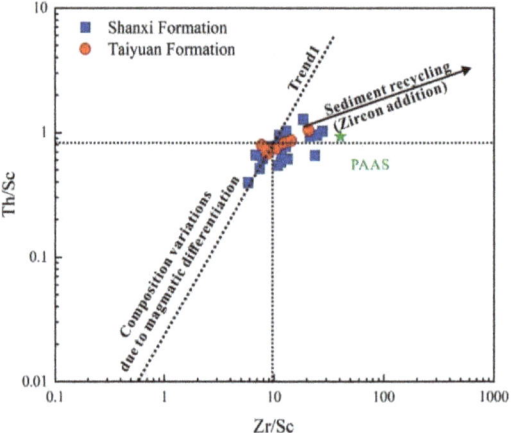

Figure 7. Th/Sc–Zr/Sc bivariate diagram of shales and carbonaceous marls to show their sediment recycling.

REE and trace elements (such as Sc, Hf, and Zr) play directional roles in tracing the source area of hydrocarbon source rocks and distinguishing the tectonic background [13,43,48]. The results of the La/Th–Hf provenance discrimination showed that most shale samples from the Shanxi Formation originated in a source area with felsic volcanic rocks. This indicates that the shale source area may be related to acidic magmatic activity. The mud shale samples from the Taiyuan Formation basically fell into the mixed area of felsic and basic rocks and the felsic source area (Figure 8a). In addition, the La/Yb–REE binary diagram is an effective index for distinguishing the source characteristics of source rocks [49]. According to Figure 8b, the samples from the Taiyuan and Shanxi Formations originated from a granite (felsic) source area, and only a single sample originated from the basic basalt source area. In summary, the source rocks of the Taiyuan and Shanxi Formations of Well KP1 chiefly originated from felsic (granite) source areas.

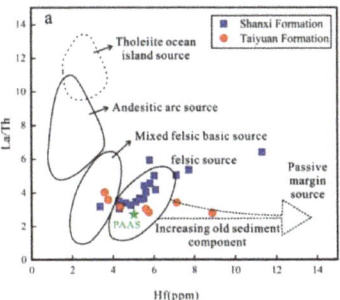
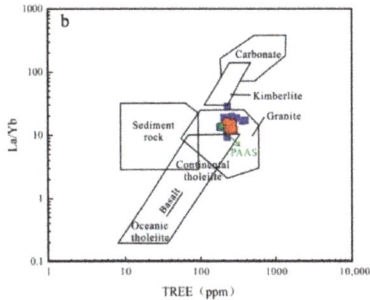

Figure 8. Provenance identification plots for the samples of KP1 well: (**a**) La/Th vs. Hf and (**b**) La/Yb vs. ∑REE.

The tectonic background of the source area can be determined using trace and rare earth element discriminant diagrams (La–Th–Sc, Th–Co–Zr/10, and Th–Sc–Zr/10) [3,50]. As shown in Figure 9, the shale samples from Well KP1 were located in or near the continental island arc area. This is consistent with the previous studies showing that the North China Basin was in a stable platform subsidence area during the Late Carboniferous–Permian [50].

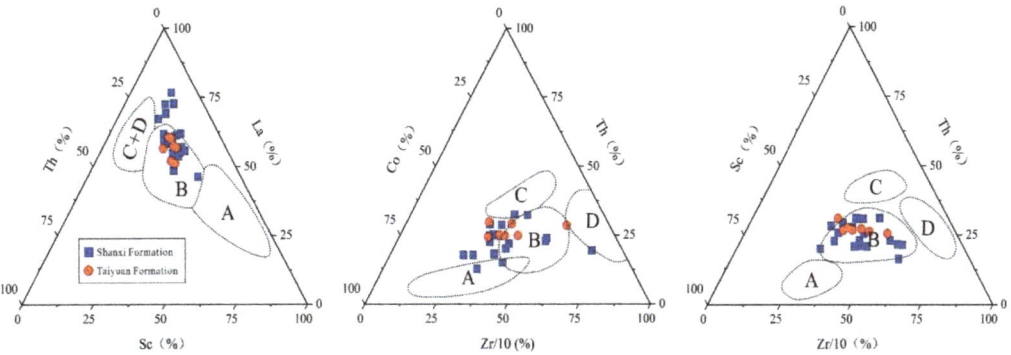

Figure 9. La–Th–Sc, Th–Co–Zr/10, and Th–Sc–Zr/10 plots of the shales from the Taiyuan and Shanxi Formations in the Well KP1 for tectonic discrimination: (A) oceanic island arc; (B) continental island arc; (C) active continental margin; and (D) passive margin.

5.2. Paleoclimatic Conditions

Paleoclimatic conditions not only regulate the weathering and denudation degree of source rocks and the growth and prosperity of paleontology but also affect the enrichment and preservation of organic matter in sediments [13,28].

Trace elements Sr, Cu, and Rb are sensitive to climatic conditions. Under warm and humid climatic conditions, Sr is preferentially lost, whereas Cu and Rb remain stable. Therefore, the concentration ratios of Sr/Cu and Rb/Sr can indicate the characteristics of paleoclimatic conditions [28,32]. When the Sr/Cu ratio is between 1 and 10, it represents warm and humid climatic conditions. When the Sr/Cu ratio is greater than 10, it indicates hot and dry climatic conditions. In addition, the Rb/Sr ratio reflects the characteristics of precipitation during the sedimentary stage. The larger the ratio, the more abundant the precipitation and the stronger the hydrodynamic conditions. The smaller the ratio, the lower the precipitation and the worse the hydrodynamic conditions [51].

According to the correlation analysis of Rb/Sr and Sr/Cu in the shale samples, the curve-fitting equation of the shale samples in Well KP1 was $y = 0.28 + 0.28/(x - 1.14)$ ($R^2 = 0.59$) (Figure 10). When the Sr/Cu ratio was less than 10, the Rb/Sr ratio increased rapidly. When the Sr/Cu ratio exceeded 10, the Rb/Sr ratio decreased slowly. This shows that the Sr content is sensitive to climate. When rainfall is rich, the Sr content is lost rapidly, which causes the Rb/Sr ratio to increase rapidly. When the climate was dry, the Sr content remained stable, causing the Rb/Sr ratio to decrease slowly (Figure 10). This indicates that paleoclimatic conditions correlate well with precipitation. The Sr/Cu ratio in the shale samples from Well KP1 in the study area was between 1.78 and 17.25, and the Rb/Sr ratio was between 0.19 and 0.82 (Table 4). The average value of Sr/Cu in the Taiyuan Formation to the Shanxi Formation is on the rise (increased from 2.18 to 5.13). The average Rb/Sr ratio was maintained at approximately 0.4. This suggests that the warm, humid paleoclimate gave way to a hot, dry one. From the Taiyuan Formation to the middle and lower parts of the Shanxi Formation, the Sr/Cu ratio was less than 10, and the Rb/Sr ratio remained low (close to 0.4). This indicates that during this sedimentary stage, the climate changed from warm and humid to hot and dry, precipitation decreased, and hydrodynamic conditions weakened. Subsequently, precipitation increased, and warm and humid conditions were experienced again.

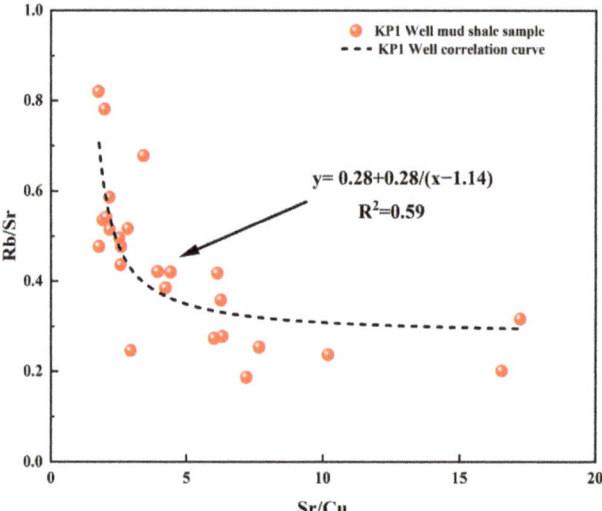

Figure 10. Relationship diagram between the proxies of paleoclimatic conditions of shale samples in northern North China Basin.

Table 4. Paleoclimatic conditions analysis by trace-element ratio.

Ratios	Index	KP1 Well Formation		Wet and Warm	Dry and Hot
		C_2t	P_1s		
Paleoclimatic	Sr/Cu	1.78 [a]–17.25 [b] (5.13) [c]	2.18 [a]–7.70 [b] (2.18) [c]	1–10	>10
	Rb/Sr	0.19 [a]–0.82 [b] (0.44) [c]	0.25 [a]–0.68 [b] (0.42) [c]	-	

[a] Maximum; [b] minimum; [c] average.

The Taiyuan Formation has a constant warm and humid climate. The climate of the Shanxi Formation shows a reciprocating change with increasing and decreasing rainfall, and the paleoclimatic conditions were even worse.

5.3. Paleoredox Conditions

According to Figure 11a, the Mo/TOC values of the Taiyuan and Shanxi Formations in the northern North China Basin are between 0.06 and 10.1 (average = 1.04). Only a single sample with medium retention intensity was in the Cariaco Basin model, and the remaining samples were in the Black Sea model. This indicated that the sedimentary environment was a strong water-retention environment [19,32,52] (Figure 11a). According to the U-EF–Mo-EF covariation diagram, the samples of the Taiyuan and Shanxi Formations were consistent with the comprehensive evaluation index, and all data points in the study area were mainly near the anoxic area. The samples from the Taiyuan Formation had a high degree of hypoxia and were in a strong water-retention environment, indicating that the sea level was relatively stable. The samples from the Shanxi Formation were in an oxygen-poor area as a whole, and the sedimentary environment was mainly a medium–strong water-retention environment. It is speculated that the stranded environment will change owing to changes in sea level (Figure 11b).

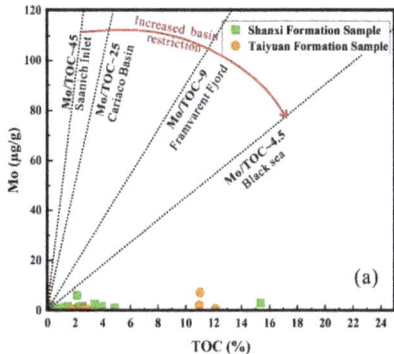 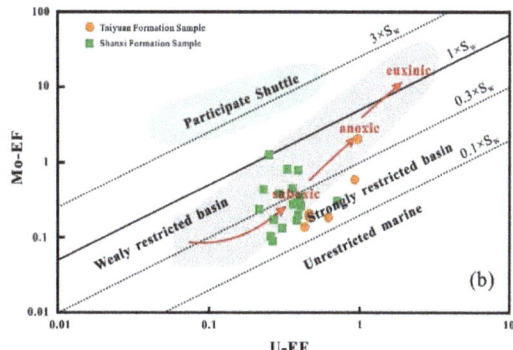

Figure 11. (a) Relationship between Mo and TOC in the shale of the Taiyuan and Shanxi Formations in the northern part of North China Basin. (b) U-EF and Mo-EF covariation diagram of shale in Taiyuan and Shanxi Formations in northern North China Basin.

In general, the paleoredox conditions of the sedimentary water in the study area were mainly anoxic reduction conditions in a medium–strong water-retention environment. The sedimentary basin of Well KP1 in the northern part of the North China Basin was in the Taiyuan Formation sedimentary stage during the Late Carboniferous. The sedimentary water body had an anoxic reduction and a strong water-retention environment. During the sedimentary stage of the Permian Shanxi Formation, the paleoredox level and retention environment of the sedimentary water fluctuated with changes in sea level; however, it was still an anoxic reduction and medium–strong water-retention environment.

5.4. Multivariate Statistical Analysis of Organic Matter Enrichment Control Factors

Organic matter enrichment may result from interactions between multiple environmental factors. However, a single environmental factor cannot explain the enrichment of organic matter. Therefore, grey correlation and multiple linear regression analyses of organic matter enrichment can systematically characterize the organic matter content under the combined effects of various environmental factors.

The principle of a grey correlation analysis is that the degree of correlation between curves can be reflected by determining the geometric similarity between the reference sequence (the TOC content) and several comparative data columns (paleoproxy water columns) [53,54] (Table 5). The grey correlation method can be used to explore the correlation between the organic matter content and paleoenvironmental factors (such as paleoclimate, paleosalinity, paleoproductivity, terrigenous clastic input, sedimentation rate, and paleoredox). The primary and secondary relationships between the paleoenvironmental factors and TOC content were determined by comparing the grey correlation degree of each paleoenvironmental factor in the system.

Table 5. Data table of organic matter content and paleoenvironmental factors of mud shale samples in Well KP1.

Stratum	Sample No.	TOC (%)	Sr/Cu	Ti (%)	Ba_{bio}	Paleo-	Mo-EF	U-EF
Shanxi Formation	KP1-1	2.15%	2.88	1.25%	657.89	0.32	1.28	0.25
	KP1-2	1.22%	1.94	1.17%	422.59	0.38	0.10	0.25
	KP1-3	2.52%	2.07	1.15%	476.39	0.30	0.26	0.40
	KP1-4	4.89%	16.56	0.84%	637.86	0.52	0.30	0.70
	KP1-5	0.87%	1.78	0.81%	496.57	0.22	0.17	0.27
	KP1-6	1.47%	2.00	0.83%	498.26	0.27	0.38	0.29
	KP1-7	0.20%	17.25	1.00%	870.33	0.48	0.44	0.23
	KP1-8	3.43%	4.26	1.11%	408.84	0.36	0.79	0.39
	KP1-9	0.30%	2.61	0.89%	376.11	0.50	0.17	0.38
	KP1-10	0.86%	1.80	1.32%	357.08	0.34	0.09	0.27
	KP1-11	1.72%	2.63	1.17%	406.56	0.44	0.20	0.39
	KP1-12	2.52%	2.56	1.26%	355.96	0.24	0.45	0.35
	KP1-13	1.03%	2.20	1.12%	377.08	0.32	0.13	0.30
	KP1-14	2.27%	3.98	1.14%	384.51	0.28	0.28	0.36
	KP1-15	2.31%	3.00	1.21%	302.89	0.47	0.24	0.21
	KP1-16	3.95%	10.21	1.31%	351.33	0.39	0.29	0.38
	KP1-17	1.04%	6.14	0.95%	505.49	0.37	0.24	0.40
	KP1-18	1.24%	6.28	0.88%	503.29	0.46	0.34	0.40
	KP1-19	15.39%	7.22	1.00%	267.82	0.48	0.81	0.33
Taiyuan Formation	KP1-20	2.36%	7.70	0.84%	507.34	0.48	0.14	0.43
	KP1-21	11.05%	6.34	0.77%	594.56	0.43	2.05	0.96
	KP1-22	11%	6.03	0.88%	466.01	0.48	0.59	0.92
	KP1-23	2.82%	4.44	0.98%	526.94	0.48	0.20	0.46
	KP1-24	1.71%	3.47	1.00%	560.33	0.24	0.19	0.46
	KP1-25	12.15%	2.19	1.31%	305.17	0.37	0.18	0.61

The correlation coefficient heat map between various influencing factors and the TOC content of the shale samples from the Shanxi Formation shows that the correlation from strong to weak is paleosalinity (Sr/Ba), paleoproductivity (Ba_{bio}), paleoredox (Paleo-), paleoclimate (Sr/Cu), debris flow (Ti, %), and deposition rate (La/Yb) $_N$. The grey correlation coefficient between the paleosalinity index and the TOC was the highest (0.99), and there was no significant correlation between the other influencing factors and the TOC content. This shows that sea-level fluctuations play a decisive role in the enrichment of shale organic matter in the Shanxi Formation in the study area (Figure 12).

According to Figure 11b, the grey correlation degree of each influencing factor of Taiyuan Formation shale from strong to weak is paleosalinity (Sr/Ba), paleoproductivity (Ba_{bio}), sedimentation rate (La/Yb) $_N$, debris flux (Ti, %), paleoredox (Paleo-), and paleoclimate (Sr/Cu). The correlation coefficient between the paleosalinity index and the TOC was the highest (0.99), followed by paleoproductivity, and the coefficient was −0.48. This indicates that sea-level fluctuations have a significant effect on the enrichment of organic matter in Taiyuan Formation shale, and the ancient production conditions also play a certain role in controlling the enrichment of organic matter. The remaining control factors have little effect on the enrichment of organic matter in the Taiyuan Formation in the study area (Figure 12).

Moreover, a multiple linear regression analysis can be used to effectively study the uncertain interdependence and restrictive relationship between organic matter enrichment and various paleoenvironmental factors. This relationship can be mathematically expressed. The purposes of this are to determine the image-specific performance of the organic matter enrichment process and to use unknown variables to predict or test the accuracy of the change [55,56].

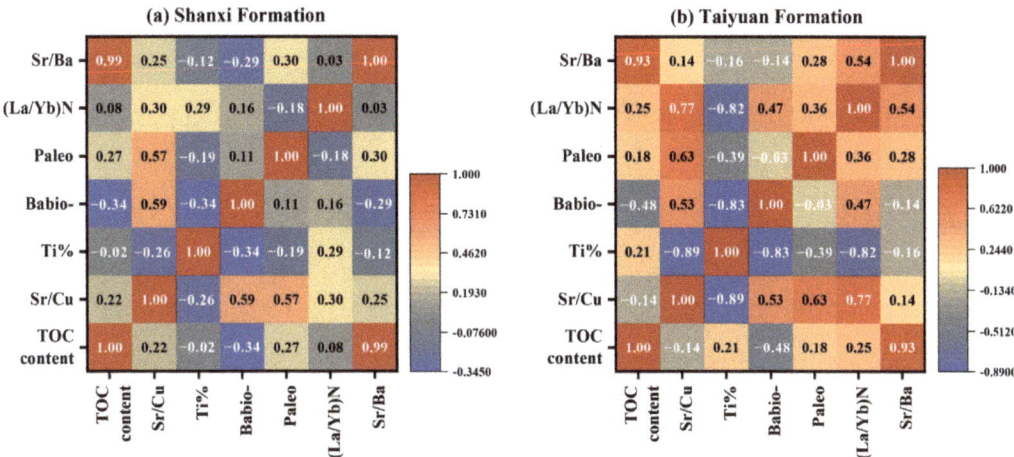

Figure 12. The correlation coefficient heat maps of paleoenvironmental factors of mud shale samples from Well KP1 in the northern part of North China Basin. (**a**) Shanxi Formation correlation coefficient heat map. (**b**) Taiyuan Formation correlation coefficient heat map.

A regression analysis must first establish a regression equation. Using the data of dependent variable Y (the TOC content) and independent variable X (paleoproxy water column), the regression parameters ($a_1, a_2, \cdots, a_m$) were estimated based on the classical least-squares principle ($V^T PV = min$). However, there were errors in both the dependent and independent variables. The classical least-squares principle was not robust. When data, especially geochemical data, have gross errors, they have a destructive effect on the parameters of the entire regression equation. In a robust regression analysis, the variance–covariance of gross error observations is continuously increased through successive iterative adjustments to automatically eliminate gross errors. Thus, a robust regression is recommended for geochemical data with gross errors. Therefore, the TOC content ensured the accuracy of the regression parameters as much as possible under the premise of gross errors in the paleoenvironmental parameters.

According to the shale data of Well KP1 in the study area, the dependent variable was the TOC content, and the independent variables were the paleoenvironmental parameters that may have a greater impact on the TOC content, namely, paleosalinity (Sr/Ba), deposition rate (La/Yb)$_N$, paleoclimate (Sr/Cu), detrital flux (Ti, %), paleoproductivity (Ba$_{bio}$), and paleoredox (Paleo). A robust regression method was used for the calculations.

The fitting parameters of the shale samples from the Shanxi Formation in Well KP1 were determined using the robust least-squares method. When the significant p-value of the paleoenvironmental factor parameter was closer to zero, the parameter had a significant effect on organic matter enrichment. In the process of organic matter enrichment in the Shanxi Formation, the p-values of detrital flux (Ti, %), deposition rate (La/Yb)$_N$, and paleosalinity (Sr/Ba) were all <0.01 (Table 6). This indicates that these three factors had a significant effect on the enrichment of organic matter in the Shanxi Formation, whereas the other factors had little effect. Furthermore, a multiple linear regression model was developed based on robust regression as follows:

$$\text{TOCcontent} = -0.017 - 0.0 \times \text{Paleoclimate (Sr/Cu)} + 1.563 \times \text{Detrital flux (Ti\%)} + 0.001 \times \text{Paleoredox}$$
$$\text{(Paleo-)} + 0.0 \times \text{Deposition rate (La/Yb)}_N + 0.01 \times \text{Paleosalinity (Sr/Ba)} \qquad (4)$$

Table 6. Robust regression solving fitted parameters for KP1 well mud shale.

Fitting Parameters	Shanxi Formation			Correlation Coefficients R^2	Taiyuan Formation			Correlation Coefficients R^2
	Parameter Value	Significance p-Value	Whether Significant		Parameter Value	Significance p-Value	Whether Significant	
Sr/Cu	0	0.993	No		−0.002	-	-	
Ti%	1.563	0.0001	Yes		2.95	-	-	
Ba$_{bio}$	0	0.095	No		0	-	-	
Paleo-	0.001	0.603	No	0.997	−0.001	-	-	0.987
(La/Yb)$_N$	0	0.006	Yes		0.003	-	-	
Sr/Ba	0.01	0.0001	Yes		0.007	-	-	
Constant value		−0.017				0.025		

The error between the fitted and actual values was small, and the correlation coefficient was $R^2 = 0.997$ (Figure 13). The two curves were well-fitted and showed strong accuracy. The sample size of the Taiyuan Formation was small, and a robust regression analysis could not be performed. The samples from the Taiyuan Formation were analyzed using the ridge regression method in least-squares regression. According to ridge regression, a multiple linear regression model was obtained as follows:

$$\text{TOCcontent} = 0.025 - 0.002 \times \text{Paleoclimate (Sr/Cu)} + 2.95 \times \text{Detrital flux (Ti\%)} - 0.001 \times \text{Paleoredox (Paleo-)} + 0.003 \times \text{Deposition rate (La/Yb)}_N + 0.007 \times \text{Paleosalinity (Sr/Ba)} \tag{5}$$

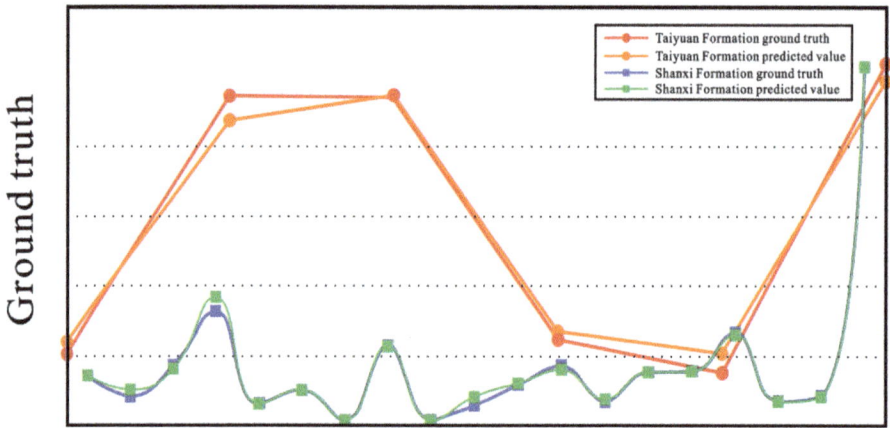

Figure 13. Curves of fitted and true values of regression analysis of mud shale samples from KP1 well.

Because of the small number of samples, the significance of each factor at that level was low, and it was impossible to determine the factors that had a decisive influence on organic matter. However, the error between the fitted and true values was small (Figure 13), and the correlation coefficient was $R^2 = 0.987$ (Table 6).

Based on these results, it is considered that the enrichment of organic matter in the Taiyuan Formation in the study area is controlled by paleosalinity and paleoproductivity. During the sedimentary stage of the Shanxi Formation, paleosalinity was the main factor controlling the enrichment of organic matter, while the remaining paleoenvironmental conditions had little effect.

5.5. Organic Matter Accumulation Model

Based on the comprehensive research results, an organic matter enrichment model of the Carboniferous–Permian Taiyuan and Shanxi Formations is discussed. During the sedimentary stage of the Taiyuan Formation, a tidal flat–lagoon environment developed in the study area. A stable and gentle transgression phase brought large amounts of deep-marine nutrients to the sea surface. Moreover, warm and humid climatic conditions promoted the growth of plankton, such as marine algae. This laid a material foundation for the mechanism-rich shale deposition of the Taiyuan Formation. The transgression process moved the sedimentary basin far away from the land. This made it difficult for terrigenous debris to reach, and the input of terrigenous debris decreased. Therefore, the input of terrigenous debris had little effect on organic matter enrichment (Figure 14a).

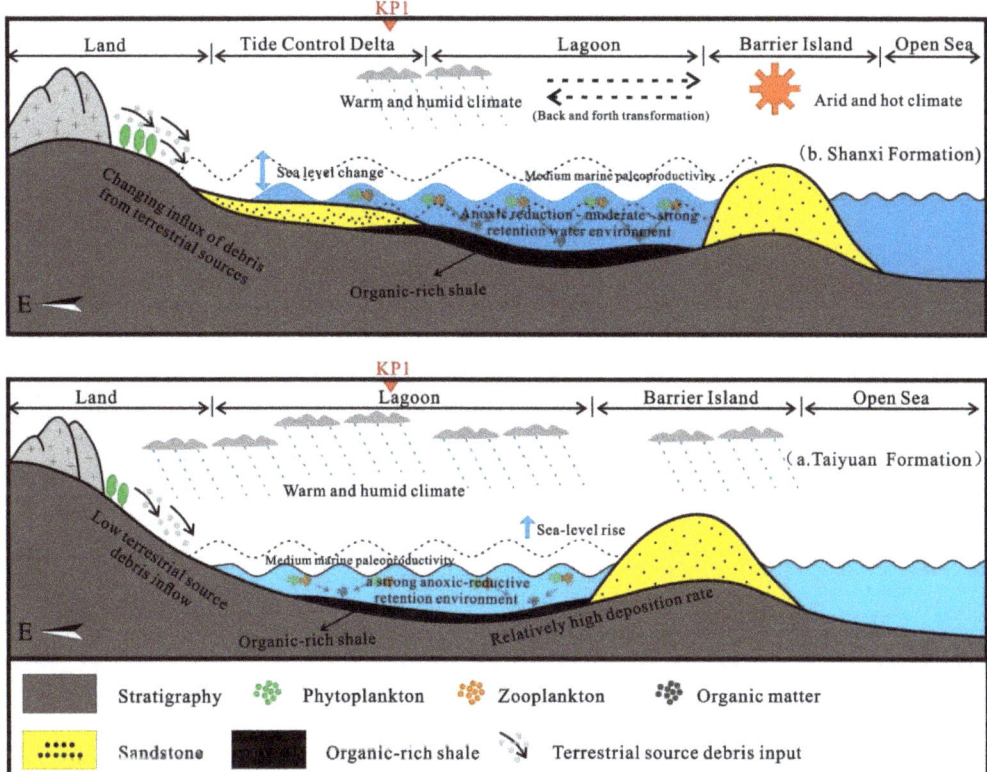

Figure 14. Organic matter enrichment pattern maps of Taiyuan and Shanxi Formations marine–continental transitional shale in northern North China Basin. (**b**) Shanxi Formation sedimentary stage. (**a**) Taiyuan Formation sedimentary stage.

During the sedimentary stage of the Shanxi Formation, a tidal-dominated delta–lagoon environment was developed in the study area. The sea level was in a stage of change, which caused the input of terrigenous debris to fluctuate and changes in the deposition rate. Moreover, the processes of transgression and regression changed the degree of retention of sedimentary water. The Shanxi Formation continues to remain in an anoxic reduction and medium–strong water-retention environment, providing better preservation conditions for organic matter. However, frequent regressions and transgressions caused the degree of enrichment of shale organic matter to differ. Therefore, the Shanxi Formation forms a set of shales with a lower organic matter content (Figure 14b).

Based on this research and analysis, the enrichment of organic matter in marine–continental transitional shales is more dependent on ancient water salinity conditions. The processes of transgression and regression not only control the amount of source debris inflow but also affect the deposition rate. In addition, they affect the redox state of the sedimentary water and the degree of water retention, thereby affecting the enrichment of organic matter. It was further proven that the enrichment of organic matter was controlled by various paleoenvironmental factors, and the optimal conditions for the generation and preservation of organic matter in shale were determined. This study reveals the mechanism of organic matter enrichment in shale and provides a theoretical basis for the exploration of transitional shale gas.

6. Conclusions

The brittle mineral content in the shales of the Taiyuan and Shanxi Formations was relatively low (<40%), and the clay mineral content was high (>50%). The average TOC content was 3.68%. The organic matter was mainly mixed and sapropelic. The felsic (granite) source area was the main source area of shale in the Taiyuan and Shanxi Formations of Well KP1, indicating that the samples were deposited in a continental island arc area.

According to the paleoclimate index and paleoredox index, the source rocks of the Taiyuan Formation in Well KP1 were mainly formed under warm and humid paleoclimate conditions and in an anoxic reduction–strong water retention environment. The source rocks of the Shanxi Formation were mainly formed under complex and changeable paleoclimate conditions and in an anoxic reduction–medium-to-strong water retention environment. The factors influencing organic matter enrichment in the shales of the Taiyuan and Shanxi Formations of Well KP1 in the northern North China Basin were determined using the grey correlation method and a robust regression analysis. The analysis showed that the enrichment of organic matter in Taiyuan Formation shale is controlled by paleosalinity and paleoproductivity. During the sedimentary stage of the Shanxi Formation, paleosalinity was the main factor controlling the enrichment of organic matter. The remaining ancient environmental conditions had little effect.

During the sedimentary stage of the Late Carboniferous–Early Permian in the northern North China Basin, paleosalinity conditions controlled the input of terrigenous debris, changes in deposition rate, redox of sedimentary water, and the degree of water retention in the study area. This provided better preservation conditions for organic matter enrichment and formed organic-rich shale in the marine–continental transitional facies.

Author Contributions: Conceptualization, H.Z. and Y.W.; methodology, H.Z.; software, H.Z., Y.W. and H.C.; validation, H.Z., Y.W. and H.C.; formal analysis, H.Z. and Y.W.; investigation, H.Z.; resources, Y.W.; data curation, Y.W.; writing—original draft preparation, H.Z.; writing—review and editing, H.Z., Y.W., H.C., Y.Z. (Yanming Zhu), J.Y., Y.Z. (Yunsheng Zhang), K.D. and Z.W.; visualization, H.Z.; supervision, Y.W.; project administration, H.C.; funding acquisition, Y.W. All authors have read and agreed to the published version of the manuscript.

Funding: This research was funded by the National Natural Science Foundation of China (42172156), the Fundamental Research Funds for the Central Universities (2022YCPY0201), and the National Key R&D Program of China (No. 2020YFA 0711800).

Institutional Review Board Statement: Not applicable.

Informed Consent Statement: Not applicable.

Data Availability Statement: Data are contained within this article.

Acknowledgments: We would like to thank the Key Laboratory of Coalbed Methane Resources and Reservoir Formation Process for all the support provided in this research.

Conflicts of Interest: The authors declare that they have no known competing financial interests or personal relationships that could have appeared to influence the work reported in this paper.

References

1. Feng, Z.J.; Dong, D.Z.; Tian, J.Q.; Qiu, Z.; Wu, W.; Zhang, C. Geochemical characteristics of longmaxi formation shale gas in the weiyuan area, Sichuan Basin, China. *J. Petrol. Sci. Eng.* **2018**, *167*, 538–548. [CrossRef]
2. Wang, Z.; Zhao, J.Z.; Chen, Y.H. Analysis of palaeo–sedimentary environment and characteristics of Shanxi formation in Shenfu area on the eastern margin of Ordos basin. *J. Xi'an Shiyou Univ. Nat. Sci. Ed.* **2019**, *34*, 24–30. (In Chinese)
3. Cao, Y.T.; Liu, L.; Chen, D.L.; Wang, C.; Yang, W.Q.; Kang, L.; Zhu, X.H. Partial melting during exhumation of Paleozoic retrograde eclogite in North Qaidam, western China. *J. Asian Earth Sci.* **2017**, *148*, 223–240. [CrossRef]
4. Li, P.; Zhang, J.C.; Tang, X.; Huo, Z.P.; Li, Z.; Luo, K.Y.; Li, Z.M. Assessment of shale gas potential of the lower Permian transitional Shanxi-Taiyuan shales in the southern North China Basin. *Aust. J. Earth Sci.* **2021**, *68*, 262–284. [CrossRef]
5. Guo, W.; Gao, J.L.; Li, H.; Kang, L.X.; Zhang, J.W.; Liu, G.H.; Liu, Y.Y. The geological and production characteristics of marine-continental transitional shale gas in China: Taking the example of shale gas from Shanxi Formation in Ordos Basin and Longtan Formation in Sichuan Basin. *Miner. Explor.* **2023**, *14*, 448–458.
6. Zheng, D.Z.; Miska, S.; Ozbayoglu, E.; Zhang, J.G. Combined Experimental and Well Log Study of Anisotropic Strength of Shale. In Proceedings of the SPE Annual Technical Conference and Exhibition, San Antonio, TX, USA, 16–18 October 2023; p. D031S046R003.
7. Liu, S.X.; Wu, C.F.; Li, T.; Wang, H.C. Multiple geochemical proxies controlling the organic matter accumulation of the marine-continental transitional shale: A case study of the Upper Permian Longtan Formation, western Guizhou, China. *J. Nat. Gas. Sci. Eng.* **2018**, *56*, 152–165. [CrossRef]
8. Luo, W.; Hou, M.C.; Liu, X.C.; Huang, S.G.; Chao, H.; Zhang, R.; Deng, X. Geological and geochemical characteristics of marine-continental transitional shale from the Upper Permian Longtan formation, Northwestern Guizhou, China. *Mar. Petrol. Geol.* **2018**, *89*, 58–67. [CrossRef]
9. Xiao, H.; Wang, T.G.; Li, M.J.; Lai, H.F.; Liu, J.G.; Mao, F.J.; Tang, Y.J. Geochemical characteristics of Cretaceous Yogou Formation source rocks and oil-source correlation within a sequence stratigraphic framework in the Termit Basin, Niger. *J. Petrol. Sci. Eng.* **2019**, *172*, 360–372. [CrossRef]
10. Zhang, S.H.; Liu, C.Y.; Liang, H.; Wang, J.Q.; Bai, J.K.; Yang, M.H.; Liu, G.H.; Huang, H.X.; Guan, Y.Z. Paleoenvironmental conditions, organic matter accumulation, and unconventional hydrocarbon potential for the Permian Lucaogou Formation organicrich rocks in Santanghu Basin, NW China. *Int. J. Coal Geol.* **2018**, *185*, 44–60. [CrossRef]
11. Lai, H.F.; Li, M.J.; Liu, J.G.; Mao, F.J.; Xiao, H.; He, W.X.; Yang, L. Organic geochemical characteristics and depositional models of Upper Cretaceous marine source rocks in the Termit Basin, Niger. *Palaeogeogr. Palaeoclimatol. Palaeoecol.* **2018**, *495*, 292–308. [CrossRef]
12. Liang, Q.S.; Zhang, X.; Tian, J.C.; Sun, X.; Chang, H.L. Geological and geochemical characteristics of marine-continental transitional shale from the lower permian Taiyuan formation, taikang uplift, southern North China basin. *Mar. Petrol. Geol.* **2018**, *98*, 229–242. [CrossRef]
13. Chen, Y.H.; Wang, Y.B.; Guo, M.Q.; Wu, H.Y.; Li, J.; Wu, W.T.; Zhao, J.Z. Differential enrichment mechanism of organic matters in the marine-continental transitional shale in northeastern Ordos Basin, China: Control of sedimentary environments. *J. Nat. Gas. Sci. Eng.* **2020**, *83*, 103625. [CrossRef]
14. Peng, Y.X.; Guo, S.B. Lithofacies analysis and paleosedimentary evolution of Taiyuan Formation in Southern North China Basin. *J. Petrol. Sci. Eng.* **2023**, *220*, 111127. [CrossRef]
15. Liu, P.; Zhang, T.Q.; Xu, C.; Wang, X.F.; Liu, C.J.; Guo, R.L.; Lin, H.F.; Yan, M.; Qin, L.; Li, Y. Organic matter inputs and depositional palaeoenvironment recorded by biomarkers of marine-terrestrial transitional shale in the Southern North China Basin. *Geol. J.* **2022**, *57*, 1617–1627. [CrossRef]
16. Nie, H.K.; Chen, Q.; Li, P.; Dang, W.; Zhang, J.C. Shale gas potential of Ordovician marine Pingliang shale and Carboniferous–Permian transitional Taiyuan-Shanxi shales in the Ordos Basin, China. *Aust. J. Earth Sci.* **2023**, *70*, 411–422. [CrossRef]
17. Lash, G.G.; Blood, D.R. Organic matter accumulation, redox, and diagenetic history of the Marcellus Formation, southwestern Pennsylvania, Appalachian basin. *Mar. Petrol. Geol.* **2014**, *57*, 244–263. [CrossRef]
18. Liu, W.Q.; Yao, J.X.; Tong, J.N.; Qiao, Y.; Chen, Y. Organic matter accumulation on the Dalong Formation (Upper Permian) in western Hubei, South China: Constraints from multiple geochemical proxies and pyrite morphology. *Palaeogeogr. Palaeocl.* **2019**, *514*, 677–689. [CrossRef]
19. Shang, F.H.; Zhu, Y.M.; Hu, Q.H.; Wang, Y.; Li, W.; Liu, R.Y.; Gao, H.T. Factors controlling organic-matter accumulation in the Upper Ordovician-Lower Silurian organic-rich shale on the northeast margin of the Upper Yangtze platform: Evidence from petrographic and geochemical proxies. *Mar. Petrol. Geol.* **2020**, *121*, 104597. [CrossRef]
20. Li, Y.; Wang, Z.S.; Gan, Q.; Niu, X.L.; Xu, W.K. Paleoenvironmental conditions and organic matter accumulation in Upper Paleozoic organic-rich rocks in the east margin of the Ordos Basin, China. *Fuel* **2019**, *252*, 172–187. [CrossRef]
21. Qi, Y.; Ju, Y.W.; Tan, J.Q.; Bowen, L.; Cai, C.F.; Yu, K.; Zhu, H.J.; Huang, C.; Zhang, W.L. Organic matter provenance and depositional environment of marine-to-continental mudstones and coals in eastern Ordos Basin, China—Evidence from molecular geochemistry and petrology. *Int. J. Coal Geol.* **2020**, *217*, 103345. [CrossRef]
22. Algeo, T.J.; Liu, J.S. A re-assessment of elemental proxies for paleoredox analysis. *Chem. Geol.* **2020**, *540*, 119549. [CrossRef]
23. Wu, B. The sedimentary geochemical characteristics and geological significance of the Wufeng–Longmaxi Formation accumulation of organic matter black shale on the Southeastern Sichuan Basin, China. *Geofluids* **2022**, *22*, 1900158. [CrossRef]

24. Bai, J.K.; Zhang, S.H.; Liu, C.Y.; Jia, L.B.; Luo, K.Y.; Jiang, T.; Peng, H. Mineralogy and geochemistry of the Middle Permian Pingdiquan Formation black shales on the eastern margin of the Junggar Basin, north–west China: Implications for palaeoenvironmental and organic matter accumulation analyses. *Geo. J.* **2022**, *57*, 1989–2006. [CrossRef]
25. Zou, C.N.; Zhu, R.K.; Chen, Z.Q.; Ogg, J.G.; Wu, S.T.; Dong, D.Z.; Qiu, Z.; Wang, Y.M.; Wang, L.; Lin, S.H.; et al. Organic-matter-rich shales of China. *Earth Sci. Rev.* **2019**, *189*, 51–78. [CrossRef]
26. Jia, S.X.; Zhang, C.K.; Zhao, J.R.; Fang, S.M.; Liu, Z.; Zhao, J.M. Crustal Structure of the Rift-Depression Basin and Yanshan Uplift in the Northeast Part of North China. *Chin. J. Geophys.* **2009**, *52*, 51–63.
27. Chen, Y.H.; Zhu, Z.W.; Zhang, L. Control actions of sedimentary environments and sedimentation rates on lacustrine oil shale distribution, an example of the oil shale in the Upper Triassic Yanchang Formation, southeastern Ordos Basin (NW China). *Mar. Petrol. Geol.* **2019**, *102*, 508–520. [CrossRef]
28. Li, Y.; Yang, J.H.; Pan, Z.J.; Meng, S.Z.; Wang, K.; Niu, X.L. Unconventional Natural Gas Accumulations in Stacked Deposits: A Discussion of Upper Paleozoic Coal-Bearing Strata in the East Margin of the Ordos Basin, China. *Acta Geol. Sin.-Engl.* **2019**, *93*, 111–129. [CrossRef]
29. Ding, J.H.; Zhang, J.C.; Huo, Z.P.; Shen, B.J.; Shi, G.; Yang, Z.H.; Li, X.Q.; Li, C.X. Controlling factors and formation models of organic matter accumulation for the Upper Permian Dalong Formation black shale in the Lower Yangtze region, South China: Constraints from geochemical evidence. *ACS Omega* **2021**, *6*, 3681–3692. [CrossRef]
30. Yan, M.; Feng, J.L. Depositional environment variations and organic matter accumulation of the first member of the Qingshankou formation in the southern Songliao Basin, China. *Front. Earth Sci.* **2023**, *11*, 1249787.
31. Wei, W.; Algeo, T.J. Elemental proxies for paleosalinity analysis of ancient shales and mudrocks. *Geochim. Cosmochim. Acta* **2020**, *287*, 341–366. [CrossRef]
32. Zhang, L.F.; Dong, D.Z.; Qiu, Z.; Wu, C.J.; Zhang, Q.; Wang, Y.M.; Liu, D.X.; Deng, Z.; Zhou, S.W.; Pan, S.Q. Sedimentology and geochemistry of Carboniferous-Permian marine-continental transitional shales in the eastern Ordos Basin, North China. *Palaeogeogr. Palaeocl.* **2021**, *571*, 110389. [CrossRef]
33. Yu, K.; Ju, Y.W.; Qian, J.; Qu, Z.H.; Shao, C.J.; Yu, K.L.; Shi, Y. Burial and thermal evolution of coal-bearing strata and its mechanisms in the southern North China Basin since the late Paleozoic. *Int. J. Coal Geol.* **2018**, *198*, 100–115. [CrossRef]
34. Qi, J.F.; Yang, Q. Cenozoic structural deformation and dynamic processes of the Bohai Bay basin province, China. *Mar. Petrol. Geol.* **2009**, *27*, 757–771. [CrossRef]
35. GB/T21114-2019; Chemical Analysis of Refractory Products by XRF-Fused Cast Bead Method. China Standard Press: Beijing, China, 2019.
36. SY/T5163-2018; Analysis Method for Clay Minerals and Ordinary Non-Clay Minerals in Sedimentary Rocks by the X-ray Diffraction. China Standard Press: Beijing, China, 2018.
37. GB/T 18602-2012; Rock pyrolysis analysis. China Standard Press: Beijing, China, 2012.
38. SY/T5125-2014; Method of Identification Microscopically the Macerals of Kerogen and Indivision the Kerogen Type by Transmitted-Light and Fluorescence. China Standard Press: Beijing, China, 2014.
39. Taylor, S.R.; McLennan, S.M. The geochemical evolution of the continental crust. *Rev. Geophys.* **1995**, *33*, 241–265. [CrossRef]
40. Francois, R.; Honjo, S.; Manganini, S.J.; Ravizza, G.E. Biogenic barium fluxes to the deep sea: Implications for paleoproductivity reconstruction. *Global Biogeochem. Cycles* **1995**, *9*, 289–303. [CrossRef]
41. Liu, J.; Yao, Y.; Liu, D.; Pan, Z.; Cai, Y. Comparison of three key marine shale reservoirs in the southeastern margin of the Sichuan Basin, SW China. *Minerals* **2017**, *7*, 179. [CrossRef]
42. He, J.H.; Ding, W.L.; Jiang, Z.X.; Li, A.; Wang, R.Y.; Sun, Y.X. Logging identification and characteristic analysis of the lacustrine organic-rich shale lithofacies: A case study from the Es3L shale in the Jiyang Depression, Bohai Bay Basin, eastern China. *J. Petrol. Sci. Eng.* **2016**, *145*, 238–255. [CrossRef]
43. Khaled, A.; Li, R.X.; Xi, S.L.; Zhao, B.S.; Wu, X.L.; Yu, Q.; Zhang, Y.N.; Li, D.L. Paleoenvironmental conditions and organic matter enrichment of the Late Paleoproterozoic Cuizhuang Formation dark shale in the Yuncheng Basin, North China. *J. Petrol. Sci. Eng.* **2022**, *208*, 109627. [CrossRef]
44. Men, X.; Mou, C.L.; Ge, X.Y. Changes in palaeoclimate and palaeoenvironment in the Upper Yangtze area (South China) during the Ordovician–Silurian transition. *Sci. Rep.* **2022**, *12*, 13186. [CrossRef]
45. Keskin, S. Geochemistry of Çamardı Formation sediments. central Anatolia (Turkey): Implication of source area weathering. provenance. and tectonic setting. *Geosci. J.* **2011**, *15*, 185–195. [CrossRef]
46. McLennan, S.M. Relationships between the trace element composition of sedimentary rocks and upper continental crust. *Geochem. Geophys. Geosyst.* **2001**, *2*, 2000GC000109. [CrossRef]
47. Taylor, S.R.; Mclennan, S.M. *The Continental Crust: Its Composition and Evolution. An Examination of the Geochemical Record Preserved in Sedimentary Rocks*; Blackwell Scientific: Oxford, UK, 1985.
48. Roser, B.P.; Korsch, R.J. Provenance signatures of sandstoneemudstone suites determined using discriminant function analysis of major element data. *Chem. Geol.* **1988**, *67*, 119–139. [CrossRef]
49. Floyd, P.A.; Leveridge, B.E. Tectonic environments of the Devonian Gramscatho basin, south Cornwall: Framework mode and geochemical evidence from turbidite sandstones. *Jour Geol. Soc. London* **1987**, *144*, 531–542. [CrossRef]
50. Bhatia, M.R.; Crook, K.A.W. Trace element characteristics of graywackes and tectonic setting discrimination of sedimentary basins. *Contrib. Mineral. Petrol.* **1986**, *92*, 181–193. [CrossRef]

51. Wang, J.; Guo, S. Comparison of geochemical characteristics of marine facies, marine-continental transitional facies and continental facies shale in typical areas of China and their control over organic-rich shale. *Energy Source Part A* **2020**, 1–13. [CrossRef]
52. Algeo, T.J.; Tribovillard, N. Environmental analysis of paleoceanographic systems based on molybdenum-uranium covariation. *Chem. Geol.* **2009**, *268*, 211–225. [CrossRef]
53. Yin, K.; Xu, J.; Li, X. A new grey comprehensive relational model based on weighted mean distance and induced intensity and its application. *Grey Syst.* **2019**, *9*, 374–384. [CrossRef]
54. Si, S.L.; You, X.Y.; Liu, H.C.; Ping, Z. DEMATEL Technique: A Systematic Review of the State-of-the-Art Literature on Methodologies and Applications. *Math. Probl. Eng.* **2018**, *2018*, 3696457. [CrossRef]
55. González, J.; Peña, D.; Romera, R. A robust partial least squares regression method with applications. *J. Chemometr.* **2009**, *23*, 78–90. [CrossRef]
56. Fan, Y.L.; Xiang, Y.Y.; Guo, Z.J. Adaptive efficient and double-robust regression based on generalized empirical likelihood. *Commun. Stat-Simul. C* **2021**, *52*, 3079–3094.

Disclaimer/Publisher's Note: The statements, opinions and data contained in all publications are solely those of the individual author(s) and contributor(s) and not of MDPI and/or the editor(s). MDPI and/or the editor(s) disclaim responsibility for any injury to people or property resulting from any ideas, methods, instructions or products referred to in the content.

Article

Influence of Organic Matter Thermal Maturity on Rare Earth Element Distribution: A Study of Middle Devonian Black Shales from the Appalachian Basin, USA

Shailee Bhattacharya [1,*], Shikha Sharma [1], Vikas Agrawal [1], Michael C. Dix [2], Giovanni Zanoni [3], Justin E. Birdwell [4], Albert S. Wylie, Jr. [5] and Tom Wagner [6]

[1] Department of Geology and Geography, West Virginia University, 98 Beechurst Ave, Morgantown, WV 26506, USA; shikha.sharma@mail.wvu.edu (S.S.)
[2] Independent Researcher, 3414 Beauchamp Street, Houston, TX 77009, USA
[3] RohmTek, 6721 Portwest Drive, Suite 100, Houston, TX 77024, USA
[4] U.S. Geological Survey, Denver Federal Center, Box 25046 MS 977, Denver, CO 80225, USA; jbirdwell@usgs.gov
[5] Independent Researcher, P.O. Box 380, Mohawk, MI 49750, USA
[6] Coterra Energy Inc., 2000 Park Lane, Suite 200, Pittsburgh, PA 15275, USA; tom.wagner@coterra.com
* Correspondence: sb0215@mix.wvu.edu

Citation: Bhattacharya, S.; Sharma, S.; Agrawal, V.; Dix, M.C.; Zanoni, G.; Birdwell, J.E.; Wylie, A.S., Jr.; Wagner, T. Influence of Organic Matter Thermal Maturity on Rare Earth Element Distribution: A Study of Middle Devonian Black Shales from the Appalachian Basin, USA. *Energies* **2024**, *17*, 2107. https://doi.org/10.3390/en17092107

Academic Editors: Shu Tao, Dameng Liu, Wei Ju, Shida Chen, Zhengguang Zhang and Jiang Han

Received: 23 January 2024
Revised: 11 April 2024
Accepted: 14 April 2024
Published: 28 April 2024

Copyright: © 2024 by the authors. Licensee MDPI, Basel, Switzerland. This article is an open access article distributed under the terms and conditions of the Creative Commons Attribution (CC BY) license (https://creativecommons.org/licenses/by/4.0/).

Abstract: This study focuses on understanding the association of rare earth elements (REE; lanthanides + yttrium + scandium) with organic matter from the Middle Devonian black shales of the Appalachian Basin. Developing a better understanding of the role of organic matter (OM) and thermal maturity in REE partitioning may help improve current geochemical models of REE enrichment in a wide range of black shales. We studied relationships between whole rock REE content and total organic carbon (TOC) and compared the correlations with a suite of global oil shales that contain TOC as high as 60 wt.%. The sequential leaching of the Appalachian shale samples was conducted to evaluate the REE content associated with carbonates, Fe–Mn oxyhydroxides, sulfides, and organics. Finally, the residue from the leaching experiment was analyzed to assess the mineralogical changes and REE extraction efficiency. Our results show that heavier REE (HREE) have a positive correlation with TOC in our Appalachian core samples. However, data from the global oil shales display an opposite trend. We propose that although TOC controls REE enrichment, thermal maturation likely plays a critical role in HREE partitioning into refractory organic phases, such as pyrobitumen. The REE inventory from a core in the Appalachian Basin shows that (1) the total REE ranges between 180 and 270 ppm and the OM-rich samples tend to contain more REE than the calcareous shales; (2) there is a relatively higher abundance of middle REE (MREE) to HREE than lighter REE (LREE); (3) there is a disproportionate increase in Y and Tb with TOC likely due to the rocks being over-mature; and (4) the REE extraction demonstrates that although the OM has higher HREE concentration, the organic leachates contain more LREE, suggesting it is more challenging to extract HREE from OM than using traditional leaching techniques.

Keywords: rare earth elements; organic matter; thermal maturity; black shales; Appalachian Basin

1. Introduction

To meet the high demands for rare earth elements (REE) needed for energy transition technologies, recent and ongoing research has been focused on identifying and evaluating unconventional sources of REE like coal mine drainage and coal-related products such as refuse, coal-fired power plant ash, overburden, underclays, and, more recently, black shales [1]. The Appalachian Basin, situated in the eastern U.S., has been one of the largest producers of oil and natural gas in the last two decades [2,3]. This basin, which includes Middle Devonian black shales of varying total organic carbon (TOC) content and spatial maturity, is an important exploration target for REE within the volume of waste material like shale cuttings from horizontally drilled oil and natural gas extraction wells. Our study is an attempt at understanding the modes of REE occurrence in the different phases of black and calcareous shales in the Appalachian Basin. While the modes of association with carbonates, sulfides, framework silicates, and clays are well documented [4,5], there is a lack of understanding regarding the role of organic matter (OM) in REE enrichment [5,6]. Furthermore, the linkage between the thermal maturity of shales and their elemental enrichment has also not been addressed adequately. Therefore, to our knowledge, this study is one of the first that attempts to develop an REE inventory in Middle Devonian black shales, with particular focus on developing a holistic geochemical model of REE distribution in the organic fraction, and finally, establishing a relationship between the thermal maturity of shales and REE enrichment.

2. Geochemical Background
2.1. Role of Organic Matter (OM) in REE Partitioning

Black shales preserve critical mineral deposits like transition metals and rare earth elements [7]. They also act as recorders of changes such as shifts in redox conditions, diagenetic processes, and thermal influences on mineral composition. REE are categorized into light (LREE) (La–Nd, Sc), middle (MREE) (Sm–Ho, Y), and heavy (HREE) (Er–Lu) using the scheme proposed by Yang et al., 2017 [5]. Variations in REE concentrations in rocks of different thermal maturities result from processes like element mobilization during catagenesis, the segregation of REE-rich mineral phases, and sediment mixing. Typically, rivers and seawater show an enrichment of MREE relative to HREE, except in cases of acidic waters with a pH below 7 [8–10]. Estuaries are a unique zone where a significant number of lanthanides are scavenged by suspended matter or complexed with Fe–Mn oxyhydroxides, which results in a decrease in REE concentration in seawater relative to rivers (10 times higher levels) [11]. In seawater, during the early stages of diagenesis, sedimentary OM, oxyhydroxides, and trace elements like REE, Th, U, and Pb may migrate to sediment porewaters. This can be due to sediment dewatering, clay mineral transformations, and Fe-Mn reduction, leading to the preferential release of incompatible elements [12–14]. When oxic degradation in OM occurs, REE are released in the order LREE > MREE > HREE due to increasing complexing efficiency across series with carbonate species [14–17]. As depth increases, this results in an increase in HREE concentration in seawater closer to the sediment–water interface, similar to those of trace elements that behave as micronutrients, such as vanadium (refer to Discussion Section 6.2). HREE enrichment in carbonate minerals may result from higher stability of HREE complexes with carbonate ions or similarity in ionic radius between heavy lanthanides and Ca^{2+} [9]. In samples without a positive correlation between Ca and HREE, it may be due to replacement of Ca by HREE [18]. Also, the rate of OM degradation significantly impacts the association of HREE with carbonates, as seen in Haley et al. (2004) [19], that may arise from factors such as episodic OM input, slower carbon burial rates, an increase in thermally labile OM contributions, or some combination of these. The relative changes in HREE and LREE content in sediment pores can vary with depth and geochemical processes at the sediment–water interface [19]. Ratios such as HREE/LREE or MREE/LREE emphasize fractionation processes to remove the effect of individual concentrations. At shallow depths (0–2 cm), while HREE is at higher levels due to stable complexations, LREE is released from particulate organic matter due to remineralization. At greater depths (2–4 cm), the incomplete remineralization of OM may decrease LREE mobility and relatively increase HREE concentration. This phenomenon explains the source of REE in oxic, suboxic, and anoxic seawaters in the absence of Fe-oxides [20]. However, OM has also been proposed to play a role in the absence of carbonate complexation. The REE profiles show a good correlation with redox-sensitive elements, such as Cu, that are known to be strongly associated with OM and organic ligands [21]. This suggests that OM might play a key role in concentrating REEs in mudstones (see Discussion Section 6.2).

Variations in OM sources, OM composition (Figure 1), and lithology can also affect the REE distribution prior to thermal maturation [22–30]. During the burial, diagenesis, and formation of authigenic minerals such as apatite, clay transformations can significantly change the source rock REE signatures. For example, apatite has a strong affinity for MREE and therefore acts as a sink for those elements only when the P_2O_5 content is greater than 0.5% [31]. Other minerals like authigenic clays and Fe–Mn oxyhydroxides can scavenge REE from highly enriched pore water. It is assumed that shale diagenesis occurs in closed systems and the overall rock chemistry remains the same even though the elemental compositions in individual mineralogical phases may differ from the source rock [18,32]. However, in nature, mudrocks or black shales can also act as open systems, leading to a significant loss or gain of old and new material, respectively [12,33,34]. At mildly high temperatures, humic acid production in the pores may increase, affecting the REE mobility and complicating the reconstruction of paleo-depositional environments ([18], and references therein).

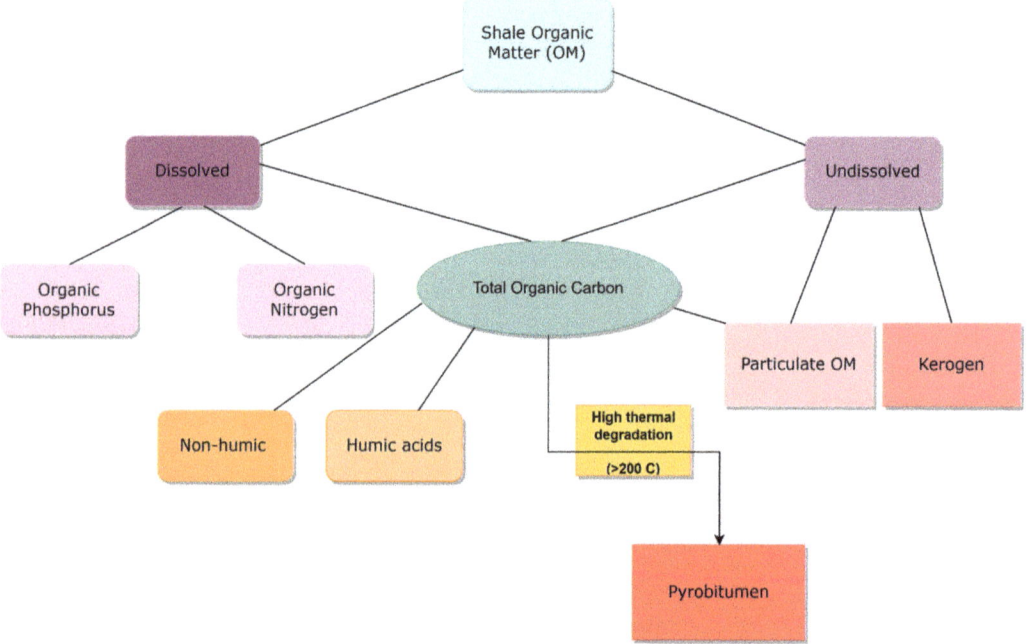

Figure 1. Constituents of shale organic matter.

2.2. Role of Thermal Maturity in HREE Partitioning

The process of thermal maturation significantly affects OM structure and can impact REE distribution. When thermal maturation progresses, OM loss may be significantly high [35], particularly for source rocks with high aliphatic carbon content [36]. The extent to which OM may be retained or lost during catagenesis was addressed in one of the pioneer studies conducted to explain the preservation of alkanes (or soluble OM) in highly mature Canadian black shales [37]. It is one of the first studies to have reported higher HREE abundance in mature black shales. The OM is found in three different forms: Insoluble remobilized bitumenite, insoluble in situ bitumenite, and soluble OM. Pyrobitumen is the type of insoluble solid bitumen (Figure 1) that represents the residue from oil, and it is generated at a later stage along with gas during the cracking of oil [38,39]. Soluble OM is expected to be eliminated from the shale at the onset of regional metamorphism [37]. The physical properties of the shale, such as the porosity, permeability, and alignment of the foliated minerals seem to be linked to controlling the amount of soluble OM retained in the formation [37]. Photomicrographs of highly mature black shales demonstrate that the mobility of previously soluble bitumen and OM that is now insoluble pyrobitumen is due to the alignment of maximum permeability along foliated phyllosilicates in the direction perpendicular to that of the increasing diagenesis and metamorphism. Consequently, some alkanes were incorporated into recrystallizing minerals such as phyllosilicates, during the generation of pyrobitumen associated with increased maturation. While OM characterization is beyond the scope of this work, it is plausible that the rocks in our study area have been subjected to similar geological processes due to their high maturation (see Discussion Section 6.1).

In xenotime, Y is known to occur at the interstices of pyrobitumen nodules and may also be present either in the structure or occur as nanoparticles [40]. The HREE in general tend to be more abundant in pyrobitumen than LREE. A detailed micro-particle induced X-ray emission (μPIXE) study of massive pyrobitumen isolates indicate the presence of HREE in the structure or as nanoparticles [40]. Recent advances in research on tetrapyrrolic and porphyrin complexes demonstrate that the heavier lanthanides and Y have strong hydrophilic properties [41]. These compounds are critical in the context of our study as it has been long known that high concentrations of Ni and V in crude oils are because of their ability to form tetrapyrrole complexes [42] that have high thermal stability [43,44], resistance to strong acids [45,46], and inertness to cation exchange reactions [47]. Furthermore, the high tenacity of tetrapyrrolic compounds is enhanced by metalation, bivalency, and small ionic radii, and all three factors are conveniently satisfied by Ni and V [43,45,48–51]. Conversely, owing to the large ionic radii and high coordination numbers, REE are also well-suited to form sandwich double- and triple-decker complexes with porphyrins, phthalocyanines, and related macrocycles [40,41,52–56]. The most favorable conditions in accumulating sediments for such bonding to occur are a reducing (Eh < 0) and anaerobic environment that supports tetrapyrrole preservation [57–59].

3. Geology of Study Area

The Appalachian Basin represents a continuous stratigraphic sequence in the eastern U.S. spanning a large area across several states [60–66]. The Middle Devonian sequence was deposited in an asymmetric foreland basin (Figure 2) [67] during the Acadian orogeny. During deposition, there were intermittent stages of active volcanism and quiescence, characterized by layers of volcanic materials and trace elemental proxies, such as Th, Rb, Cs, Ta, and LREE [68]. Organic-rich black shales were deposited during periods of reduced sediment influx into the Appalachian Basin due to its distal location. Besides the prevailing tectonic influences, rising sea level is considered the dominant factor controlling Middle Devonian black shale deposition. Three major tectophases occurred in the Devonian. The second tectophase (Eifelian–early Givetian) triggered the deposition of the Marcellus Shale, which marked an abrupt transition from the Onondaga Limestone platform [69]. This lithological transition is attributed to the combined effect of rapid subsidence due to tectonic loading and eustatic sea level rise [61]. Furthermore, various studies using proxies such as concentrations of Mo and U, degree of pyritization, Mo and S isotopes, rare earth patterns in organic-rich and organic-lean sections, and I/Ca ratio in sediments have concluded that these black shales or mudrocks were deposited under fluctuating redox conditions [60,63,70–78]. As the orogeny advanced in the craton-ward direction, the organic-rich sediment underwent clastic dilution, and the deposits became successively poorer in carbonates and OM.

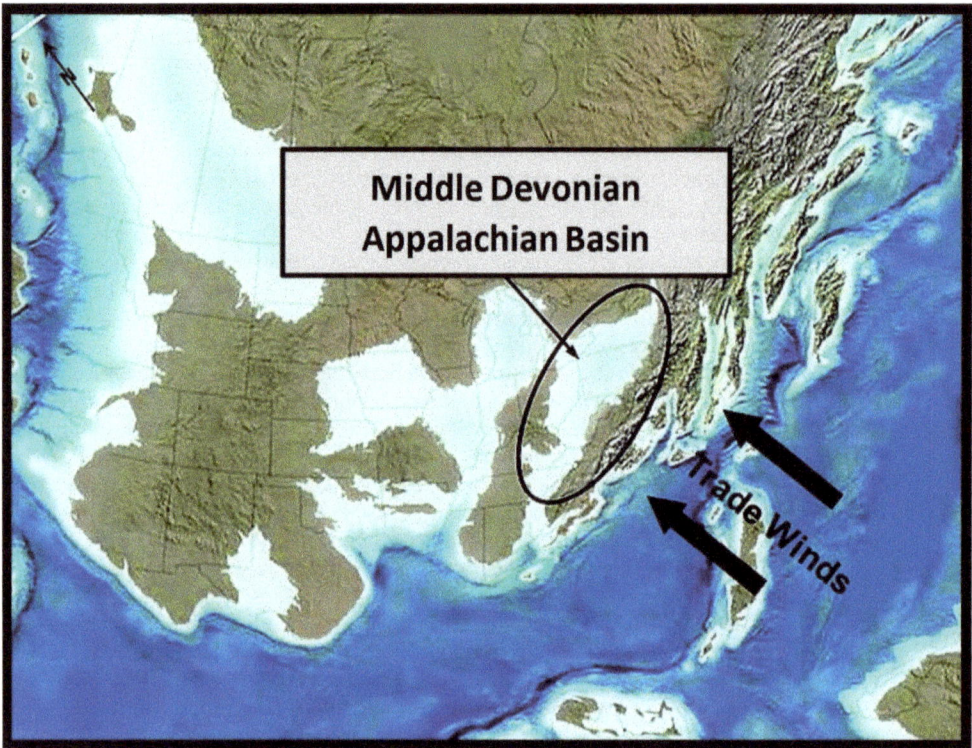

Figure 2. Paleogeographic map during the Middle Devonian (385 Ma) Acadian orogeny. Black oval indicates the general location of the study area; black arrows indicate paleo-wind direction. Modified from Blakey, 2005 [79].

The current stratigraphic divisions link the fine-grained Marcellus subgroup of the distal western margin of the basin with the proximal eastern basin such that the Marcellus Shale is defined as a part of a generally shallowing upwards trend of basinal black shale to nearshore sandstone and alluvial deposits [80–82]. The Purcell member of the Marcellus Shale comprises bedded and nodular fine-grained limestone [80,81]. The limestone occurs between the upper and lower Marcellus subgroups above the Union Springs member of the Marcellus Shale but is relatively thinner than the black shales. The Union Springs member at the base of the lower Marcellus subgroup above the Onondaga platform is characterized by the black shale of the Shamokin member of the Marcellus Shale [81]. The Union Spring's basal section exhibits exceptionally high radioactivity and low density, and is composed of quartz, pyrite, TOC, and lower clay content [81]. In contrast, the upper part of the formation reports a less prominent radioactive response and increased bulk density due to relatively lower OM content and a higher content of framework silicates [83].

4. Materials and Methods

The samples given in Table 1 were collected from a cored well in the Appalachian Basin (exact well location and stratigraphic unit names and depth information are proprietary). The general location is represented in Figure 2. These samples are labeled from A1 (bottom-most) to D1 (topmost) in decreasing order of depth. A few samples have the same identifier suggesting that those were collected from different depths of a particular formation. For example, the first A2 sample at the bottom of Table 1 is a deeper sample than the overlying A2. To screen the samples and identify an appropriate set for this study, preliminary elemental concentrations were obtained using energy dispersive-X-ray fluorescence (ED-XRF;

hand-held Bruker Tracer 5i) from the slab core face at 1.0-inch intervals. TOC and total sulfur values were determined using Leco analyzer from the powders drilled from the back of the cores and from the same depths as the ED-XRF measurements. All data thus obtained in this study correspond to the same depths and intervals. This study comprises ten core samples (plugs and segments) that were stratigraphically above, below, and from the Marcellus Shale (Table 1) and belong to the dry gas window. The samples were homogenized to 75 μm at the Premier Corex Laboratory and sent to West Virginia University (Morgantown, WV, USA) for whole rock analyses and sequential leaching experiments.

Table 1. General description of the core samples.

Sample Identifier	Lithofacies	Sample Type
D1	Clay-rich	1.5-inch plug, center depth (Upper)
D1	Clay-rich	1.5-inch plug, center depth (Lower)
C3	Clay-rich	1.5-inch plug and laterally adjacent core segment
C2	Carbonate-rich	1.5-inch plug, center depth
C1	Carbonate-rich	1.5-inch plug, center depth
B	Organic-rich	1.5-inch plug, center depth (Upper)
B	Organic-rich	1.5-inch plug, center depth (Lower)
A2	Organic-rich	1.5-inch plug, center depth (Upper)
A2	Organic-rich	core segment (Lower)
A1	Carbonate-rich	core segment

Mineralogical characterization of the samples was conducted at the Premier Corex Laboratory through X-ray diffraction (XRD). The XRD analysis was carried out on the bulk-rock fraction using a Bruker D8 Advance instrument. Initial preparation of bulk-rock samples involved powdering the material in a McCrone mill and side-loading before conducting bulk-rock measurements. Subsequently, clay analysis was performed following the separation of the clay fraction using centrifugation, adhering to the company's proprietary workflow. The measurement parameters included a step scan in the Bragg–Brentano geometry employing CuKα radiation (40 kV and 30 mA). For both bulk-rock and clay fraction, samples were scanned at a counting time of 1.8 s per $0.02°$ 2θ, from 3 to $70°$ 2θ, and 1 s per $0.02°$ 2θ from 3 to $30°$ 2θ, for bulk-rock and clay fraction, respectively. Mineral phase interpretation and quantification were achieved using the Reference Intensity Ratio method, calibrated with in-house artificial mixes.

For the Sequential leaching procedure, 10 g of a sample (75 μm) was first washed with 150 mL of deionized (DI) water in a 400 mL borosilicate beaker and rolled for 18 h on an orbital shaker. The fluid was filtered using a 0.45 μm Millipore membrane filter. Following this, reagents were added to sequentially extract the inorganic and organic fractions from the shale. First, 80 mL of 1 M magnesium chloride was used to dissolve the exchangeable fraction. Second, the carbonates and phosphates were targeted using 150 mL of 1 N acetic acid, and constant shaking for 6 h at room temperature. Next, the Fe–Mn oxyhydroxides were dissolved using 150 mL of 0.05 M hydroxylamine hydrochloride in 25% acetic acid for 6 h at pH 2. Following the oxyhydroxides, pyrite was dissolved in 150 mL of 2 M nitric acid by constant shaking for 18 h at room temperature. The final step was designed to target the organically associated particles. First, the sample residue remaining after pyrite dissolution was combusted in a furnace at 650 °C in porcelain crucibles for 3 h to oxidize the organics. This step was performed to ensure the REE would be concentrated as rare earth oxides in the burnt residue that is predominantly a refractory material, i.e., silicates. After combustion, the sample was washed in 150 mL of 0.1 M HCl and shaken for 4 h to separate the REE from the residue. The supernatant fluids formed after every leaching step were collected by vacuum filtering using 0.45 μm membrane filters and subsequently acidified with 1% conc. nitric acid to prevent chemical deterioration of the sample. Additionally, the sample residue after each leaching step was washed in 150 mL of DI water, collected, and acidified for future analyses (if necessary) to account for the elemental loss in between the leaching steps.

About 5 g of each powdered shale sample was used for whole-rock digestion by the sodium peroxide fusion method. Major and trace element concentrations were measured in acidified leachates and whole-rock concentrations using EPA methods 200.7 and a modified 200.8, respectively [84,85]. These steps were taken to ensure analytical accuracy and reproducibility following QA/QC protocol. The instrument was run daily with four points and a blank along with a positive and a negative check. The positive check was to ensure recoveries were proper. The negative check (or the continuing blank) was performed to ensure that the blank was below the method detection limit (MDL). Check standards were run after every 10 injections to ensure that the analytical uncertainty was within $\pm 10\%$ for the calibration verification standard, and calibration blank standards were run every 10 samples. Each batch of 20 samples had a batch blank and laboratory control spike in addition to one sample run in duplicate. The MDLs for the elements are provided in the Supplementary Materials.

Kerogen isolation was performed on pulverized aliquots of the oil shale samples using a sequential acid treatment method [86]. Initially, 18% w/w hydrochloric acid was added to the samples to remove carbonate minerals, followed by 52% w/w hydrofluoric acid to remove aluminosilicates, and then boiling 37% w/w hydrochloric acid was used to remove fluorosilicates produced by the hydrofluoric acid treatment. After removal of residual acid with multiple washes using deionized water, a heavy-liquid separation was performed in a zinc bromide solution to reduce pyrite and other heavy acid-resistant minerals. Additional hot HCl treatments to remove minerals like ralstonite (evaluated by qualitative X-ray diffraction analysis) were employed when required. Finally, a Soxhlet extraction of the organic isolate using a 60:40 (wt%) benzene-methanol azeotrope was employed to remove any residual $ZnBr_2$ and extractable OM. Kerogen samples were then dried in a vacuum oven overnight (60 °C). Major and trace element analyses were performed on whole rock powders and kerogen isolates (after ashing) by SGS Laboratories (Toronto, Canada) using an inductively coupled plasma-optical emission spectroscopy-mass spectroscopy (ICP-OES-MS) method following preparation of samples by sodium peroxide fusion [87].

5. Results

5.1. General Mineralogical Information of the Shale Samples

In our investigation, the mineralogical results in Table 2 and Figure 3 show that samples are typically organic-rich mudrocks and have a moderate to high amount of quartz (19.3–34.0%), feldspar (Plag + Kfsp) (4.7–7.4%), total clay (illite/mica + chlorite) (13.3–60.1%), carbonates (Cal + Dol + Fe-Dol) (1.2–60.7%), pyrite (1.4–6.5%), and TOC (0.41–8.73%). Other minerals were anatase (0.5–1.1%) and barite (trace amount). The TOC increases in the interval between D1 and C2, decreasing sharply at C1, before increasing substantially in the lower horizons until reaching 4.90% at A1, which is a carbonate platform with 53.5% calcite. Other horizons that are carbonate-rich but relatively less in TOC are C2 and C1 ($CaCO_3$ = 19.5% and 35.8% respectively). The TOC-rich zones are concentrated in the lower portion of the core found in three horizons, namely B and A2.

Table 3 provides the major elemental contents as oxides that constitute the major shale-forming minerals. The siliciclastic indicators, namely, SiO_2 ranges from 34.26 wt% in A1 to 60.14 wt% in D1, TiO_2 from 0.24 wt% to 0.78 wt% in D1, Al_2O_3 from 4.42 wt% in A1 to 16.35 wt% in D1, Na_2O 0.63 wt% in A1 to 1.12 wt% in upper B, and K_2O from 0.82 wt% in A1 to 4.15 wt% in the upper D1 sample. Pyrite is represented by Fe_2O_3 ranging from 1.61 wt% in A1 to 7.52 wt% in lower D1 and total sulfur from 0.83 wt% in C2 to 4.06 wt% in lower B, which is also the sample with highest pyrite content. Carbonates are constituted of CaO, ranging from 0.97 wt% in D1 to 25.19 wt% in A1, and MgO from 0.92 wt% in A2 to 3.04 wt% in C1.

Table 2. XRD-determined mineralogical content and Leco-TOC of core samples.

ID	Qtz	Plag	Kfsp	Ill/Mic	Chl	Cal	Dol	Fe-Dol	Pyr	Bar	Anat	TOC
	%	%	%	%	%	%	%	%	%	%	%	%
D1	27.9	5.6	0.5	52.4	7.7	0.6	0.6	0.0	2.5	0.0	1.1	1.90
D1	30.6	6.0	0.4	45.7	5.6	1.4	0.3	1.1	5.9	0.0	1.1	3.37
C3	29.0	6.8	0.3	40.1	3.1	13.6	1.4	4.5	1.8	0.0	1.1	4.18
C2	19.3	5.4	0.4	38.2	0.9	30.5	3.7	1.6	1.4	0.0	0.7	3.27
C1	24.6	6.0	0.3	37.7	1.0	26.7	8.8	11.0	2.5	0.0	0.8	0.41
B	27.2	7.4	0.0	42.1	1.6	11.0	4.2	1.7	4.5	0.0	0.6	5.63
B	30.3	6.7	0.0	34.8	0.0	12.3	1.3	0.6	6.5	0.0	0.7	8.69
A2	26.8	6.5	0.0	22.5	0.0	30.0	1.4	0.3	4.7	0.0	0.7	8.73
A2	34.0	6.7	0.0	16.9	0.0	29.4	5.8	0.3	5.5	0.0	0.7	6.75
A1	21.1	4.7	0.0	13.3	0.0	53.5	6.2	1.0	2.0	0.0	0.5	4.90

Qtz: Quartz; Plag: Plagioclase feldspar; Kfsp: K-Feldspar; IllMic: IlliteMica; Chl: Chlorite; Dol: Dolomite; Pyr: Pyrite; Bar: Barite; Anat: Anatase; %: weight percent.

Table 3. Major elemental content of the whole rock.

Sample ID	SiO_2 %	TiO_2 %	Al_2O_3 %	Fe_2O_3 %	MnO %	MgO %	CaO %	Na_2O %	K_2O %	P_2O_5 %	Leco S %	BaO %
D1	59.02	0.78	16.35	6.37	0.03	1.70	0.97	1.05	4.15	0.08	1.59	0.25
D1	60.14	0.69	14.24	7.52	0.03	1.45	1.24	1.07	3.70	0.09	3.90	0.37
C3	53.10	0.54	11.91	3.64	0.03	1.87	6.99	1.10	3.09	0.05	1.10	0.24
C2	40.76	0.57	11.22	2.63	0.03	1.60	16.77	0.89	2.91	0.04	0.83	0.26
C1	46.35	0.63	11.21	5.74	0.07	3.04	10.63	1.00	2.65	0.13	1.87	0.36
B	55.14	0.63	13.08	5.24	0.03	1.43	4.81	1.12	3.43	0.09	2.55	0.36
B	52.42	0.49	10.37	5.57	0.02	1.06	6.67	1.08	2.65	0.10	4.06	0.29
A2	41.48	0.32	6.92	3.80	0.02	0.92	16.64	1.01	1.90	0.09	2.84	0.16
A2	45.75	0.28	5.45	5.45	0.02	1.89	15.42	0.78	1.21	0.10	3.91	0.12
A1	34.26	0.24	4.42	1.61	0.01	2.07	25.19	0.63	0.82	0.11	1.05	0.14

5.2. Rare Earth Elements in the Whole Rock

The individual REE contents in the whole rock are given in Table 4. We further categorized these elements into light (LREE) (La–Nd, Sc), middle (MREE) (Sm–Ho, Y), and heavy (HREE) (Er–Lu) using the scheme proposed by [5]. The Total REE (TREE) content is consistently higher in the black shales than in the calcareous shales. The increasing trend in REE content from the base to the top section coincides with an increase in Al_2O_3 and TiO_2 as well as SiO_2 and K_2O. However, when grouped into light, middle, and heavy, there are more distinct relationships observed as given in the heat map in Table 5.

Table 4. Individual and total REE content in the whole rock and PAAS.

ID	Sc ppm	Y ppm	La ppm	Ce ppm	Pr ppm	Nd ppm	Sm ppm	Eu ppm	Gd ppm	Tb ppm	Dy ppm	Ho ppm	Er ppm	Tm ppm	Yb ppm	Lu ppm	TREE ppm
D1	16.7	25.0	44.5	90.2	10.6	40.9	7.5	1.63	7.5	1.1	6.8	1.40	3.9	0.6	3.7	0.6	262.5
D1	13.2	24.9	45.3	89.6	11.3	44.9	9.4	2.07	9.2	1.4	7.5	1.55	4.4	0.7	4.3	0.6	270.3
C3	13.4	18.7	37.0	72.1	9.2	35.6	6.7	1.39	6.5	0.9	5.1	1.08	3.2	0.4	2.9	0.6	214.7
C2	11.9	18.9	38.6	73.3	9.3	35.1	5.9	1.28	6.1	0.8	5.1	0.99	2.9	0.4	2.9	0.4	213.8
C1	10.5	15.2	33.0	65.5	7.8	29.7	5.8	1.40	6.1	0.9	5.2	1.02	2.8	0.4	2.7	0.4	188.5
B	13.6	24.4	40.4	78.8	9.8	38.6	7.7	1.97	8.1	1.2	7.2	1.47	4.2	0.6	3.8	0.4	242.2
B	13.5	29.8	39.2	73.7	9.8	39.5	8.6	2.03	9.4	1.4	8.2	1.68	4.7	0.7	4.3	0.6	247.1
A2	11.4	56.5	39.1	58.9	9.2	39.1	8.9	2.24	11.2	1.7	10.0	2.20	6.2	0.9	5.6	0.6	264.1
A2	11.9	34.7	28.9	46.7	7.9	32.5	6.7	1.54	7.6	1.1	6.8	1.35	4.0	0.6	3.8	0.9	196.7
A1	7.7	26.6	23.9	33.0	5.3	21.3	4.7	1.07	6.0	0.9	5.2	1.08	2.9	0.4	2.6	0.5	143.2
PAAS	15.9	27.3	44.6	88.3	10.2	37.3	6.9	1.2	6.0	0.9	5.3	1.1	3.1	0.5	3.0	0.4	252.0

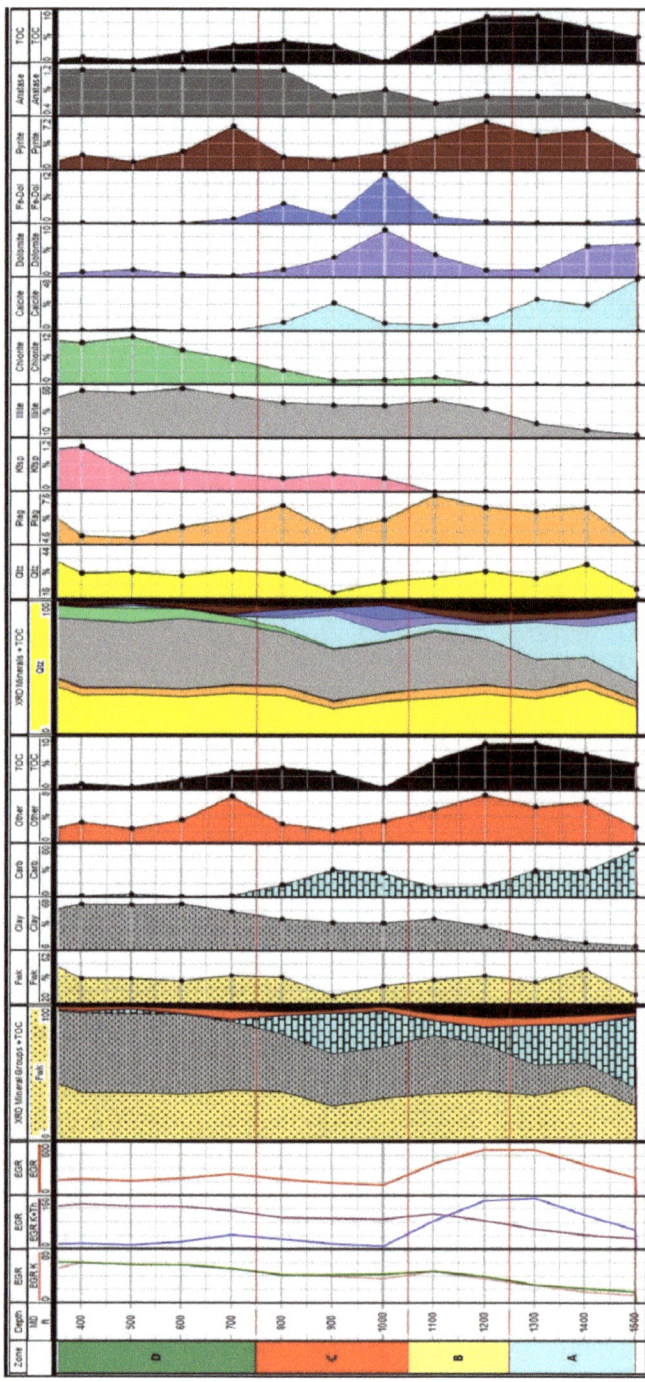

Figure 3. XRD-determined mineralogical proportions of the lithologic units of our study. The dominant groups comprising the whole shale are the siliciclastic, carbonates, mica clays, and pyrite and TOC.

Table 5. Heat map illustrating the relationship between the major shale phases and the REE groups.

ID	SiO$_2$	TiO$_2$	Al$_2$O$_3$	K$_2$O	CaO	Chlorite	LREE (La–Nd)	Leco S	Leco Bulk TOC	HREE (Tb–Lu)
	%	%	%	%	%	%	ppm	%	%	ppm
D1	59.02	0.78	16.35	4.15	0.97	7.7	186.11	1.59	1.9	18.04
D1	60.14	0.69	14.24	3.70	1.24	5.6	191.09	3.90	3.37	20.35
C3	53.10	0.54	11.91	3.09	6.99	3.1	153.95	1.10	4.18	14.09
C2	40.76	0.57	11.22	2.91	16.77	0.9	156.25	0.83	3.27	13.48
C1	46.35	0.63	11.21	2.65	10.63	1.0	136.03	1.87	0.413	13.37
B	55.14	0.63	13.08	3.43	4.81	1.6	167.55	2.55	5.63	18.95
B	52.42	0.49	10.37	2.65	6.67	0.0	162.13	4.06	8.69	21.61
A2	41.48	0.32	6.92	1.90	16.64	0.0	146.24	2.84	8.73	27.66
A2	45.75	0.28	5.45	1.21	15.42	0.0	116.03	3.91	6.75	18.21
A1	34.26	0.24	4.42	0.82	25.19	0.0	83.50	1.05	4.9	13.34

Min. values indicated in white. Max. values indicated in red.

The heat map in Table 5 demonstrates the covariation of the major inorganic and organic phases in our samples with LREE (La–Nd) and HREE (Tb–Lu), respectively. The numbers are assigned a white-red color gradient to represent the relative concentration of the sample in SiO$_2$, K$_2$O, TiO$_2$ (indicative of siliciclastic minerals), chlorite, CaO (calcite), S (sulfides), and TOC. The reason for excluding the Sc and Y from the grouping was to evaluate if these elements behaved similar to or different than a particular REE group. The two uppermost samples (D1) have the highest siliciclastic contents which corresponds with the highest LREE contents, followed by B and C3. The least silicate input is found in the carbonaceous shales that are C2, C1, A2, and A1. We found that the siliciclastic indicators for minerals like quartz, feldspar, and total clays show an affinity for LREE. On the other hand, S and TOC show a clear association with elevated Y and HREE from C1 down. This illustrates that although the REE have a preferential mode of enrichment into different inorganic species, the classical grouping needs further modification to better constrain the behavior of individual REE. This classification can be defined by chemical properties such as the nature of electronic orbitals and the specific electron interactions between particular types of REE and the phases in which they are enriched.

Figure 4 shows the Post-Archean Australian Shale (PAAS)-normalized [88] rare earth distribution in the whole rock with a moderately MREE to HREE-enriched pattern. Using the REE classification from [5], the MREE category includes six lanthanides (Sm, Eu, Gd, Tb, Dy, and Ho). However, in this study, it is evident that a transition in REE abundance starts at Tb. Hence, for a better understanding of the behavior of the heavy lanthanides, we hereby group the HREE from Tb to Lu. The Ce and Eu anomalies [9,89] were calculated as follows:

$$Ce/Ce^* = 2Ce_{SN}/(La_{SN} + Pr_{SN}) \qquad (1)$$

$$Eu/Eu^* = 2Eu_{SN}/(Sm_{SN} + Gd_{SN}) \qquad (2)$$

The anomalies in our samples occur as expected in black shales formed in deep ocean anoxic to suboxic basins (see Supplementary Materials, Table S1). There is significantly higher levels of Y relative to PAAS in the four deepest samples, i.e., one A1, two A2, and one B. The rest of the samples show a depletion in Y relative to PAAS, indicating that the source of Y higher up in the lithologic sections was either absent or these rocks may have undergone post-depositional changes to remove Y. We also observed a similar Tb enrichment, but, in this case, it is present in all samples relative to PAAS. It is likely that the modes of Y and Tb enrichment are similar, but during diagenesis, these elements can respond differently to fluid conditions.

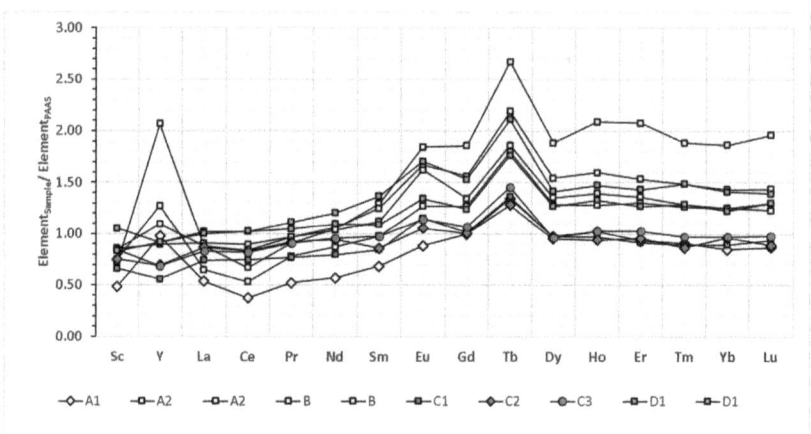

Figure 4. PAAS-normalized REE abundance in the 10 Middle Devonian shales.

5.3. Rare Earth Elements in the Organic Leachate

The ICP-determined concentration of individual rare earth elements extracted from organic leachate fraction and the total REE are given in Table 6. The total concentration extracted varies from only 0.62 to 6 ppm, indicating that a significant amount remained in the silicate residue. The LREE had higher levels in the leachates, followed by the MREE and the HREE. However, the multi-element graphical representation (Discussion 6.3 helps us develop key insights into the plausible mechanisms for such observations. Furthermore, the whole rock REE content also provides an understanding of the key factors that control REE enrichment in black shale.

Table 6. Individual REE and Total REE concentrations in the organic leachate fraction.

ID	Sc	Y	La	Ce	Pr	Nd	Sm	Eu	Gd	Tb	Dy	Ho	Er	Tm	Yb	Lu	TREY
	mg/L	mg/L	mg/L	mg/L	mg/L	mg/L	mg/L	mg/L	mg/L	mg/L	mg/L	mg/L	mg/L	mg/L	mg/L	mg/L	mg/L
D1	0.396	0.162	1.036	2.160	0.327	1.432	0.188	0.028	0.112	0.013	0.065	0.013	0.038	0.005	0.026	0.004	6.004
D1	0.303	0.221	0.556	1.344	0.219	1.053	0.137	0.022	0.087	0.013	0.065	0.013	0.039	0.005	0.028	0.005	4.110
C3	0.177	0.083	0.431	1.051	0.178	0.698	0.069	0.013	0.055	0.007	0.036	0.007	0.019	0.002	0.014	0.002	2.842
C2	0.071	0.076	0.404	1.133	0.178	0.610	0.056	0.011	0.047	0.006	0.040	0.008	0.022	0.003	0.014	0.002	2.682
C1	0.127	0.148	0.285	0.777	0.125	0.482	0.068	0.016	0.072	0.010	0.061	0.012	0.032	0.004	0.021	0.002	2.241
B	0.206	0.247	0.361	0.963	0.167	0.664	0.078	0.019	0.081	0.014	0.078	0.015	0.044	0.005	0.035	0.005	2.980
B	0.129	0.161	0.534	1.294	0.208	0.729	0.080	0.016	0.080	0.000	0.058	0.010	0.031	0.004	0.024	0.003	3.359
A2	0.049	0.087	0.403	0.706	0.084	0.249	0.037	0.010	0.041	0.000	0.030	0.006	0.015	0.002	0.012	0.001	1.735
A2	0.061	0.074	0.279	0.409	0.050	0.157	0.027	0.008	0.030	0.000	0.024	0.005	0.015	0.002	0.014	0.002	1.158
A1	0.018	0.055	0.154	0.169	0.025	0.103	0.025	0.007	0.029	0.000	0.020	0.003	0.009	0.001	0.005	0.000	0.623

6. Discussion

6.1. HREE Enrichment in Black Shale

The mineralogical changes down the stratigraphic column (Figure 3) clearly demonstrate that two lithological end members are present based on the silicate and carbonate contents in the chosen sequence with varying proportions of other minerals and TOC. The clay content is dominated by illite, which is as high as 60% in the upper D1 sample and gradually decreases as the units become increasingly calcareous. We see a similar trend with chlorite, although it disappears completely at the B–A2 transition. As for the carbonates (calcite, dolomite, and Fe-dolomite), they are present in low concentrations in the higher units and increase towards A1. We observe a sharp increase in calcite and dolomite in the C1 and C2 samples, indicating a change in the siliciclastic and clay input into the marine waters. Pyrite content varies concomitantly with OM from C1 to the base of the sequence. This implies that higher OM productivity combined with anoxia primarily controls the mineralogical composition of the deeper shales.

Since the chemical fractionation of REE occurs in a predictable fashion that corresponds to their increasing atomic number, it is assumed that individual REE will behave similarly when partitioning into the various shale-forming minerals. This concept will help us substantiate the relationships we observe between HREE and TOC from the Middle Devonian rocks and kerogen isolates of global oil shales. Kerogen isolates, which are a concentrated form of organic carbon, can be useful to develop a mechanistic understanding of HREE partitioning in pure kerogen. Although the comparison is between two different kinds of samples, we argue that in both cases the focus lies on HREE association with organic matter components in black shales. Furthermore, Figure 5a,b also draw our attention to whether the thermal maturity of the shale has an influence on partitioning, as we see opposite HREE trends with increasing TOC contents. Therefore, it is reasonable to consider comparing the HREE data of the highly mature Middle Devonian samples with the relatively more immature kerogen samples of global oil shale. We note that there is a moderately positive trend in HREE with an increase in TOC in Figure 5a. This observation is in agreement with a few previous studies that propose that the HREE enrichment in the whole rock is contributed by organics [21]. On the contrary, the HREE content in kerogen isolates extracted from global oil shales (see Table 7) show an inverse relationship with TOC (Figure 5b). This opposite trend suggests that other factors besides TOC might be controlling the HREE partitioning; for example, it is possible that hydrocarbon-generating aliphatic organic moieties are not a repository for HREE. To corroborate the data, we assessed the relationship between TOC and certain redox-sensitive elements, such as V, Ni, Cu, Mo, and U in our Appalachian Basin samples.

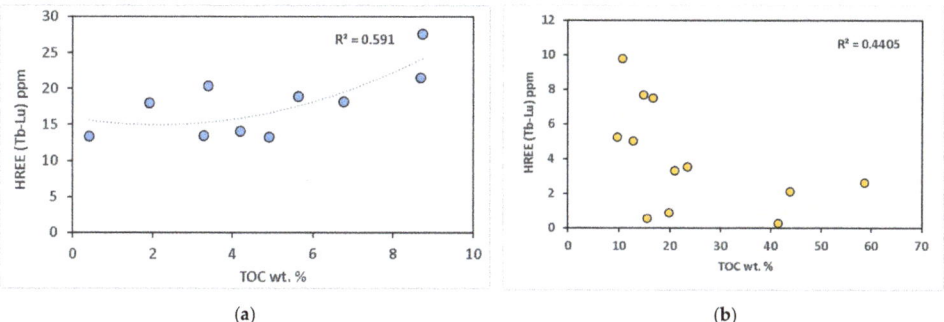

Figure 5. (**a**) HREE in overmature Middle Devonian shales. (**b**) HREE in kerogen isolate of immature global oil shales.

6.2. REE Distributions in Immature Oil Shales and Kerogen Isolates

The oil shale samples examined as part of this work represent a range of sedimentary rock formations from various basins around the world. All samples are thermally immature, based on maceral reflectance (vitrinite or solid bitumen), programmed pyrolysis parameters, and kerogen elemental ratios. The samples include examples from the Mahogany oil-shale zone of the Eocene Green River Formation (GR; Piceance Basin, CO, USA); Cretaceous Timahdit oil shale (TM; Morocco); Cretaceous Ghareb Formation (GI and GJ) shales from two locations (Israel and Jordan; sometimes referred to as the Muwaqqar Formation in Jordan); Jurassic Kimmeridgian Blackstone (KB; England); Permian Irati Formation marinite (IF; Brazil); Permian Glen Davis torbanite (GD; Australia); Permian shale of the Phosphoria Formation (PR; MT, USA); Carboniferous Pumpherston torbanite (PO; Scotland); Mississippian–Devonian New Albany shale (NA; IN, USA); Ordovician Narva-E mine kukersite (EK; Estonia); and the Cambrian Alum Shale Formation (AS; Sweden).

Table 7. HREE content in global oil shales kerogen isolates (BRL = below reporting limit).

Sample ID	Mineralogy	Y	La	Ce	Pr	Nd	Sm	Eu	Gd	Tb	Dy	Ho	Er	Tm	Yb	Lu	TREE	HREE
		ppm	ppm	ppm	ppm	ppm	ppm	ppm	ppm	ppm	ppm	ppm	ppm	ppm	ppm	ppm	ppm	ppm
GR	Carbonate-quartz/feldspar	7.5	8.9	14.4	1.56	4.6	0.7	0.13	0.6	0.16	1.03	0.24	0.79	0.15	1	0.18	34.46	2.12
TM	ML clays–carbonate	10.5	3	5.8	0.61	1.9	0.5	0.13	0.9	0.19	1.56	0.34	1.21	0.21	1.5	0.27	18.08	3.19
GI	Carbonate	1.4	0.8	1.7	0.19	0.6	0.1	BRL	0.1	BRL	0.21	BRL	0.15	BRL	0.2	BRL	4.05	0.35
GJ	Carbonate–clay	2	1	1.9	0.2	0.6	0.1	BRL	0.1	BRL	0.26	0.07	0.2	BRL	0.3	0.06	4.82	0.56
KB	Claystone	5	1.9	3.6	0.44	1.5	0.3	0.09	0.5	0.1	0.73	0.16	0.49	0.08	0.5	0.08	10.47	1.15
IF	Quartz–illite	22.9	19	35.1	4.22	15.4	3.4	0.66	3.8	0.7	4.05	0.76	2.02	0.31	1.7	0.29	91.39	4.32
GD	Quartz	7.3	0.5	0.9	0.11	0.5	0.1	BRL	0.1	BRL	0.54	0.18	0.72	0.14	0.9	0.16	4.88	1.92
PR	Quartz–ML clays	8.9	19	7.3	1.08	3.1	0.5	0.13	0.6	0.15	1.04	0.22	0.73	0.14	0.9	0.15	35.05	1.92
PO	Quartz–kaolinite	14.6	6.9	11.8	1.37	4.3	0.8	0.22	1.1	0.26	2.01	0.46	1.72	0.33	2.3	0.43	33.96	4.78
NA	Quartz–illite	18.3	6.5	11.3	1.22	4	0.9	0.27	1.8	0.41	2.86	0.6	1.73	0.27	1.6	0.23	33.73	3.83
EK	Carbonate–clay	1.3	0.6	1	0.16	0.6	0.1	BRL	0.2	BRL	0.18	BRL	0.09	BRL	BRL	BRL	2.91	0.09
AS	Quartz–illite	12.4	11.4	17.2	1.68	5.3	1.1	0.27	1.5	0.28	1.89	0.4	1.15	0.16	1	0.16	43.47	2.47

These elements demonstrate a distinct mineralogical dependence as the rock TOC content increases downward from C1 (Figure 6). Between B and A1, the sharp increase in the trace metal concentration coincides with an increase in TOC from 3–9.5 wt.%. V, Ni, Cu, Mo, and U vary from about 80–280 ppm, 0–150 ppm, 0–130 ppm, 20–100 ppm, and 0–15 ppm in the upper sections, respectively, and to about 280–640 ppm, 150–370 ppm, 130–280 ppm, 60–150 ppm, and 30–60 ppm in the lower sections, respectively. This increase in elemental concentrations relates directly to the associated reduction in dissolved oxygen concentrations with depth [19,21,32,42,90]. As the REE demonstrate a similar response to a change in redox with depth (Figure 6), we therefore use the HREE data reliably to understand its relationship with TOC. A study on Utica Shale magnafacies in Quebec, Ontario, and New York shows that REE in the organic fraction represents up to almost 20% of the whole rock content [18]. Abanda and Hannigan (2006) [18] reported that the elemental association with OM was much higher than with the sulfide or carbonate fraction. The trend of HREE distribution in our samples also supports that organics can play a key role in controlling the whole rock REE content, in addition to the silicate fraction. To verify the trend, it is necessary to find visible HREE enrichment patterns, which is evident in Figure 4.

It is important to consider other factors in addition to the role of OM that could influence the LREE-HREE enrichment pattern in the whole rocks. We see a variation in the mildly negative Ce anomalies and that can likely be attributed to differences in the extent of biologically mediated activities at particular depths (Figure 4). We hypothesize that local disturbances in the pore waters may have affected the redox conditions and influenced the oxidation of Ce and the growth of microbial communities. More negative anomalies in the deeper stratigraphic horizon would suggest persistent anoxia during the deposition of A1 [20] that prevented the formation of discrete CeO_2 grains. However, it is difficult to resolve if the same biological mediation was responsible for simultaneous Mn-oxide formation or if the degrees of oxidation are controlled by a common process [20]. It is also important to note that although we have a reasonable understanding of the modes of REE partitioning during sediment–fluid interactions in marine settings, most of these studies have been conducted on immature shales. Therefore, the published geochemical models do not consider the effects of thermal alteration on REE distribution in black shales and this is why it is challenging to adequately interpret the REE patterns of the highly mature samples in our study.

The cross plot of the Y concentration against the TOC content in our samples shows that there is a non-linear positive trend in Y with an increase in TOC (Figure 7). This observation aligns with a study by Fuchs et al. (2016) [40] that focused on metal and REE occurrences in pyrobitumen. They reported significantly higher amounts of Y in pyrobitumen (mean concentration of 600 ppm) relative to the mineral matrix of black shales. It is known that Y is an abundant mineral in xenotime and in the interstices of pyrobitumen nodules, and it may also be present either in the structure or occur as nanoparticles. The higher Y content may invariably be a function of pyrobitumen formation during thermal maturation. This suggests that HREE in general tend to be more concentrated in pyrobitumen than the LREE, regardless of the variations in lithology.

Based on the whole rock pattern and moderately strong positive correlation between Y and TOC, we argue that pyrobitumen is potentially the primary host of the heavier lanthanides. Several studies have reported that an increase in shale porosity with increasing thermal maturity is associated strongly with changes in the porosity within OM [91–97]. The lanthanides can hence be incorporated either in OM-hosted pores or in the pyrobitumen structure. A study by Chen and Xiao (2014) [98] evaluated the evolutionary characteristics of OM-hosted nanoporosity in artificially matured shales. It was found that vitrinite reflectance has a strong control on the meso-, micro- and nanopore development within the thermally evolving OM. Therefore, careful evaluation is warranted to determine the nature of association of HREE with pyrobitumen, as that has implications for the development of extraction techniques.

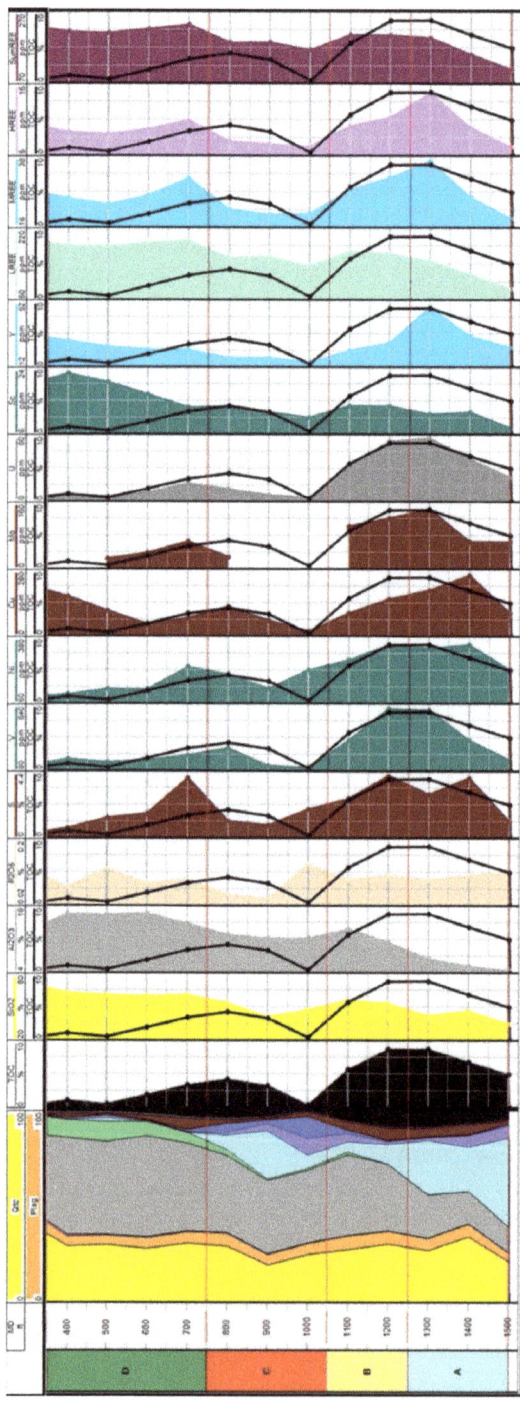

Figure 6. A comparative illustration to track the covariance of redox sensitive parameters, namely, sulfur, TOC.

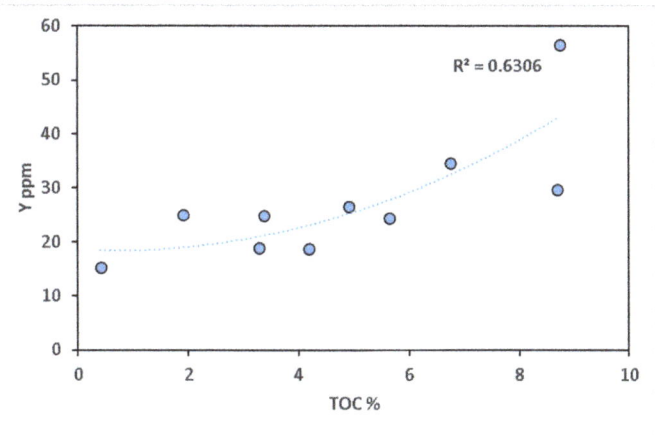

Figure 7. Increasing Y with TOC in whole rock.

Devonian black shales in the Appalachian Basin have been extensively studied with respect to the thermal maturity, mineral assemblages, and nature of OM. In highly mature and overmature rocks (>3% VRo), similar to our Appalachian samples studied here, the dominating OM identified is pyrobitumen [99–101]. Photomicrographs and the 3D modeling of the Marcellus Formation from our Appalachian samples used in this study confirm the presence of thermally altered organics in the form of highly aromatic pyrobitumen (cannot be published for proprietary reasons). Furthermore, the presence of sub-greenschist metamorphic facies and prehnite-pumpellyite assemblage allude to post-mature Devonian shales in highly mature parts of the basin. This corroborates that the highly mature shales contain higher OM porosity compared with those at lower maturities in the wet-gas window. These highly mature regions of the Marcellus basin contain an organic network that may also preserve conditions favorable for the existence of tetrapyrrolic and porphyrin complexes of HREE, including Y. This supports our hypothesis that overmature OM is the primary repository of heavy lanthanides. Therefore, the HREE-enriched pattern in the whole rock (Figure 4) and the positive correlation between TOC and HREE (Figure 5a) demonstrate the role of OM in the incorporation of REE distribution.

6.3. REE Partitioning during Chemical Leaching of OM

To ensure that all the REE were effectively leached from the organic fraction, the sample was combusted at 650 °C for 3.5 h to concentrate them as rare earth oxides prior to rinsing with diluted acid. Subsequently, a Leco TOC analysis was performed to verify if all the OM had been removed at the combustion step, which showed that OM was efficiently removed during the experiment (see Figure S1 in Supplementary Materials). The PAAS-normalized REE pattern indicates that the REE are significantly depleted in the organic leachate fraction. OM was completely removed except for two A samples (Figure 8). Hence, the REE volume in the organic leachate is a true representation of the total REE likely associated with the pyrobitumen solubilized with an acid in the whole rock.

This is, however, contradictory to our expectation that the organic leachates from our extractions show higher relative concentrations of lighter REEs instead of the heavier ones that are known to be present in higher concentrations in OM in the whole rock (Figures 5a and 6). This observation suggests that while OM has an affinity for HREE, it is also more challenging to decouple them using traditional leaching techniques. We hypothesize that nanoscopic specks of apatite, and potentially, xenotime, embedded in OM-hosted pores could be responsible for generating positive Y anomalies in the whole rock but negative anomalies in the organic leachates. The REE abundance also drastically decreased from Tb, indicating that Dy and Ho, which are conventionally grouped into MREE, are observed to behave similar to the HREE. We also saw similar distribution patterns in the whole rock, indicating that there is some similarity in the MREE and HREE distribution patterns (Figure 4). We propose that this warrants a reevaluation of REE classifications based on the objectives of the study. For the purpose of understanding the cause of poor recoveries from our samples, grouping Tb, Y, Dy, Ho, Er, Tm, Yb, and Lu as HREE was determined to be a reasonable way to evaluate HREE behavior in mature black shales. Furthermore, Tb and Y could not be extracted from the high-TOC samples, suggesting these two elements behave similarly as the other HREE, and therefore the PAAS-normalized Y and Tb ratios could be reliable indicators of the behavior of HREE in overmatured rocks. A similar anomalous behavior of Y and Tb has been reported previously [6]. The soluble OM, which was extracted via the traditional leaching procedure, mobilized the lighter rare earth elements, because LREE are known to be incompatible elements. It is important to reiterate that the samples are overmatured and that the likelihood of residual soluble OM is expected to be very low. Since our OM removal was successful, as seen from the Leco TOC analysis of the residue, we propose a few mechanisms that corroborate our data. First, some fraction of the TOC is composed of labile hydrocarbons that were not expelled at the time of thermal maturation and had retained the LREE volume that was generated during leaching. We also assume that the soluble OM may have been trapped by association with refractory silicates during remineralization [37] and that is plausibly the primary source of the remnant LREE in organic leachates. The second mechanism explains the presence of low LREE by accounting for significant loss during processes such as kerogen cracking, the migration of oil, and oil cracking. These three processes are progressive stages of thermal maturation [102]. Therefore, a highly mature shale is expected to be significantly depleted in LREE. Thirdly, to account for the HREE enrichment, we attribute it to the formation of thermally resistant stable organic compounds (discussed in detail previously), which made LREE less likely to be preserved in OM. Therefore, the whole rock and leachate REE patterns agree with our hypothesis that although OM has a higher affinity for HREE, thermal maturity plays a more critical role in determining the modes of REE occurrence.

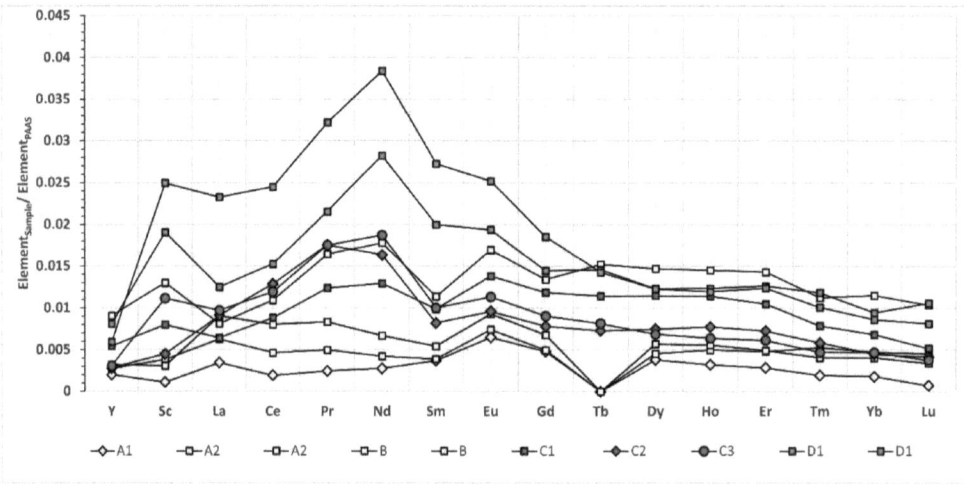

Figure 8. PAAS-normalized REE abundances in the organic leachate fraction.

6.4. Geochemical Model for REE Incorporation in Mature Organic-Rich Shale

The current understanding of REE partitioning during the thermal maturation of black shales is limited. There can be several modes of occurrence depending not only on the source but also on the post-burial diagenetic processes and the presence of authigenic minerals or lack thereof. Since we do not find the occurrence of minerals that have an affinity for rare earth elements, we focused our geochemical model to role of OM and siliciclastics, excluding carbonates and sulfides.

The hypothetical model can be divided into depositional and post-depositional processes and can be described as follows:

Depositional stage: During a high influx of organic material from various sources into the marine waters, dissolved REE in seawater [20,103,104] is scavenged by carboxyl groups of humic acids during the early stages of OM deposition. As the depth in the water column increases, there is a relative increase in HREE over LREE due to the increasing stability of complexes with an increase in the atomic number of the lanthanides. Therefore, depositing OM preferentially takes up more HREE. With compaction, there is a release of REE into pore waters due to the reductive dissolution of Fe–Mn oxyhydroxides [105–108] that may further increase the possibility of the REE to be incorporated into OM.

Post-depositional stage: After compaction is complete, and when thermal maturation reaches a point that initiates a release of aggressive pore fluids from OM and the dewatering of smectite to forms illitic clays, there is a further preferential partitioning of HREE into the solid OM. Other processes that can mobilize HREE are through the formation of stable tetrapyrrolic and porphyrin complexes which are tenacious at high temperatures and resistant to cation exchange reactions. When oil is released due to maturation, these compounds migrate into nano- and micropores from the origin along the temperature gradient. As maturation progresses from the oil to wet gas and then dry gas, there is significant modification of OM-hosted shale porosity. While fluids remain in the evolving shale system, certain organic networks develop which may host a fraction of the rare earths, preferentially retaining the heavy REE, simultaneously depleting the host rock in LREE. At further stages of maturation, there is cracking of oil, leading to the generation of gas and pyrobitumen. With the expulsion of oil, a volume of LREE may be lost and the stable tetrapyrroles are preserved in thermally recalcitrant pyrobitumen, which may later become partially graphitized at higher temperatures. This sequence of events can ultimately give rise to a HREE-enriched pattern that we observe in all the high TOC-rich samples. On the other hand, the negative correlation between HREE and TOC in the relatively immature oil shales may be attributed to the lack of significant thermal maturation resulting in the preferential retention of HREEs in organic residues not occurring.

7. Conclusions

We conclude that organic matter in highly matured black shales plays a significant role in partitioning and remobilizing heavy rare earths. The opposite trends between global oil shales and over-matured Middle Devonian shales with respect to the relationship between TOC and HREE help develop the idea that thermal maturation process is an important factor to consider when studying geochemical behavior of REE association. It is therefore likely that although TOC has a control on REE enrichment relative to PAAS, thermal maturity is the most important factor that governs HREE partitioning into refractory materials, such as pyrobitumen.

- We established an REE inventory from the D1 through A1. The total REE ranges from 180 to 270 ppm and the OM-rich samples tend to contain more REE than the calcareous shales.
- The samples show a relatively higher abundance of middle and heavy REEs than light REEs.
- There is a disproportionate increase in Y and Tb with TOC, suggesting these elements are more strongly bound to OM and thus can be used in tracer studies.
- The organic leachates from our experiments contained more LREE than HREE despite the HREE concentration being higher in OM. This observation suggests that while OM has an affinity for HREE, it also is more challenging to decouple them using traditional leaching techniques. The high Y and Tb content in the whole rock reflects that the HREE are incorporated under high thermal conditions and these elements can be used

as reliable proxies for determining nature of refractory OM in highly mature black shales.

The REE in pore waters and their marine sediment counterparts can exhibit a wide range of REE concentrations controlled by different redox conditions, diagenetic environments, and thermal maturation. Therefore, this study shows the need to further explore shales from other basins and the effects of other maturation processes on the geochemical behavior of REE. Our preliminary findings contribute to the development of a better understanding of the role of maturation history, the scavenging capacity of pyrobitumen, and the potential role of clays in the remobilization of REE from the silicates. Finally, building an REE inventory in other major shale plays could help tap into unconventional REE resources to meet the global demands of these relevant elements.

Supplementary Materials: The following supporting information can be downloaded at: https://www.mdpi.com/article/10.3390/en17092107/s1. Figure S1: Comparative illustration to track the mineralogical changes in the residual samples after leaching. Figure S2: Ternary plots for mineralogical proportions of the samples. Table S1: Method Detection limits; Table S2: Cerium and Europium anomalies in the whole rock; Table S3: Formation location of global oil shales.

Author Contributions: Conceptualization: S.B., V.A. and S.S.; Supervision: S.S.; Formal Analysis: S.S., M.C.D. and G.Z.; Conducted the experiments: S.B. and V.A.; Visualization: M.C.D. and S.B.; Writing original draft: S.B.; Edited the manuscript: S.B., S.S., V.A., M.C.D., J.E.B., G.Z., T.W. and A.S.W.J.; Resources: S.S. and T.W. All authors have read and agreed to the published version of the manuscript.

Funding: This research was funded by Coterra Energy Inc. Oil shale and kerogen work was funded by the U.S. Geological Survey Energy Resources Program. Publication of this article was supported by the John C. Ludlum Geology Endowment, Department of Geology and Geography, West Virginia University, Morgantown.

Data Availability Statement: Concentration data (lanthanide and related elements) for oil shale whole rock and kerogen samples available in a U.S. Geological survey data release [109] on ScienceBase.gov.

Conflicts of Interest: The authors declare no conflicts of interest. M.C.D., G.Z. were employed by Premier Corex Laboratories and W.T. were employed by the company Coterra Energy Inc. The remaining authors declare that the research was conducted in the absence of any commercial or financial relationships that could be construed as a potential conflict of interest.

References

1. Arnold, B.J. A review of element partitioning in coal preparation. *Int. J. Coal Geol.* **2023**, *274*, 104296. [CrossRef]
2. Wang, Q.; Chen, X.; Jha, A.N.; Rogers, H. Natural gas from shale formation–the evolution, evidences and challenges of shale gas revolution in United States. *Renew. Sustain. Energy Rev.* **2014**, *30*, 1–28. [CrossRef]
3. Considine, T.; Watson, R.; Blumsack, S. *The Economic Impacts of the Pennsylvania Marcellus Shale Natural Gas Play: An Update*; The Pennsylvania State University, Department of Energy and Mineral Engineering: State College, PA, USA, 2010.
4. Phan, T.T.; Hakala, J.A.; Lopano, C.L.; Sharma, S. Rare earth elements and radiogenic strontium isotopes in carbonate minerals reveal diagenetic influence in shales and limestones in the Appalachian Basin. *Chem. Geol.* **2019**, *509*, 194–212. [CrossRef]
5. Yang, J.; Torres, M.; McManus, J.; Algeo, T.J.; Hakala, J.A.; Verba, C. Controls on rare earth element distributions in ancient organic-rich sedimentary sequences: Role of post-depositional diagenesis of phosphorus phases. *Chem. Geol.* **2017**, *466*, 533–544. [CrossRef]
6. Bhattacharya, S.; Agrawal, V.; Sharma, S. Association of Rare Earths in Different Phases of Marcellus and Haynesville Shale: Implications on Release and Recovery Strategies. *Minerals* **2022**, *12*, 1120. [CrossRef]
7. Scott, C.; Slack, J.F.; Kelley, K.D. The hyper-enrichment of V and Zn in black shales of the Late Devonian-Early Mississippian Bakken Formation (USA). *Chem. Geol.* **2017**, *452*, 24–33. [CrossRef]
8. Freslon, N.; Bayon, G.; Toucanne, S.; Bermell, S.; Bollinger, C.; Chéron, S.; Etoubleau, J.; Germain, Y.; Khripounoff, A.; Ponzevera, E.; et al. Rare earth elements and neodymium isotopes in sedimentary organic matter. *Geochim. Cosmochim. Acta* **2014**, *140*, 177–198. [CrossRef]
9. Elderfield, H.; Greaves, M.J. The rare earth elements in seawater. *Nature* **1982**, *296*, 214–219. [CrossRef]
10. Elderfield, H. The oceanic chemistry of the rare-earth elements. *Philos. Trans. R. Soc. Lond. Ser. A Math. Phys. Sci.* **1988**, *325*, 105–126.

11. Sholkovitz, E.; Ronald, S. The estuarine chemistry of rare earth elements: Comparison of the Amazon, Fly, Sepik and the Gulf of Papua systems. *Earth Planet. Sci. Lett.* **2000**, *179*, 299–309. [CrossRef]
12. Milodowski, A.E.; Zalasiewicz, J.A. Redistribution of rare earth elements during diagenesis of turbidite/hemipelagite mudrock sequences of Llandovery age from central Wales. *Geol. Soc. Lond. Spec. Publ.* **1991**, *57*, 101–124. [CrossRef]
13. Goldberg, E.D.; Koide, M.; Schmitt, R.A.; Smith, R.H. Rare-Earth distributions in the marine environment. *J. Geophys. Res.* **1963**, *68*, 4209–4217. [CrossRef]
14. Byrne, R.H.; Sholkovitz, E.R. Marine chemistry and geochemistry of the lanthanides. *Handb. Phys. Chem. Rare Earths* **1996**, *23*, 497–593.
15. Wood, S.A. The aqueous geochemistry of the rare-earth elements and yttrium: 1. Review of available low-temperature data for inorganic complexes and the inorganic REE speciation of natural waters. *Chem. Geol.* **1990**, *82*, 159–186. [CrossRef]
16. Millero, F.J. Stability constants for the formation of rare-earth-inorganic complexes as a function of ionic strength. *Geochim. Cosmochim. Acta* **1992**, *56*, 3123–3132. [CrossRef]
17. Lee, J.H.; Byrne, R.H. Examination of comparative rare earth element complexation behavior using linear free-energy relationships. *Geochim. Cosmochim. Acta* **1992**, *56*, 1127–1137. [CrossRef]
18. Abanda, P.A.; Hannigan, R.E. Effect of diagenesis on trace element partitioning in shales. *Chem. Geol.* **2006**, *230*, 42–59. [CrossRef]
19. Haley, B.A.; Klinkhammer, G.P.; McManus, J. Rare earth elements in pore waters of marine sediments. *Geochim. Cosmochim. Acta* **2004**, *68*, 1265–1279. [CrossRef]
20. Moffett, J.W. Microbially mediated cerium oxidation in sea water. *Nature* **1990**, *345*, 421–423. [CrossRef]
21. Klinkhammer, G.; Heggie, D.T.; Graham, D.W. Metal diagenesis in oxic marine sediments. *Earth Planet. Sci. Lett.* **1982**, *61*, 211–219. [CrossRef]
22. Moldowan, J.M.; Sundararaman, P.; Schoell, M. Sensitivity of biomarker properties to depositional environment and/or source input in the Lower Toarcian of SW-Germany. *Org. Geochem.* **1986**, *10*, 915–926. [CrossRef]
23. Ten Haven, H.L.; De Leeuw, J.W.; Peakman, T.M.; Maxwell, J.R. Anomalies in steroid and hopanoid maturity indices. *Geochim. Cosmochim. Acta* **1986**, *50*, 853–855. [CrossRef]
24. Ten Haven, H.L.; De Leeuw, J.W.; Damsté, J.S.S.; Schenck, P.A.; Palmer, S.E.; Zumberge, J.E. Application of biological markers in the recognition of palaeohypersaline environments. *Geol. Soc. Lond. Spec. Publ.* **1988**, *40*, 123–130. [CrossRef]
25. Curiale, J.A.; Odermatt, J.R. Short-term biomarker variability in the Monterey Formation, Santa Maria basin. *Org. Geochem.* **1989**, *14*, 1–13. [CrossRef]
26. Strachan, R.A.; Smith, M.; Harris, A.L.; Fettes, D.J. *The Northern Highland and Grampian Terranes*; Geological Society of London: London, UK, 2002.
27. Dahl, I.M.; Kolboe, S. On the reaction mechanism for propene formation in the MTO reaction over SAPO-34. *Catal. Lett.* **1993**, *20*, 329–336. [CrossRef]
28. Snowdon, L.R. Rock-Eval Tmax suppression: Documentation and amelioration. *AAPG Bull.* **1995**, *79*, 1337–1348.
29. French, K.L.; Hallmann, C.; Hope, J.M.; Schoon, P.L.; Zumberge, J.A.; Hoshino, Y.; Peters, C.A.; George, S.C.; Love, G.D.; Brocks, J.J.; et al. Reappraisal of hydrocarbon biomarkers in Archean rocks. *Proc. Natl. Acad. Sci. USA* **2015**, *112*, 5915–5920. [CrossRef] [PubMed]
30. French, K.; Birdwell, J.; Berg, V. Biomarker similarities between the saline lacustrine eocene green river and the paleoproterozoic Barney Creek Formations. *Geochim. Cosmochim. Acta* **2020**, *274*, 228–245. [CrossRef]
31. Kidder, D.L.; Krishnaswamy, R.; Mapes, R.H. Elemental mobility in phosphatic shales during concretion growth and implications for provenance analysis. *Chem. Geol.* **2003**, *198*, 335–353. [CrossRef]
32. Lev, S.M.; McLennan, S.M.; Hanson, G.N. Mineralogic controls on REE mobility during black-shale diagenesis. *J. Sediment. Res.* **1999**, *69*, 1071–1082. [CrossRef]
33. Awwiller, D.N. Illite/smectite formation and potassium mass transfer during burial diagenesis of mudrocks; a study from the Texas Gulf Coast Paleocene-Eocene. *J. Sediment. Res.* **1993**, *63*, 501–512.
34. Bloch, J.; Hutcheon, I.E. Shale Diagenesis: A Case Study from the Albian Harmon Member (Peace River Formation), Western Canada. *Clays Clay Miner.* **1992**, *40*, 682–699. [CrossRef]
35. Raiswell, R.; Berner, R.A. Organic carbon losses during burial and thermal maturation of normal marine shales. *Geology* **1987**, *15*, 853–856. [CrossRef]
36. Miknis, F.P.; Jiao, Z.S.; MacGowan, D.B.; Surdam, R.C. Solid-state NMR characterization of Mowry shale from the Powder River Basin. *Org. Geochem.* **1993**, *20*, 339–347. [CrossRef]
37. Tait, L. *The Character of Organic Matter and the Partitioning of Trace and Rare Earth Elements in Black Shales; Blondeau Formation, Chibougamau, Québec*; Université du Québec à Chicoutimi: Chicoutimi, QC, Canada, 1987.
38. Chen, Z.; Simoneit, B.R.; Wang, T.-G.; Ni, Z.; Yuan, G.; Chang, X. Molecular markers, carbon isotopes, and rare earth elements of highly mature reservoir pyrobitumens from Sichuan Basin, southwestern China: Implications for PreCambrian-Lower Cambrian petroleum systems. *Precambrian Res.* **2018**, *317*, 33–56. [CrossRef]
39. Mastalerz, M.; Glikson, M. In-situ analysis of solid bitumen in coal: Examples from the Bowen Basin and the Illinois Basin. *Int. J. Coal Geol.* **2000**, *42*, 207–220. [CrossRef]
40. Fuchs, S.; Williams-Jones, A.E.; Jackson, S.E.; Przybylowicz, W.J. Metal distribution in pyrobitumen of the Carbon Leader Reef, Witwatersrand Supergroup, South Africa: Evidence for liquid hydrocarbon ore fluids. *Chem. Geol.* **2016**, *426*, 45–59. [CrossRef]

41. Martynov, A.G.; Horii, Y.; Katoh, K.; Bian, Y.; Jiang, J.; Yamashita, M.; Gorbunova, Y.G. Rare-earth based tetrapyrrolic sandwiches: Chemistry, materials and applications. *Chem. Soc. Rev.* **2022**, *51*, 9262–9339. [CrossRef] [PubMed]
42. Lewan, M.D.; Maynard, J.B. Factors controlling enrichment of vanadium and nickel in the bitumen of organic sedimentary rocks. *Geochim. Cosmochim. Acta* **1982**, *46*, 2547–2560. [CrossRef]
43. Hodgson, G.W.; Baker, B.L. Vanadium, nickel, and porphyrins in thermal geochemistry of petroleum. *AAPG Bull.* **1957**, *41*, 2413–2426.
44. Rosscup, R.J.; Bowman, D.H. Thermal Stabilities of Vanadium and Nickel Porphyrins. *Div. Pet. Chem. Am. Chem. Soc.* **1967**, *12*, 77.
45. Caughey, W.S.; Corwin, A.H. The Stability of Metalloetioporphyrins toward Acids1. *J. Am. Chem. Soc.* **1955**, *77*, 1509–1513. [CrossRef]
46. Dean, R.A.; Girdler, R.B. Reaction of metal etioporphyrins on dissolution in sulfuric acid. *Chem. Indust.* **1960**, 100–101.
47. Barnes, J.W.; Dorough, G.D. Exchange and Replacement Reactions of α, β, γ, δ-Tetraphenyl-metalloporphins1. *J. Am. Chem. Soc.* **1950**, *72*, 4045–4050. [CrossRef]
48. Corwin, A.H. Petroporphyrins. In Proceedings of the 5th World Petroleum Congress, New York, NY, USA, 30 May–5 June 1959; Section V. pp. 120–129.
49. Dunning, H.N.; Moore, J.W.; Denekas, M.O. Interfacial activities and porphyrin contents of petroleum extracts. *Ind. Eng. Chem.* **1953**, *45*, 1759–1765. [CrossRef]
50. Erdman, J.G.; Walter, J.W.; Hanson, W.E. The stability of the porphyrin metallo complexes. *Amer. Chem. Soc. Div. Petrol. Chem. Prepr.* **1957**, *2*, 259–267.
51. Fleischer, E.B. The structure of nickel etioporphyrin-I. *J. Am. Chem. Soc.* **1963**, *85*, 146–148. [CrossRef]
52. Jiang, J.; Ng, D.K.P. A decade journey in the chemistry of sandwich-type tetrapyrrolato– rare earth complexes. *Acc. Chem. Res.* **2009**, *42*, 79–88. [CrossRef]
53. Jiang, J.; Bian, Y.; Furuya, F.; Liu, W.; Choi, M.T.M.; Kobayashi, N.; Li, H.-W.; Yang, Q.; Mak, T.C.W.; Ng, D.K.P. Synthesis, Structure, Spectroscopic Properties, and Electrochemistry of Rare Earth Sandwich Compounds with Mixed 2, 3-Naphthalocyaninato and Octaethylporphyrinato Ligands. *Chem. A Eur. J.* **2001**, *7*, 5059–5069. [CrossRef]
54. Pushkarev, V.E.; Tomilova, L.G.; Nemykin, V.N. Historic overview and new developments in synthetic methods for preparation of the rare-earth tetrapyrrolic complexes. *Coord. Chem. Rev.* **2016**, *319*, 110–179. [CrossRef]
55. Lysenko, A.B.; Malinovskii, V.L.; Padmaja, K.; Wei, L.; Diers, J.R.; Bocian, D.F.; Lindsey, J.S. Multistate molecular information storage using S-acetylthio-derivatized dyads of triple-decker sandwich coordination compounds. *J. Porphyr. Phthalocyanines* **2005**, *9*, 491–508. [CrossRef]
56. Ali, M.F.; Abbas, S. A review of methods for the demetallization of residual fuel oils. *Fuel Process. Technol.* **2006**, *87*, 573–584. [CrossRef]
57. Brongersma-Sanders, M. On conditions favouring the preservation of chlorophyll in marine sediments. In Proceedings of the World Petroleum Congress, The Hague, The Netherlands, 28 May–6 June 1951; p. WPC-4027.
58. Gorham, E.; Sanger, J. Plant pigments in woodland soils. *Ecology* **1967**, *48*, 306–308. [CrossRef]
59. Drozdova, T.V.; Gorskiy, Y.N. Conditions of preservation of chlorophyll, pheophytin and humic matter in Black Sea sediments. *Geokhimiya* **1972**, *3*, 323–334.
60. He, R.; Lu, W.; Junium, C.K.; Straeten, C.A.V.; Lu, Z. Paleo-redox context of the Mid-Devonian Appalachian Basin and its relevance to biocrises. *Geochim. Cosmochim. Acta* **2020**, *287*, 328–340. [CrossRef]
61. Brett, C.E.; Baird, G.C.; Bartholomew, A.J.; DeSantis, M.K.; Straeten, C.A.V. Sequence stratigraphy and a revised sea-level curve for the Middle Devonian of eastern North America. *Palaeogeogr. Palaeoclimatol. Palaeoecol.* **2011**, *304*, 21–53. [CrossRef]
62. Ver Straeten, C.A.; Brett, C.E.; Sageman, B.B. Mudrock sequence stratigraphy: A multi-proxy (sedimentological, paleobiological and geochemical) approach, Devonian Appalachian Basin. *Palaeogeogr. Palaeoclimatol. Palaeoecol.* **2011**, *304*, 54–73. [CrossRef]
63. Ver Straeten, C.; Baird, G.; Brett, C.; Lash, G.; Over, J.; Karaca, C.; Jordan Blood, R. The Marcellus Subgroup in its type area, Finger Lakes area of New York, New York State Geological Association Field Guide. In Proceedings of the 83rd Annual Meeting, 14–16 October 2011; pp. 23–86.
64. Stein, W.E.; Mannolini, F.; Hernick, L.V.; Landing, E.; Berry, C.M. Giant cladoxylopsid trees resolve the enigma of the Earth's earliest forest stumps at Gilboa. *Nature* **2007**, *446*, 904–907. [CrossRef] [PubMed]
65. Ettensohn, F.R.; Miller, M.L.; Dillman, S.B.; Elam, T.D.; Geller, K.L.; Swager, D.R.; Markowitz, G.; Woock, R.D.; Barron, L.S. *Characterization and Implications of the Devonian-Mississippian Black Shale Sequence, Eastern and Central Kentucky, USA: Pycnoclines, Transgression, Regression, and Tectonism*; AAPG: Tulsa, OK, USA, 1988; pp. 323–345.
66. Woodrow, D.L.; Dennison, J.M.; Ettensohn, F.R.; Sevon, W.T.; Kirchgasser, W.T. *Middle and Upper Devonian Stratigraphy and Paleogeography of the Central and Southern Appalachians and Eastern Midcontinent, USA*; AAPG: Tulsa, OK, USA, 1988; pp. 277–301.
67. Ettensohn, F.R.; Barron, L.S. Tectono-climatic model for origin of Devonian-Mississippian black gas shales of east-central United States. *AAPG Bull.* **1981**, *65*, 923.
68. Chen, R.; Sharma, S. Linking the Acadian Orogeny with organic-rich black shale deposition: Evidence from the Marcellus Shale. *Mar. Pet. Geol.* **2017**, *79*, 149–158. [CrossRef]
69. Ettensohn, F.R. Modeling the nature and development of major Paleozoic clastic wedges in the Appalachian Basin, USA. *J. Geodyn.* **2004**, *37*, 657–681. [CrossRef]

70. Murphy, A.E.; Sageman, B.B.; Hollander, D.J.; Lyons, T.W.; Brett, C.E. Black shale deposition and faunal overturn in the Devonian Appalachian Basin: Clastic starvation, seasonal water-column mixing, and efficient biolimiting nutrient recycling. *Paleoceanography* **2000**, *15*, 280–291. [CrossRef]
71. Werne, J.P.; Sageman, B.B.; Lyons, T.W.; Hollander, D.J. An integrated assessment of a "type euxinic" deposit: Evidence for multiple controls on black shale deposition in the Middle Devonian Oatka Creek Formation. *Am. J. Sci.* **2002**, *302*, 110–143. [CrossRef]
72. Sageman, B.B.; Murphy, A.E.; Werne, J.P.; Straeten, C.A.V.; Hollander, D.J.; Lyons, T.W. A tale of shales: The relative roles of production, decomposition, and dilution in the accumulation of organic-rich strata, Middle–Upper Devonian, Appalachian basin. *Chem. Geol.* **2003**, *195*, 229–273. [CrossRef]
73. Algeo, T.J. Can marine anoxic events draw down the trace element inventory of seawater? *Geology* **2004**, *32*, 1057–1060. [CrossRef]
74. Rimmer, S.M. Geochemical paleoredox indicators in Devonian–Mississippian black shales, central Appalachian Basin (USA). *Chem. Geol.* **2004**, *206*, 373–391. [CrossRef]
75. Gordon, G.W.; Lyons, T.W.; Arnold, G.L.; Roe, J.; Sageman, B.B.; Anbar, A.D. When do black shales tell molybdenum isotope tales? *Geology* **2009**, *37*, 535–538. [CrossRef]
76. Lash, G.G.; Blood, D.R. Organic matter accumulation, redox, and diagenetic history of the Marcellus Formation, southwestern Pennsylvania, Appalachian basin. *Mar. Pet. Geol.* **2014**, *57*, 244–263. [CrossRef]
77. Blood, D.R.; Lash, G.G.; Larsen, D.; Egenhoff, S.O.; Fishman, N.S. Dynamic redox conditions in the Marcellus Shale as recorded by pyrite framboid size distributions. *Paying Atten. Mudrocks Priceless* **2015**, *515*, 153–168.
78. Chen, R.; Sharma, S. Role of alternating redox conditions in the formation of organic-rich interval in the Middle Devonian Marcellus Shale, Appalachian Basin, USA. *Palaeogeogr. Palaeoclimatol. Palaeoecol.* **2016**, *446*, 85–97. [CrossRef]
79. Blakey, R. Global Paleogeography. 2005. Available online: http://jan.ucc.nau.edu/~rcb7/globaltext2.html (accessed on 24 September 2009).
80. Ver Straeten, C.A. Microstratigraphy and depositional environments of a middle Devonian foreland basin: Berne and Otsego members, Mount Marion formation, eastern New York state. *Stud. Stratigr. Paleontol. Honor Donald W. Fish. N. Y. State Mus. Bull.* **1994**, *481*, 367–380.
81. Straeten, C.A.V.; Brett, C.E. Pragian to Eifelian strata (middle Lower to lower Middle Devonian), northern Appalachian Basin-stratigraphic nomenclatural changes. *Northeast. Geol. Environ. Sci.* **2006**, *28*, 80.
82. Ver Straeten, C.A. Basinwide stratigraphic synthesis and sequence stratigraphy, upper Pragian, Emsian and Eifelian stages (Lower to Middle Devonian), Appalachian Basin. *Geol. Soc. Lond. Spec. Publ.* **2007**, *278*, 39–81. [CrossRef]
83. Engelder, T.; Lash, G.G.; Uzcátegui, R.S. Joint sets that enhance production from Middle and Upper Devonian gas shales of the Appalachian Basin. *AAPG Bull.* **2009**, *93*, 857–889. [CrossRef]
84. U.S. EPA. *Method 200.7: Determination of Metals and Trace Elements in Water and Wastes by Inductively Coupled Plasma-Atomic Emission Spectrometry*; U.S. EPA: Cincinnati, OH, USA, 1994; Revision 4.4.
85. U.S. EPA. *Method 200.8: Determination of Trace Elements in Waters and Wastes by Inductively Coupled Plasma-Mass Spectrometry*; U.S. EPA: Cincinnati, OH, USA, 1994; Revision 5.4.
86. Lewan, M.D.; Bjorøy, M.; Dolcater, D.L. Effects of thermal maturation on steroid hydrocarbons as determined by hydrous pyrolysis of Phosphoria Retort Shale. *Geochim. Cosmochim. Acta* **1986**, *50*, 1977–1987. [CrossRef]
87. USGS Mineral Resources Program. Method 18—Sixty Elements by Inductively Coupled Plasma-Optical Emission Spectroscopy-Mass Spectroscopy (ICP-OES/MS), Sodium Peroxide Fusion (ICP-60). 2018. Available online: https://www.usgs.gov/media/files/60-elements-icp-oes-ms-na2o-fusion-method (accessed on 15 April 2022).
88. Pourmand, A.; Dauphas, N.; Ireland, T.J. A novel extraction chromatography and MC-ICP-MS technique for rapid analysis of REE, Sc and Y: Revising CI-chondrite and Post-Archean Australian Shale (PAAS) abundances. *Chem. Geol.* **2012**, *291*, 38–54. [CrossRef]
89. De Baar, H.J.W.; Bacon, M.P.; Brewer, P.G. Rare-earth distributions with a positive Ce anomaly in the Western North Atlantic Ocean. *Nature* **1983**, *301*, 324–327. [CrossRef]
90. Kim, J.-H.; Torres, M.E.; Haley, B.A.; Kastner, M.; Pohlman, J.W.; Riedel, M.; Lee, Y.-J. The effect of diagenesis and fluid migration on rare earth element distribution in pore fluids of the northern Cascadia accretionary margin. *Chem. Geol.* **2012**, *291*, 152–165. [CrossRef]
91. Ross, D.J.K.; Bustin, R.M. Shale gas potential of the lower Jurassic Gordondale member, northeastern British Columbia, Canada. *Bull. Can. Pet. Geol.* **2007**, *55*, 51–75. [CrossRef]
92. Ross, D.J.K.; Bustin, R.M. Characterizing the shale gas resource potential of Devonian–Mississippian strata in the Western Canada sedimentary basin: Application of an integrated formation evaluation. *AAPG Bull.* **2008**, *92*, 87–125. [CrossRef]
93. Chalmers, G.R.L.; Bustin, R.M. The organic matter distribution and methane capacity of the Lower Cretaceous strata of North-eastern British Columbia, Canada. *Int. J. Coal Geol.* **2007**, *70*, 223–239. [CrossRef]
94. Chalmers, G.R.L.; Bustin, R.M. Lower Cretaceous gas shales in northeastern British Columbia, Part I: Geological controls on methane sorption capacity. *Bull. Can. Pet. Geol.* **2008**, *56*, 1–21. [CrossRef]
95. Valenza, J.J.; Drenzek, N.; Marques, F.; Pagels, M.; Mastalerz, M. Geochemical controls on shale microstructure. *Geology* **2013**, *41*, 611–614. [CrossRef]

96. Carroll, A.R. Upper Permian lacustrine organic facies evolution, southern Junggar Basin, NW China. *Org. Geochem.* **1998**, *28*, 649–667. [CrossRef]
97. Milliken, K.L.; Rudnicki, M.; Awwiller, D.N.; Zhang, T. Organic matter–hosted pore system, Marcellus formation (Devonian), Pennsylvania. *AAPG Bull.* **2013**, *97*, 177–200. [CrossRef]
98. Chen, J.; Xiao, X. Evolution of nanoporosity in organic-rich shales during thermal maturation. *Fuel* **2014**, *129*, 173–181. [CrossRef]
99. Laughrey, C.D. Produced Gas and Condensate Geochemistry of the Marcellus Formation in the Appalachian Basin: Insights into Petroleum Maturity, Migration, and Alteration in an Unconventional Shale Reservoir. *Minerals* **2022**, *12*, 1222. [CrossRef]
100. Laughrey, C.D.; Lemmens, H.; Ruble, T.E.; Butcher, A.R.; Walker, G.; Kostelnik, J.; Barnes, J.; Knowles, W. Black shale diagenesis: Insights from integrated high-definition analyses of post-mature Marcellus Formation rocks, northeastern Pennsylvania. In *Critical Assessment of Shale Resource Plays*; AAPG: Tulsa, OK, USA, 2013; AAPG Memoir 103.
101. Delle Piane, C.; Bourdet, J.; Josh, M.; Clennell, M.B.; Rickard, W.D.A.; Saunders, M.; Sherwood, N.; Li, Z.; Dewhurst, D.N.; Raven, M.D. Organic matter network in post-mature Marcellus Shale: Effects on petrophysical properties. *AAPG Bull.* **2018**, *102*, 2305–2332. [CrossRef]
102. Tissot, B.P.; Welte, D.H. *Petroleum Formation and Occurrence*, 2nd ed.; Springer: Berlin/Heidelberg, Germany, 1984; p. 699.
103. Sholkovitz, E.R.; Landing, W.M.; Lewis, B.L. Ocean particle chemistry: The fractionation of rare earth elements between suspended particles and seawater. *Geochim. Cosmochim. Acta* **1994**, *58*, 1567–1579. [CrossRef]
104. Hathorne, E.C.; Stichel, T.; Brück, B.; Frank, M. Rare earth element distribution in the Atlantic sector of the Southern Ocean: The balance between particle scavenging and vertical supply. *Mar. Chem.* **2015**, *177*, 157–171. [CrossRef]
105. Klinkhammer, G.P.; Elderfield, H.; Edmond, J.M.; Mitra, A. Geochemical implications of rare earth element patterns in hydrothermal fluids from mid-ocean ridges. *Geochim. Cosmochim. Acta* **1994**, *58*, 5105–5113. [CrossRef]
106. Douville, E.; Charlou, J.L.; Oelkers, E.H.; Bienvenu, P.; Colon, C.F.J.; Donval, J.P.; Fouquet, Y.; Prieur, D.; Appriou, P. The rainbow vent fluids (36 14′ N, MAR): The influence of ultramafic rocks and phase separation on trace metal content in Mid-Atlantic Ridge hydrothermal fluids. *Chem. Geol.* **2002**, *184*, 37–48. [CrossRef]
107. Tostevin, R.; Wood, R.A.; Shields, G.A.; Poulton, S.W.; Guilbaud, R.; Bowyer, F.; Penny, A.M.; He, T.; Curtis, A.; Hoffmann, K.H.; et al. Low-oxygen waters limited habitable space for early animals. *Nat. Commun.* **2016**, *7*, 12818. [CrossRef]
108. Sensarma, S.; Saha, A.; Hazra, A. Implications of REE incorporation and host sediment influence on the origin and growth processes of ferromanganese nodules from Central Indian Ocean Basin. *Geosci. Front.* **2021**, *12*, 101123. [CrossRef]
109. Birdwell, J.E. *Rare Earth Element Concentrations for Oil Shales and Isolated Kerogens from around the world: U.S. Geological Survey Data Release*; U.S. Geological Survey: Reston, VA, USA, 2024. [CrossRef]

Disclaimer/Publisher's Note: The statements, opinions and data contained in all publications are solely those of the individual author(s) and contributor(s) and not of MDPI and/or the editor(s). MDPI and/or the editor(s) disclaim responsibility for any injury to people or property resulting from any ideas, methods, instructions or products referred to in the content.

Article

Solvent Exsolution and Liberation from Different Heavy Oil–Solvent Systems in Bulk Phases and Porous Media: A Comparison Study

Wei Zou and Yongan Gu *

Petroleum Technology Research Centre (PTRC), Petroleum Systems Engineering, Faculty of Engineering and Applied Science, University of Regina, Regina, SK S4S 0A2, Canada; wzh349@uregina.ca
* Correspondence: peter.gu@uregina.ca; Tel.: +1-(306)-585-4630

Abstract: In this paper, experimental and numerical studies were conducted to differentiate solvent exsolution and liberation processes from different heavy oil–solvent systems in bulk phases and porous media. Experimentally, two series of constant-composition-expansion (CCE) tests in a PVT cell and differential fluid production (DFP) tests in a sandpacked model were performed and compared in the heavy oil–CO_2, heavy oil–CH_4, and heavy oil–C_3H_8 systems. The experimental results showed that the solvent exsolution from each heavy oil–solvent system in the porous media occurred at a higher pressure. The measured bubble-nucleation pressures (P_n) of the heavy oil–CO_2 system, heavy oil–CH_4 system, and heavy oil–C_3H_8 system in the porous media were 0.24 MPa, 0.90 MPa, and 0.02 MPa higher than those in the bulk phases, respectively. In addition, the nucleation of CH_4 bubbles was found to be more instantaneous than that of CO_2 or C_3H_8 bubbles. Numerically, a robust kinetic reaction model in the commercial CMG-STARS module was utilized to simulate the gas exsolution and liberation processes of the CCE and DFP tests. The respective reaction frequency factors for gas exsolution (rff_e) and liberation (rff_l) were obtained in the numerical simulations. Higher values of rff_e were found for the tests in the porous media in comparison with those in the bulk phases, suggesting that the presence of the porous media facilitated the gas exsolution. The magnitudes of rff_e for the three different heavy oil–solvent systems followed the order of CO_2 > CH_4 > C_3H_8 in the bulk phases and CH_4 > CO_2 > C_3H_8 in the porous media. Hence, CO_2 was exsolved from the heavy oil most readily in the bulk phases, whereas CH_4 was exsolved from the heavy oil most easily in the porous media. Among the three solvents, CH_4 was also found most difficult to be liberated from the heavy oil in the DFP test with the lowest rff_l of 0.00019 min^{-1}. This study indicates that foamy-oil evolution processes in the heavy oil reservoirs are rather different from those observed from the bulk-phase tests, such as the PVT tests.

Keywords: solvent exsolution and liberation; non-equilibrium phase behaviour; foamy-oil formation and flow; heavy oil–solvent systems; heavy oil reservoirs

1. Introduction

The primary productions in some heavy oil reservoirs exhibit anomalous characteristics, such as low producing gas–oil ratios (GORs), high oil production rates and recovery factors (RFs) [1]. These abnormal characteristics could be attributed to foamy-oil flow, which is an unusual two-phase flow of oil with dispersed gas [2]. After the primary production, solvent-based enhanced oil recovery (EOR) methods can be applied by injecting one or several solvents into the heavy oil reservoirs to reduce the heavy oil viscosity and continue the foamy-oil production. Two major non-equilibrium processes greatly affect the foamy-oil production: solvent exsolution and liberation, which are referred to as the foamy-oil formation and evolution. Therefore, it is of practical importance to study the foamy-oil formation and evolution in the heavy oil reservoirs.

Physically, the dissolved gas in the oil is exsolved in the form of microbubbles when the pressure is reduced below the so-called bubble-point pressure. The bubbles could be dispersed in the heavy oil for a long time due to its high viscosity. As the dispersed bubbles grow and coalesce, they gradually liberate from the foamy oil to form the free gas. Extensive experimental, numerical, and theoretical studies have been conducted to understand the processes and mechanisms involved in the foamy-oil formation and evolution. Several factors influencing the foamy-oil formation and evolution have been experimentally studied in bulk phases and porous media, including the dead heavy oil viscosity, solvent type, and concentration as well as pressure drawdown method and rate.

The experimental results in the bulk-phase tests show that foamy-oil stability increases as the viscosity of the dead heavy oil increases [2,3]. First, the high viscosity of the heavy oil reduces the gas diffusion rate, causing the microbubbles to grow slowly through mass transfer. Second, high viscosity also hinders bubbles from coalescing and liberating from the foamy oil. Thus, oil recovery can even increase with the increase in the heavy oil viscosity [4]. From an engineering point of view, however, an optimum heavy oil viscosity could exist to achieve the highest oil RF. Wu et al. conducted pressure depletion tests in a sandpacked model by using three heavy oil samples with different viscosities but at the same CO_2 concentration [5]. They found that the heavy oil sample with an intermediate viscosity gave the highest oil recovery. This was because the heavy oil with a lower viscosity had less stable foamy oil, whereas the heavy oil with a higher viscosity had a lower mobility. These counteracting effects on the oil recovery were also studied by utilizing a viscosity reducer in heavy oil [6,7]. It was found that the oil recovery factors were increased markedly with the increase of viscosity-reducer concentration in the low concentration range. This was attributed to the decreased oil viscosity and increased oil mobility. At higher concentrations of the viscosity reducer, however, the increase in the oil recovery was more gradual because the foamy oil was less stable [6].

In addition to the heavy oil viscosity, solvent type and concentrations also play critical roles in foamy-oil formation and evolution. Three common solvents, CH_4, C_3H_8, and CO_2 as well as their mixtures, are mostly used in the heavy oil–solvent systems. Sun et al. conducted a series of tests in a PVT cell to study the different foamy-oil evolution processes in the heavy oil–CH_4/C_3H_8/CO_2 systems [8]. They found that more gas was evolved in the heavy oil–CH_4 system than in the heavy oil–C_3H_8/CO_2 system at the same solvent molar concentration, which indicated that CH_4 bubbles were nucleated more readily. The larger difference between the two pressures, at which bubbles started to nucleate and liberate, was found in the heavy oil–CH_4 system, which indicated that CH_4 bubbles were more difficult to liberate from foamy oil. Zhou et al. performed pressure depletion tests in a sandpacked model by using CH_4, C_3H_8, and a mixture of CH_4 and C_3H_8 [9]. The foamy oil in the heavy oil–CH_4 system was also found more stable than those in the other two heavy oil–solvent systems. Furthermore, the amount of the dissolved gas in heavy oil also affects foamy-oil evolution. PVT tests show that the higher the initial solution GOR in a heavy oil–solvent system is, the longer the bubbles can be dispersed in the heavy oil [2]. This was attributed to a higher supersaturation in the system at a higher solvent concentration. A higher supersaturation can lead to more and smaller bubbles being generated. Experimental results from sandpacked tests also reveal the same trend between initial solution GOR and supersaturation as found in the PVT tests. In addition, it was found that a lower limit of the initial solution GOR exists in order to have the foamy oil [10].

Another essential factor in foamy-oil formation and evolution is pressure drawdown. Three pressure drawdown methods are typically adopted in the experimental studies with bulk phase and porous media: (a) constant pressure depletion rate [11]; (b) constant volume withdrawal rate [12,13]; and (c) constant pressure depletion stepsize [14]. The measured pressure vs. volume (P–V) data in the PVT tests could show different characteristics by using the different pressure drawdown methods. In particular, the measured P–V data with the volume withdrawal rate method could show a rebound region when the pressure reached the bubble-nucleation pressure due to a high bubble growth rate [15]. In contrast, smooth P–V data are usually obtained by using the constant pressure depletion rate method [11]. The P–V data obtained by using the constant pressure depletion stepsize method could have one or two turning points, at which the bubble-point pressure or pseudo bubble-point pressure was achieved [14]. How fast the pressure was reduced affects the gas exsolution and liberation processes. Experimental results show that as the pressure depletion rate was increased, the volume of dispersed gas and the duration in which gas bubbles were dispersed in the oil also were increased [11]. This is attributed to two reasons. First, a higher depletion rate leads to an insufficient time for bubbles to nucleate and thus lowers the pressure at which bubbles start to liberate to form the free gas. Second, a higher depletion rate results in a higher supersaturation and a higher bubble nucleation rate so that the bubbles nucleated can also be smaller [2].

Many theoretical models have been proposed to describe the foamy-oil formation and/or evolution processes. These models can be broadly categorized into two types: physics-based models [16,17] and kinetic reaction models [18–20]. In the physics-based models, bubble nucleation was considered as either progressive nucleation or instantaneous nucleation. Progressive nucleation has been modeled by using the classic bubble nucleation theory or the pre-existing bubble theory [21,22]. In instantaneous nucleation, bubbles are nucleated instantaneously and no more new bubbles are nucleated [23]. The bubble number at the onset of bubble nucleation is assigned an initial guessing value and then adjusted to match the calculated data with the measured data. The bubble growth was controlled by mechanic expansion and mass transfer [4,23]. For the bubble liberation, some empirical correlations are made available in the literature and have been verified by limited experimental data. The kinetic reaction models are developed from chemical kinetics theories and assume that three or more gas components exist in the foamy-oil formation and evolution. Typically, the gas components include the solution gas, dispersed gas, and free gas [24–26]. The dispersed gas and free gas together are also referred to as the evolved gas. The reaction rate of a gas component transferring to another gas component is assumed to be proportional to the concentration of the reactant gas component. The kinetics model proposed by Coombe and Maini [27] has been used in the CMG-STARS module since it is predictive and can simulate the time-dependent phenomena [28].

Although numerous studies have investigated the foamy-oil formation and/or evolution in the porous media or bulk phases alone, few studies have focused on studying the effects of the bulk phases and porous media on the foamy-oil formation and evolution processes in different heavy oil–solvent systems. The research results obtained in the bulk phases and porous media could be different and cause confusion in the heavy oil industry. Hence, it is important to understand their differences prior to the oil field applications. In this paper, three constant-composition-expansion (CCE) tests in a PVT cell and three differential fluid production (DFP) tests in a sandpacked model were conducted in the heavy oil–CO_2, heavy oil–CH_4, and heavy oil–C_3H_8 systems. The total isothermal compressibility (c_t) vs. test pressure (P) data were measured in all the tests, while the oil and gas production data were also measured in the tests in the porous media. Afterwards, numerical simulations were executed to simulate the six tests. From these simulations, two reaction frequency factors were obtained to quantify the foamy-oil formation and evolution processes. The results from the experimental tests and numerical simulations in this study provide insights into the differences of foamy-oil formations and evolutions in the bulk phases and porous media.

2. Experimental

2.1. Materials

The heavy oil sample (Well No.: 16A-3-59-7) used in this study was collected from the Colony formation in the Bonnyville area, Alberta, Canada. Prior to the measurements of the heavy oil properties, the heavy oil was centrifuged to remove the possibly existing sands and/or brine from the oil. The compositions of the Colony heavy oil were measured by using the standard ASTM D86 and are given in Table 1. It can be seen from the table that the minimum carbon number of the Colony heavy oil was C_9 and that the hydrocarbons were measured up to C_{60}. The heavy oil density and viscosity were measured to be $\rho_o = 0.992$ g/cm^3 and $\mu_o = 33,876$ cP by using a densitometer (DMA 4200, Anton Paar, Graz, Austria) and a viscometer (DV-II+, Brookfield Engineering, Middleboro, MA, USA) at $P_a = 1$ atm and $T_{lab} = 21°$ C, respectively. The respective measured densities and viscosities of the Colony heavy oil at $P_a = 1$ atm and different temperatures are listed in Table 2. Both the density and viscosity of the heavy oil decrease with the increase in temperature. The molecular weight of the heavy oil was measured to be $MW_o = 547.7$ g/mol by using an automatic high-sensitivity wide-range cryoscopy (Model 5009, Precision Systems Inc., Natick, MA, USA). The asphaltene content of the heavy oil was measured to be $W_{asp} = 18.3$ wt.% (n-C_5 insoluble) by using the standard ASTM D2007-19 method. Three solvents used in this study, CO_2, CH_4, and C_3H_8, were purchased from Linde Canada Inc. (Mississauga, ON, CA) and had purities of 99.998 mol.%, 99.97 mol.%, and 99.5 wt.%, respectively.

Table 1. Compositional analysis result of the Colony heavy oil (Well No.: 16A-3-59-7) collected from the Bonnyville area, Alberta [29].

Carbon no.	mol.%	wt.%	Carbon no.	mol.%	wt.%
C_1	0.00	0.00	C_{32}	1.46	1.61
C_2	0.00	0.00	C_{33}	1.32	1.50
C_3	0.00	0.00	C_{34}	1.28	1.50
C_4	0.00	0.00	C_{35}	1.15	1.39
C_5	0.00	0.00	C_{36}	1.05	1.30
C_6	0.00	0.00	C_{37}	1.16	1.48
C_7	0.00	0.00	C_{38}	1.00	1.31
C_8	0.00	0.00	C_{39}	0.99	1.33
C_9	0.79	0.25	C_{40}	1.05	1.45
C_{10}	2.61	0.91	C_{41}	0.93	1.32
C_{11}	2.21	0.85	C_{42}	0.98	1.42
C_{12}	3.80	1.59	C_{43}	1.06	1.57
C_{13}	4.20	1.90	C_{44}	0.96	1.45
C_{14}	4.51	2.19	C_{45}	0.89	1.39
C_{15}	4.49	2.34	C_{46}	0.83	1.32
C_{16}	4.67	2.59	C_{47}	0.77	1.25
C_{17}	4.43	2.61	C_{48}	0.67	1.11
C_{18}	4.18	2.61	C_{49}	0.55	0.93
C_{19}	3.87	2.55	C_{50}	0.64	1.11
C_{20}	3.52	2.44	C_{51}	0.63	1.11
C_{21}	3.16	2.30	C_{52}	0.51	0.92
C_{22}	2.99	2.28	C_{53}	0.40	0.74
C_{23}	2.54	2.02	C_{54}	0.53	0.99
C_{24}	2.47	2.05	C_{55}	0.47	0.89
C_{25}	2.21	1.91	C_{56}	0.53	1.03
C_{26}	2.22	2.00	C_{57}	0.40	0.78
C_{27}	2.24	2.09	C_{58}	0.44	0.88
C_{28}	1.89	1.83	C_{59}	0.39	0.79
C_{29}	1.97	1.98	C_{60}	0.35	0.72
C_{30}	1.68	1.74	C_{61+}	8.53	20.85
C_{31}	1.43	1.53	Total	100.00	100.00

Table 2. The Colony heavy oil densities (ρ_o) and viscosity (μ_o) at P_a = 1 atm and different temperatures.

T (°C)	ρ_o (g/cm³)	μ_o (cP)
21	0.9920	33,876
30	0.9864	10,539
40	0.9804	3,980
50	0.9743	1,679
60	0.9681	795.9

2.2. Experimental Setup and Procedures for the CCE Tests

In this study, three CCE tests were carried out by using a mercury-free PVT system (PVT-0150-100-200-316-155, DBR, Canada), as schematically shown in Figure 1. The PVT system was mainly composed of a PVT cell, an air bath, an oil injection unit, and a digital image-based height measurement system. The see-through PVT cell was a glass tube with an inner diameter (ID) of 1.252 inches and a length of 8.005 inches. A freely movable piston was positioned in the PVT cell to pressurize the test fluids. The PVT cell can withstand a high pressure up to 69 MPa and a high temperature of 200.0 °C. A syringe pump (100 DX, Teledyne ISCO, Lincoln, NE, USA) connected to the PVT cell was used to apply the test pressure and overburden pressure to the PVT cell. The test temperature can be controlled by using an air bath, which is surrounded by the PVT cell. The oil injection unit consisted of a high-pressure transfer cylinder and another syringe pump (500 DX, Teledyne ISCO, Lincoln, NE, USA). The high-pressure transfer cylinder held the solvent-saturated live oil, and the syringe pump was used to inject the solvent-saturated live heavy oil into the PVT cell. The digital image-based height measurement system included a high-resolution color monitor, a cathetometer, and an encoder. The image of the piston or fluid interface (s) in the PVT cell can be viewed and captured by the camera and displayed on the monitor. The encoder was used to adjust and record the height of the piston or fluid interface (s). The height measurement resolution of the system was 0.001 inches.

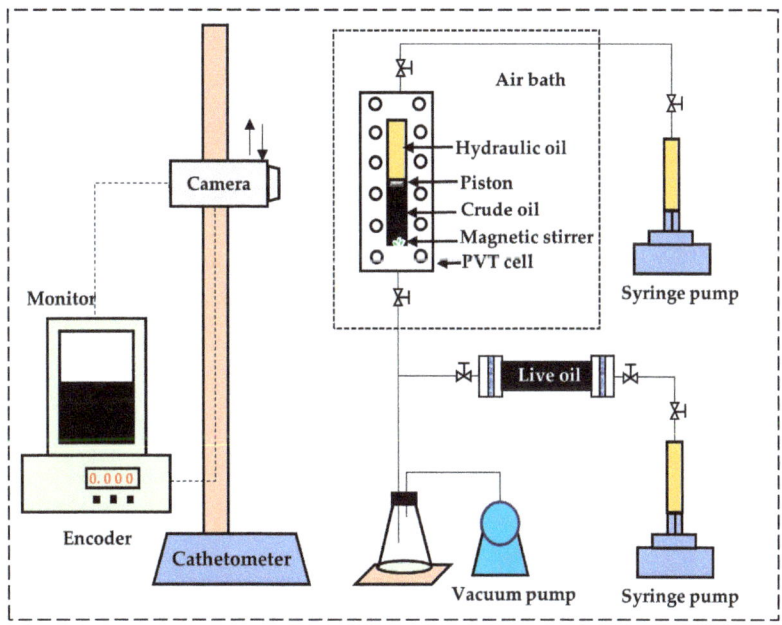

Figure 1. Schematic diagram of the experimental setup for conducting the CCE tests [30].

The three CCE tests for the three heavy oil–solvent systems were conducted by using the following experimental procedures:

(1) The $CO_2/CH_4/C_3H_8$-saturated heavy oil was prepared by mixing a solvent ($CO_2/CH_4/C_3H_8$) and the dead Colony heavy oil in two high-pressure transfer cylinders. The mixture in the two high-pressure transfer cylinders was pumped back and forth for about 15–20 days until the target GOR of the live oil was achieved. The solvent concentrations (C_s) in the prepared heavy oil–CO_2 system, heavy oil–CH_4 system and heavy oil–C_3H_8 system were measured and are listed in Table 3.

(2) The PVT cell was cleaned with kerosene and ethanol and then air-dried for 24 h. A leakage test was conducted by injecting CO_2 into the PVT cell at a constant pressure of 3.5 MPa. The PVT cell was considered leakage-free if the pressure reduction was less than 5 psi after 24 h. Then, the CO_2 in the cell was released to an exhaust ventilation hose. Afterwards, a vacuum pump was used to vacuum the PVT cell.

(3) The piston in the PVT cell was pushed to the bottom of the PVT cell. The PVT cell pressure and its overburden pressure were raised to a pressure higher than the bubble-point pressure of the live oil by using the ISCO pump, which was connected to the PVT cell and its overburden chamber. Then, a pre-specified volume of the live oil (V_{oi}) was injected into the PVT cell by using another syringe pump. The volumes of the injected three live oils in the three CCE tests (Tests #1, #3 and #5) are given in Table 3.

(4) The PVT cell pressure was reduced stepwise by using a pre-specified pressure step (ΔP) given in Table 3. The height (H) of the test fluid column in the PVT cell and the volume of the syringe pump connected to the PVT cell were monitored and recorded every 10 min.

(5) The total test-fluid volume (V_t) in the PVT cell, total isothermal compressibility (c_t) and relative volume (R_v) were measured when the volume change of the syringe pump was less than 0.12 cc within 10 min. Here, V_t, c_t and R_v are defined as

$$V_t = \frac{\pi D^2}{4} \cdot H - V_{dv} \quad (1)$$

$$c_t = -\frac{1}{V_t}\left(\frac{\Delta V_t}{\Delta P_{cell}}\right)_T \quad (2)$$

$$R_v = \frac{V_t}{V_{oi}} \quad (3)$$

where D and H are the ID of the PVT cell glass tube and the height of the test fluid column in the PVT cell; V_{dv} is the dead volume in the PVT cell; ΔV_t is the change of total test-fluid volume caused by the change of the PVT cell pressure (ΔP_{cell}); and V_{oi} is the volume of the live oil that was initially injected into the PVT cell.

(6) Steps 4 and 5 were repeated until the PVT cell pressure reached a pre-specified test termination pressure.

Table 3. Test details, measured data, and tuned reaction frequency factors of Tests #1–6.

Test No.	Test	C_s (mol.%)	V_{oi} (cm³)	ϕ (%)	k (D)	S_{oi} (%)	ΔP (MPa)	P_b (MPa)	P_n (MPa)	rff_e (min⁻¹)	rff_1 (min⁻¹)
1	CO_2-CCE	27.66	39.78	-	-	-	0.20	2.00	1.73	0.00052	-
2	CO_2-DFP		-	40	2.6	98			1.97	0.00075	0.009
3	CH_4-CCE	11.32	38.51	-	-	-	0.20	1.70	0.60	0.00009	-
4	CH_4-DFP		-	39	2.7	97			1.50	0.013	0.00019
5	C_3H_8-CCE	27.31	40.54	-	-	-	0.05	0.50	0.48	0.0001, 0.000065	-
6	C_3H_8-DFP		-	39	1.9	97			0.50	0.00025	0.0045

2.3. Experimental Setup for the DFP Tests

The experimental setup for conducting the three DFP tests mainly consisted of four operating units: a 2-D rectangular sandpacked physical model; a water/oil injection module; a fluids production system; and a data acquisition system (DAS). The schematic diagram of the experimental setup for conducting the three DFP tests is shown in Figure 2. The physical model was composed of four parts: a stainless-steel base; a thin polycarbonate plate; a thick transparent acrylic plate; and a stainless-steel cover. The stainless-steel base had a rectangular cavity with the dimensions of 40 cm × 10 cm × 2 cm for packing the sand grains. The polycarbonate plate was placed between the stainless-steel base and the acrylic plate to avoid scratching on the latter. The stainless-steel cover was used to place and secure the stainless-steel base, the polycarbonate plate, and the acrylic plate together. The physical model has two ports at its two ends. The one on the right-hand side was used as a reservoir fluids producer in each test. The other, on the left-hand side, was used to inject water/oil and establish the initial water/oil saturation.

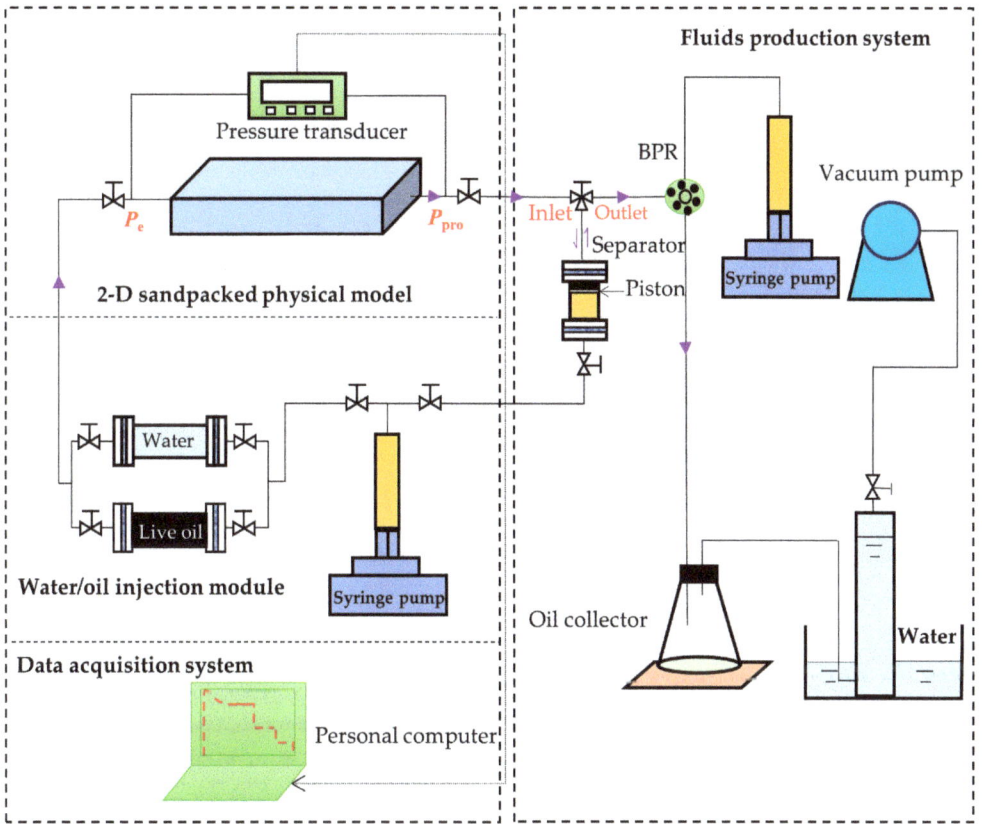

Figure 2. Schematic diagram of the experimental setup for conducting the DFP tests.

The water/oil injection module comprised two high-pressure transfer cylinders and a syringe pump (500 D, ISCO Inc., USA). One transfer cylinder held the de-ionized water and the other contained the solvent-saturated live heavy oil. Along the produced reservoir fluids flow direction, the fluids production system was composed of a high-pressure transfer cylinder with a freely movable piston used as a separator, a high-precision back-pressure regulator (BPR, LBS4 Series, Swagelok, Solon, OH, USA), a glass flask, an electronic scale, and a gas bubbler. The separator's bottom port was connected to the syringe pump (500 D, Teledyne ISCO, Lincoln, NE, USA) used in the water/oil injection module, while its top port was connected to the common port of an L-type three-way valve. The three-way valve's inlet port was connected to the sandpacked model and its outlet port was connected to the BPR. The BPR was connected to a different syringe pump (260 D, Teledyne ISCO, Lincoln, NE, USA) to maintain its back pressure. After the BPR, the glass flask, electronic scale, and bubbler were used to collect the produced dead oil, measure its mass, and measure the produced gas volume at the atmospheric conditions, respectively. The DAS mainly consisted of two high-precision pressure transducers (PXM409, Omegadyne Inc., Sunbury, OH, USA) at the two ends of the physical model to measure the two pressures at any time during each test. The pressure data were recorded and stored in a personal computer, to which the pressure transducers were connected. The respective pressures measured at the fluids production end on the right-hand side and at the other end on the left-hand side were denoted by P_{pro} and P_e.

The DFP tests were conducted by using the following experimental procedures:

(1) The sieved sand grains (Bell & Mackenzie, Toronto, ON, Canada) with sizes of 60–80 mesh were added into the physical model through one of its ports. The sandpacked model was hammered continuously by using a rubber hammer until no void space existed in the model.
(2) The leakage test was conducted by injecting CO_2 into the sandpacked model at 3 MPa for each DFP test. The sandpacked model was considered leakage-free if its pressure reduction was less than 0.2 MPa within 24 h.
(3) The porosity (ϕ) and permeability (k) measurements were performed by using the imbibition method and the Darcy's law, respectively. The measured permeabilities and porosities of the three DFP tests (Tests #2, #4, #6) are listed in Table 3.
(4) Water was injected into the 2-D physical model first to raise the reservoir pressure higher than the bubble-point pressure of the live heavy oil. Then the solvent-saturated live oil was injected to displace the water until no more water was produced from the 2-D physical model and the measured solution GOR reached the initial solution GOR. The initial oil saturations (S_{oi}) were determined by using the material balance equation and are given in Table 3.
(5) The separator (i.e., a high-pressure cylinder) was connected to the 2-D physical model after its pressure was increased to the reservoir pressure. A period of 24 h was allowed for the 2-D physical model and separator to reach an equilibrium state.
(6) The separator and reservoir pressures were reduced simultaneously by applying a pre-specified pressure depletion stepsize (ΔP). The produced fluids (heavy oil and/or gas) were collected in the separator and their total volume (V_f) at the separator pressure was monitored and recorded by using a syringe pump. The pressures at the two ends of the sandpacked physical model (P_{pro} and P_e) were monitored and recorded.
(7) The production data, including P_{pro} and P_e, were measured when the pump volume change was less than 0.12 cc within 10 min. First, the three-way valve was switched to disconnect the separator from the physical model and the produced fluids in the separator were displaced to measure the dead-oil and gas volumes (V_{do} and V_g) at the atmospheric pressure. Second, the volumes of the cumulative produced fluids at the reservoir conditions (Q_f), cumulative produced dead oil and gas (Q_o and Q_g) at the atmospheric conditions, as well as the total isothermal compressibility (c_t), were measured with the following equations:

$$Q_f = \sum V_f \qquad (4)$$

$$Q_o = \sum V_o \qquad (5)$$

$$Q_g = \sum V_g \qquad (6)$$

$$c_t = -\frac{1}{PV}\left(\frac{V_f}{\Delta P_{res}^{ave}}\right)_T \tag{7}$$

where PV and ΔP_{res}^{ave} represent the pore volume of the physical model and the change of the average reservoir pressure, $P_{res}^{ave} = (P_{pro} + P_e)/2$.

(8) Each DFP test was continued by repeating Steps 6 and 7 until the production pressure reached a pre-specified ending production pressure, e.g., P_{pro} =0.8, 0.3, and 0.3 MPa in the heavy oil–CO_2 system, heavy oil–CH_4 system, and heavy oil–C_3H_8 system, respectively.

3. Numerical Modeling

Numerical simulations were undertaken by using the CMG simulator to simulate the experimental tests in order to obtain the quantitative results of the foamy-oil formations and evolutions in the three heavy oil–solvent systems with or without the porous media. The fluid models were built in the CMG-WinProp module by matching the measured dead oil and live oil properties. The CMG-STARS module with the foamy-oil kinetic reaction model was utilized to history match the relative volume vs. test pressure data in the CCE tests and the production data in the DFP tests.

3.1. Fluid Models

In this work, the fluid models for the three heavy oil–solvent systems were generated by using the Peng–Robinson equation of state (PR-EOS) in the CMG-WinProp module. The components of the heavy oil listed in Table 1 were lumped into two pseudo-components: 9.59 mol.% C_{9-12} and 90.41 mol.% C_{13+}. The related physicochemical properties of the two pseudo-components were used as the adjustable parameters to match the measured densities and viscosities at P_a = 1 atm and different temperatures in Table 2. The tuned physicochemical properties of the two pseudo-components of the Colony heavy oil are given in Table 4. Afterwards, a solvent (CO_2/CH_4/C_3H_8) was added into the two pseudo-components and the measured live-oil bubble-point pressures in Table 3 were matched to generate the CO_2/CH_4/C_3H_8-saturated fluid models. The tuned corresponding binary interaction parameters (BIPs) between each solvent and the two pseudo-components are listed in Table 4.

Table 4. Physicochemical properties of the two pseudo-components of the Colony heavy oil and their respective binary interaction parameters (BIPs) with CO_2/CH_4/C_3H_8.

Properties and BIPs	Two Pseudo-Components	
	C_{9-12}	C_{13+}
Critical pressure P_c (MPa)	2.33	0.69
Critical temperature T_c (K)	642.86	938.31
Pitzer acentric factor ω	0.4789	0.9805
Molecular weight MW (g/mol)	146.9	596.7
BIP with CO_2	0.12	0.14
BIP with CH_4	0.045	0.085
BIP with C_3H_8	0.036	0.106

3.2. Numerical Simulations of the CCE and DFP Tests

The CMG-STARS module was employed as a thermal compositional simulator to simulate the measured relative volume data in the three CCE tests (Tests #1, #3, and #5). The numerical simulations of the CCE tests using the CMG-STARS module in this study were similar to those conducted before [31]. The cylindrical PVT cell in each CCE test was modeled by building a simulation model with $1 \times 1 \times 4$ grid blocks along the x, y and z directions, as shown in Figure 3. Initially, the two upper grid blocks (1, 1, 1) and (1, 1, 2) had the live heavy oil only, whereas the two lower grid blocks (1, 1, 3) and (1, 1, 4) had water only. In order to simulate the CCE tests in the CMG-STARS module, an extremely high porosity of $\phi = 0.99$ and an absolute permeability of $k = 10{,}000$ D were assigned to the four grid blocks. The dimensions of the simulation model in its x and y directions were 2.83×2.83 cm^2 so that its horizontal cross-sectional area was almost the same as that of the PVT cell (ID = 3.18 cm) after the porosity (ϕ) was considered. The thicknesses of the two upper grid blocks (1, 1, 1) and (1, 1, 2) were the same and were determined by the volume of the injected live oil in each CCE test. Therefore, the two upper grid blocks (1, 1, 1) and (1, 1, 2) had the same thickness of 2×2.50, 2×2.42 and 2×2.55 cm for the heavy oil–CO$_2$ system ($V_{oi} = 39.78$ cm^3), heavy oil–CH$_4$ system ($V_{oi} = 38.51$ cm^3) and heavy oil–C$_3$H$_8$ system ($V_{oi} = 40.54$ cm^3), respectively. The thicknesses of the two lower grid blocks (1, 1, 3) and (1, 1, 4) were equal to 5.00 cm for all three heavy oil–solvent systems. A vertical producer was drilled from the top grid block to the bottom grid block but only the bottom grid block (1, 1, 4) was perforated. In this way, the simulation model pressure could be controlled by the vertical producer at the top of the simulation model. The water was produced as the simulation model pressure was reduced. The volume of the produced water at the reservoir conditions was equivalent to the expanded total volume of the test fluids. Hence, the relative volume (R_v) in the CCE tests can be simulated by using the CMG-STARS module. It is worthwhile to note that the water compressibility was set to be zero to eliminate its effect on the produced water volume.

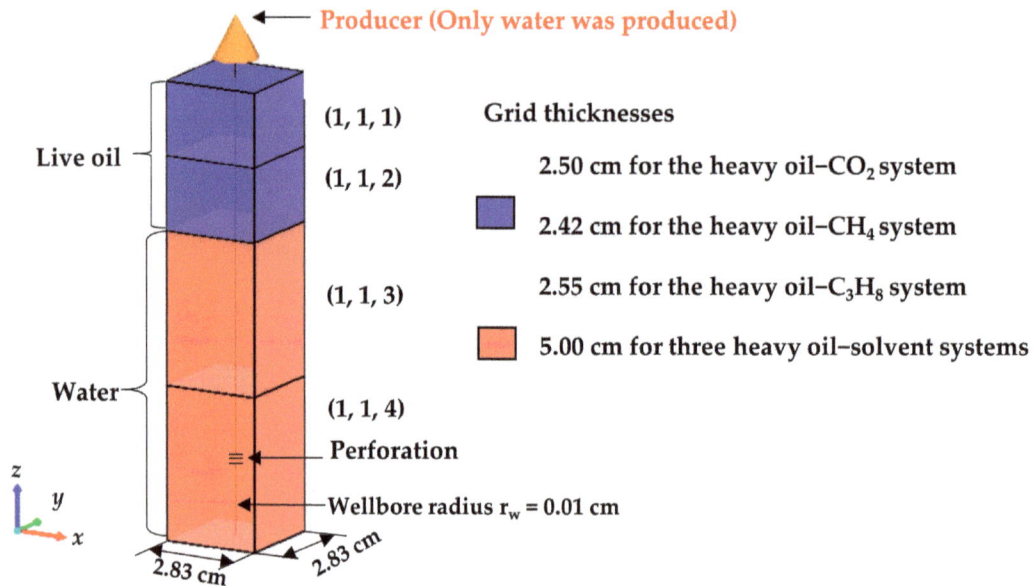

Figure 3. Three-dimensional view of the rectangular simulation model with four grid blocks for the CCE tests.

In the numerical simulations conducted by before [31], the relative permeability curves were altered to prevent oil and gas from moving between the grid blocks. In this study, however, three sets of relative permeability data were used for different grid blocks in order to improve the convergence of numerical simulation. In grid blocks (1, 1, 1) and (1, 1, 2), the oil and gas components had high relative permeabilities. Grid block (1, 1, 3) was set as a transition zone, in which all the components could move and the relative permeabilities of the oil and water were proportional to their corresponding saturations in the liquid phase. In grid block (1, 1, 4), the water relative permeability was set to be extremely high, whereas the oil and gas relative permeabilities were set to be extremely low so that only the water in the grid block (1, 1, 4) could be produced but the oil and gas could not. Therefore, the overall composition of all the test fluids (heavy oil + gas) remained constant in the numerical simulation. The three sets of relative permeability data are listed in Table 5.

Table 5. Three sets of relative permeability data used in different grid blocks.

Grid Block	Liquid Saturation and Relative Permeability		
	S_w [a]	k_{rw} [b]	k_{row} [c]
	0	0	1
	1.0×10^{-5}	0	1
	0.97999	0	1
(1, 1, 1) and (1, 1, 2)	1	1	0
	S_l [d]	k_{rg} [e]	k_{rog} [f]
	0	1	0
	1.0×10^{-5}	1	1
	0.9999	1	1
	1	0	1
	S_w	k_{rw}	k_{row}
	0	0	1
	1	1	0
	S_l	k_{rg}	k_{rog}
(1, 1, 3)	0	1	0
	1.0×10^{-5}	1	1
	0.9999	1	1
	1	0	1
	S_w	k_{rw}	k_{row}
	0	0	1.0×10^{-9}
	0.1	1	0
(1, 1, 4)	1	1	0
	S_l	k_{rg}	k_{rog}
	0	1.0×10^{-9}	0
	1	0	1.0×10^{-9}

Notes: [a] Water saturation; [b] Relative permeability to water at S_w; [c] Relative permeability to oil at S_w. [d] Liquid saturation; [e] Relative permeability to gas at S_l; [f] Relative permeability to oil at S_l.

Similar to the CCE tests, three DFP tests (Tests #2, #4, and #6) were also simulated by using the CMG-STARS module. The simulation model had $40 \times 5 \times 1$ grid blocks with the dimensions $40 \times 10 \times 2$ cm^3 in the x, y, and z directions. The model's porosities and permeabilities in the x, y, and z directions were assumed to be uniform and the same as those of the DFP tests. The initial reservoir conditions were also set to be the same as those in the DFP tests. The oil–water and liquid–gas relative permeabilities were first generated by using the modified Brooks–Corey correlations [32] and then adjusted by matching the predicted production data (i.e., P_e, Q_o, Q_g, Q_f) from the CMG-CMOST module to those measured production data in the DFP tests.

3.3. Foamy-Oil Kinetic Reaction Model

Foamy oil was simulated in the CMG-STARS module by using the entrained-gas model. The entrained-gas model includes two kinetic reactions, which describe the gas exsolution and liberation processes, respectively. The reaction rates of gas exsolution and liberation are expressed as follows [33]:

$$r_e = rff_e \cdot \exp\left(-\frac{E_a}{RT}\right) \cdot \phi \cdot S_{fo} \cdot n_{fo} \cdot (x - x_{eq}), \tag{8}$$

$$r_l = rff_l \cdot \exp\left(-\frac{E_a}{RT}\right) \cdot \phi \cdot S_{fo} \cdot n_{fo} \cdot x_{bub}, \tag{9}$$

where r denotes the reaction rate per bulk volume in mol/(min·cm^3); rff denotes the reaction frequency factor in 1/min; the subscripts e and l represent gas exsolution and liberation, respectively. E_a is the activation energy in J/mol; R is the universal gas constant of 8.3145 J/(mol· K); T is the reaction temperature in K; ϕ is the porosity; S_{fo} is the foamy-oil saturation in fraction; n_{fo} is the molar density of the foamy oil in mol/cm^3; x, x_{eq} and x_{bub} are the mole fractions of the dissolved gas, the dissolved gas at the equilibrium state and the dispersed bubble in the foamy oil. Since the DFP tests were conducted at the constant temperature of 21 °C, the activation energy E_a was set to be zero [34,35]. Hence, only the reaction frequency factors (rff_e and rff_l) in Equations (8) and (9) need to be determined.

Physically, $(x - x_{eq})$ in Equation (8) represents the deviation of the dissolved gas mole fraction from its equilibrium value at the existing test pressure. The x_{eq} can be calculated from the equilibrium constant (K-value) generated from the CMG-WinProp module. The higher $(x - x_{eq})$ is, the higher supersaturation in the live oil. The rff_e can be used to compare how fast the dissolved gas was exsolved to form dispersed gas in two systems at the same supersaturation. Similarly, rff_l in Equation (9) indicates how fast the dispersed gas is liberated from the foamy oil to form free gas. In the numerical simulations, the reaction rates of gas exsolution and liberation can be calculated with the two factors (rff_e and rff_l) by using Equations (8) and (9) once $(x - x_{eq})$ becomes greater than zero due to supersaturation. Then, the calculated reaction rates could be easily incorporated into the existing continuity equations with the mass-transfer terms in the CMG-STARS module.

4. Results and Discussion

4.1. Heavy Oil–CO$_2$ System

In this study, the equilibrium CCE data, namely the R_v vs. P_{cell} data, were predicted by using the PR-EOS in the CMG-WinProp module. The non-equilibrium R_v vs. P_{cell} data were predicted by using the CMG-STARS module. Figure 4 compares the predicted equilibrium R_v vs. P_{cell} data with the measured and predicted non-equilibrium R_v vs. P_{cell} data of Test #1 for the heavy oil–CO$_2$ system. It can be seen from the figure that the predicted equilibrium R_v vs. P_{cell} data had an obvious turning point at P_{cell} = 2.00 MPa, which was called the bubble-point pressure (P_b) of the heavy oil–CO$_2$ system. It can be seen from the predicted equilibrium R_v vs. P_{cell} data that the total volume of the test fluids (the heavy oil and gas) increased drastically once the test pressure was reduced to below P_b. In contrast, the measured non-equilibrium R_v vs. P_{cell} data did not show a noticeable change until P_{cell} = 1.73 MPa, at which a certain amount of the bubbles were nucleated. In this study, this threshold pressure was called the bubble-nucleation pressure (P_n), i.e., $P_n = P_{cell}$ = 1.73 MPa.

Here, P_n was used for the non-equilibrium tests to distinguish it from P_b, which was defined at the equilibrium state. Supersaturation started to occur in the live heavy oil at $P_{cell} < P_b$ and was the driving force for the bubble nucleation [36]. The bubbles could not be nucleated until a certain degree of supersaturation was reached, which is defined as the critical supersaturation (P_{crit}), namely, the pressure difference between P_b and P_n in this study.

Figure 4 also shows that an increased relative volume at each pressure drop was much smaller at $P_{cell} < P_n = 1.73$ MPa at the non-equilibrium state than that at $P_{cell} < P_b = 2.00$ MPa at the equilibrium state. This difference was attributed to the relatively smaller amount of the gas that was exsolved in the non-equilibrium state than that in the equilibrium. It was difficult to reach the equilibrium state in a heavy oil–solvent system during a CCE test due to the low bubble nucleation rate and CO_2 diffusion rate in the heavy oil [37]. The predicted non-equilibrium R_v vs. P_{cell} data from the CMG-STARS module are also plotted in Figure 4 after history matching the measured R_v vs. P_{cell} data. The predicted non-equilibrium relative volume data matched well with the measured data. The value of rff_e in Equation (8) was tuned to be 0.00052 min^{-1}. The rff_1 in Equation (9) was not obtained in the CMG-STARS module for the CCE tests. This was because the keyword *GASSYLIQ was used in the simulation module to give the same isothermal compressibility of the dispersed gas as that of the free gas [38]. This means that the system's total isothermal compressibility was determined by the total amount of the dispersed gas and free gas (i.e., the evolved gas) and thus was irrelevant to the individual amount of either gas. In this case, the tuned rff_e was sufficient to calculate the amount of the evolved gas and match the total volume changes measured in the CCE tests.

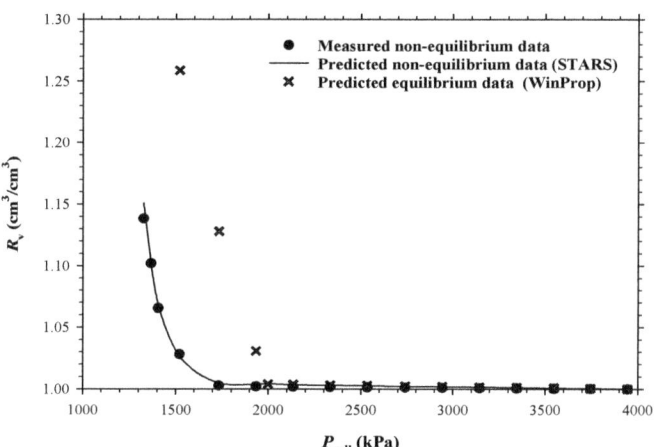

Figure 4. Measured and predicted non-equilibrium relative volume (R_v) vs. PVT cell pressure (P_{cell}) data as well as predicted equilibrium R_v vs. P_{cell} data in Test #1 for the heavy oil–CO_2 system.

Test #2 was conducted to study the gas exsolution in the porous media. Figure 5 depicts the measured and predicted production data (P_e, Q_o, Q_g, Q_f) of Test #2. As shown in the figure, P_e decreased quickly and the slopes of Q_o, Q_g, and Q_f were almost constant at $P_e > P_n$. This was because the reservoir fluids in the sandpacked model were still in the liquid phase. When P_e was slightly lower than P_n, P_e was reduced more slowly, indicating the existence of bubbles in the sandpacked model. This was because the gas phase had a much higher compressibility than that of the liquid phase and thus could much better help to maintain the reservoir pressure. The absolute slope of Q_o was decreased at $P_e < P_n$. This was because the CO_2-diluted heavy oil had a relatively lower viscosity at $P_e > P_n$. The viscosity of the CO_2-diluted heavy oil was drastically increased after the dissolved gas was exsolved from the live oil, causing a lower oil production rate. As P_e continued to decrease, the absolute slope of Q_o did not change much initially and then started to decrease due to the free-gas production.

The total isothermal compressibilities (c_t) measured in Tests #1 and #2 are compared in Figure 6. Two turning points were found and three regions (Regions I, II, and III) could be identified in each dataset. In Region I, the compressibility data were low and remained almost constant. This region represented the single liquid phase. In Region II, bubbles started to be nucleated and the total isothermal compressibilities of the two tests were increased accordingly. The measured bubble-nucleation pressures (P_n) were found to be 1.73 MPa in Test #1 and 1.97 MPa in Test #2 from Figure 6. Therefore, the supersaturation at P_n in Test #1 was higher than that in Test #2. The main reason for the lower supersaturation in Test #2 was the presence of the porous media in the test. The sand grains in Test #2 provided tremendous bubble nucleation sites for bubbles to nucleate easily and thus decreased the supersaturation needed for the bubble nucleation [39]. The numerical simulation results are consistent with the experimental results. It was found that the values of rff_e obtained from Tests #1 and #2 were 0.00052 min^{-1} and 0.00075 min^{-1}, respectively. A higher rff_e indicated that the dissolved gas was more readily exsolved, to form gas bubbles and/or diffuse into the existing gas bubbles in Test #2 with the porous media.

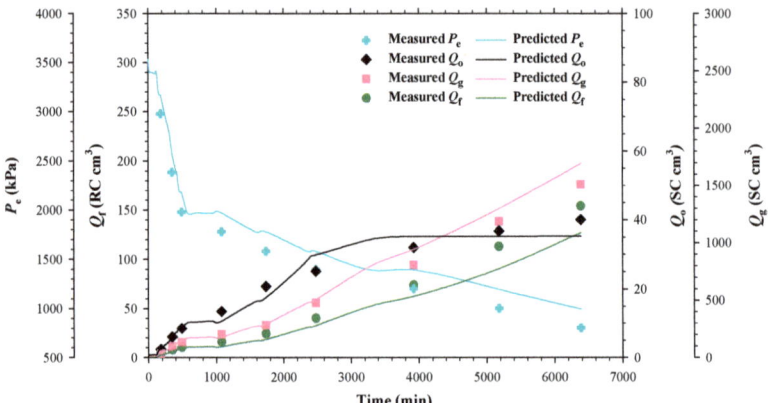

Figure 5. Measured and predicted pressures at the opposite end of the producer (P_e); cumulative total fluids production (Q_f) at the reservoir conditions; cumulative oil and gas productions (Q_o and Q_g) at the atmospheric pressure in Test #2 for the heavy oil–CO_2 system.

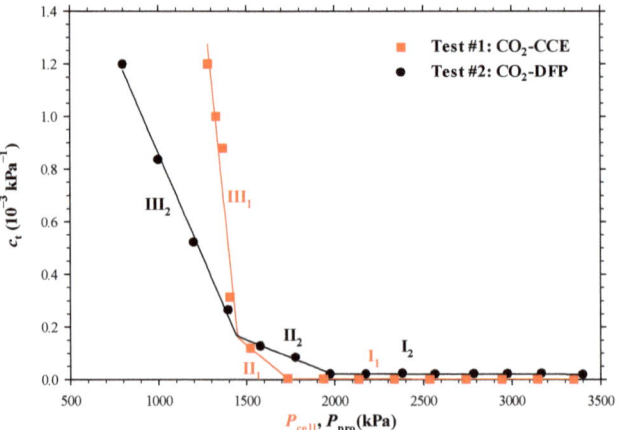

Figure 6. Measured total isothermal compressibilities (c_t) in Tests #1 and #2.

In the third regions of the two tests, the absolute slopes of the total isothermal compressibilities became much larger. This means that the system's compressibility started to be dominated by the evolved gas phase. Bubbles were liberated in the third region, though some free gas could already exist at the late stage of the second region [14]. In the numerical simulations of Test #2, the bubble liberation rate depended on the amount of bubbles in the foamy heavy oil. The reaction frequency factor (rff_l) for the bubble liberation in Test #2 was found to be 0.009 min^{-1}.

4.2. Heavy Oil–CH$_4$ System

The equilibrium R_v vs. P_{cell} data predicted by using the PR-EOS in the CMG-WinProp module, the non-equilibrium R_v vs. P_{cell} data predicted by using the CMG-STARS module and the measured non-equilibrium R_v vs. P_{cell} data in Test #3 for the heavy oil–CH$_4$ system are plotted in Figure 7. The equilibrium and non-equilibrium R_v vs. P_{cell} data were almost the same when the test pressure was higher than the bubble-point pressure P_b = 1.70 MPa of the heavy oil–CH$_4$ system since the test fluid was still in the liquid phase at $P_{cell} > P_b$. The two relative volumes started to deviate from each other at P_b. The predicted equilibrium R_v increased suddenly when the test pressure was reduced from 1.70 MPa to 1.50 MPa. With further pressure reduction, the increase in R_v with each pressure drop became larger and larger. Conversely, the measured or predicted non-equilibrium R_v was increased by less than 0.5% as the test pressure was decreased from 1.70 MPa to 1.50 MPa. As the test pressure continued to decrease, the non-equilibrium R_v was increased slightly with each pressure drop until P_{cell} = 0.70 MPa. This indicates that no gas was exsolved from the live heavy oil in the non-equilibrium CCE test until P_{cell} was 1 MPa lower than P_b. The measured or predicted non-equilibrium R_v was increased drastically when the test pressure was reduced from 0.70 MPa to 0.50 MPa. This was because a large number of bubbles were nucleated at P_{cell} = 0.50 MPa. The tuned rff_e in Test #3 was found to be 0.00009 min^{-1} in the CMG-STARTS module. The low rff_e in Test #3 indicated that the dissolved gas was exsolved from the live oil rather slowly, which was consistent with the high supersaturation measured in the test.

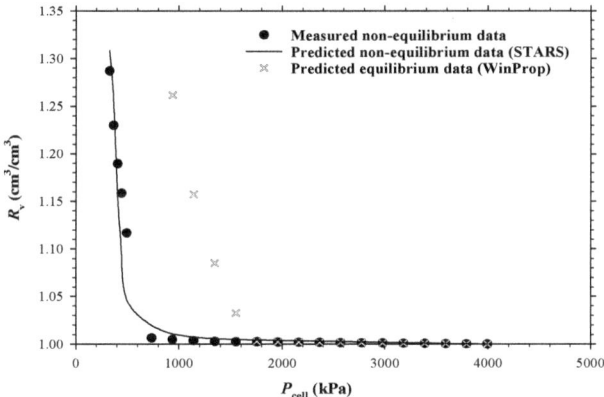

Figure 7. Measured and predicted non-equilibrium relative volume (R_v) vs. PVT cell pressure (P_{cell}) data as well as predicted equilibrium R_v vs. P_{cell} data in Test #3 for the heavy oil–CH$_4$ system.

The measured and predicted production data (P_e, Q_f, Q_o, Q_g) in Test #4 are depicted in Figure 8. It can be found from the history matching that the predicted cumulative total fluids production data had relatively larger deviations from the measured data at lower test pressures. The deviations of the predicted data from the measured data were also found at low test pressures in the CO$_2$-DFP test (Test #2). These deviations could be attributed to the measured cumulative total fluids production data (Q_f) at low pressures. The accuracy of Q_f measurements was compromised at a low pressure due to the large expansion of free gas in the separator at the low pressure. The overall production trends in the CH$_4$-DFP test (Test #4) were similar to those in the CO$_2$-DFP test (Test #2). The major difference was the cumulative oil productions (Q_o) in the two tests. In Test #2, the slope of Q_o vs. time data had an obvious decrease when P_n = 1.97 MPa was reached. In contrast, the slope of Q_o vs. time data in Test #4 was marginally increased once P_e was reduced below P_n. The slope remained almost the same for a long time before the free gas was produced. This could be because the exsolution of CH$_4$ from the live heavy oil did not significantly affect the foamy-oil viscosity [40].

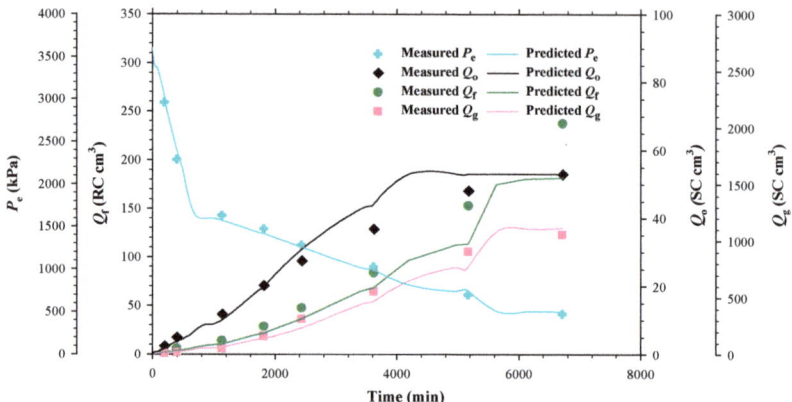

Figure 8. Measured and predicted pressures at the opposite end of the producer (P_e); cumulative total fluids production (Q_f) at the reservoir conditions; cumulative oil and gas productions (Q_o and Q_g) at the atmospheric pressure in Test #4 for the heavy oil–CH_4 system.

The measured total isothermal compressibilities (c_t) in Tests #3 and #4 are shown in Figure 9. Unlike the measured total isothermal compressibilities in Tests #1 and #2 as shown in Figure 6, the measured total isothermal compressibility data in Tests #3 and #4 only had one turning point in each dataset. This meant that CH_4 bubbles tended to be nucleated more instantaneously in comparison with CO_2 bubbles. The turning points in the two datasets in Figure 9 were $P_n = 0.60$ and 1.50 MPa in Tests #3 and #4, respectively. Therefore, the critical supersaturation at $P_n = 0.60$ MPa was $P_{crit} = 1.70 - 0.60 = 1.10$ MPa in Test #3, whereas the critical supersaturation at $P_n = 1.50$ MPa was $P_{crit} = 1.70 - 1.50 = 0.20$ MPa in Test #4. The large difference of the critical supersaturations in the two tests shows that the porous media strongly affect P_n and P_{crit} for the heavy oil–CH_4 system. In the numerical simulations for the two tests, the tuned rff_e in Test #3 (CH_4-CCE) was 0.00009 min^{-1}, which was much smaller than the tuned $rff_e = 0.013$ min^{-1} in Test #4 (CH_4-DFP). Hence, the numerical simulation results also show that the presence of the porous media can cause the dissolved gas to be exsolved more readily. The value of rff_1 in Test #4 was tuned to be 0.00019 min^{-1}. The high rff_e but low rff_1 for the heavy oil–CH_4 system in the porous media indicate that CH_4 is an excellent solvent to induce and maintain foamy-oil flow in the porous media.

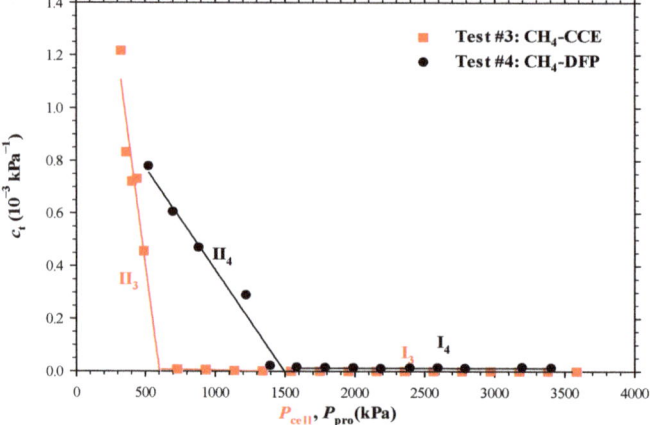

Figure 9. Measured total isothermal compressibilities (c_t) in Tests #3 and #4.

4.3. Heavy Oil–C_3H_8 System

Figure 10 compares the three different R_v vs. P_{cell} data for the heavy oil–C_3H_8 system: the predicted equilibrium data from the GMG-WinProp module; the measured non-equilibrium data from CCE test; and the predicted non-equilibrium data from the CMG-STARS module. This figure shows that, similar to the predicted equilibrium R_v vs. P_{cell} data in the other two heavy oil–solvent systems, the predicted equilibrium R_v data had a sudden increase once P_{cell} was decreased to P_b = 0.48 MPa. Both the measured and predicted non-equilibrium values of R_v were increased gradually after the pressure was decreased slightly below P_b. This fact means that the critical supersaturation required for the C_3H_8 bubbles to nucleate was lower than that for the CO_2 bubbles. As the test pressure was further reduced, the predicted equilibrium R_v increased quickly while the non-equilibrium R_v was increased rather slowly. This large deviation was caused by the low bubble nucleation rate and gas diffusion rate in the heavy oil [41]. The predicted R_v vs. P_{cell} data agreed well with the measured R_v vs. P_{cell} data. It was found in the numerical simulation results that a single rff_e did not give a satisfactory matching result. Therefore, two different rff_e values had to be used in two pressure ranges in the CMG-STARS module. Both values of rff_e were found to be rather low for the heavy oil–C_3H_8 system, 0.0001 min^{-1} in the test pressures of 0.50–0.40 MPa and 0.000065 min^{-1} in the test pressures of 0.40–0.30 MPa.

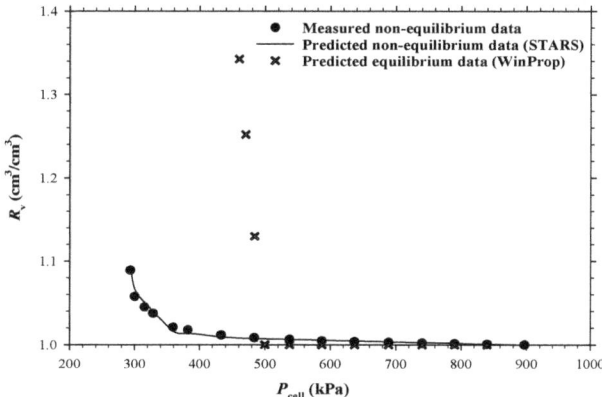

Figure 10. Measured and predicted non-equilibrium relative volume (R_v) vs. PVT cell pressure (P_{cell}) data as well as predicted equilibrium R_v vs. P_{cell} data in Test #5 for the heavy oil–C_3H_8 system.

The measured and predicted production data for Test #6 are plotted in Figure 11. The predicted data had a satisfactory agreement with the measured data with a global error of 8.2%. As shown in the figure, P_e had a similar trend to those in Tests #2 (CO_2-DFP) and #4 (CH_4-DFP). It decreased quickly at $P_e > P_n$ and much more slowly after the bubble started to nucleate. The tuned rff_e and rff_1 for Test #6 were found to be 0.00025 min^{-1} and 0.0045 min^{-1}. Figure 12 shows the total isothermal compressibilities (c_t) measured in Tests #5 and #6. Similar to the measured total isothermal compressibilities (c_t) in Tests #1 (CO_2-CCE) and Test #2 (CO_2-DFP), the measured c_t data in Tests #5 and #6 also had two turning points and three regions in each dataset. This suggests that C_3H_8 and CO_2 bubbles were nucleated more gradually than CH_4 bubbles. The pressures at the first turning points of the two datasets were the bubble-nucleation pressures (P_n). They were 480 kPa in Test #5 and 500 kPa in Test #6, which were close to and the same as P_b = 500 kPa. This means that the supersaturation in the heavy oil–C_3H_8 system is almost negligible, whether a test was conducted in the porous media or not. In comparison with the heavy oil–CO_2 system, the heavy oil–C_3H_8 system had lower rff_e values in the bulk phases and porous media. This indicates that gas was exsolved more slowly in the heavy oil–C_3H_8 system at the same supersaturation. In addition, the tuned rff_1 of 0.0045 min^{-1} in Test #6 with the porous media was also lower than rff_1 = 0.009 min^{-1} in Test #2 with the porous media.

Therefore, the heavy oil–C_3H_8 system had more stable foamy oil than that in the heavy oil–CO_2 system.

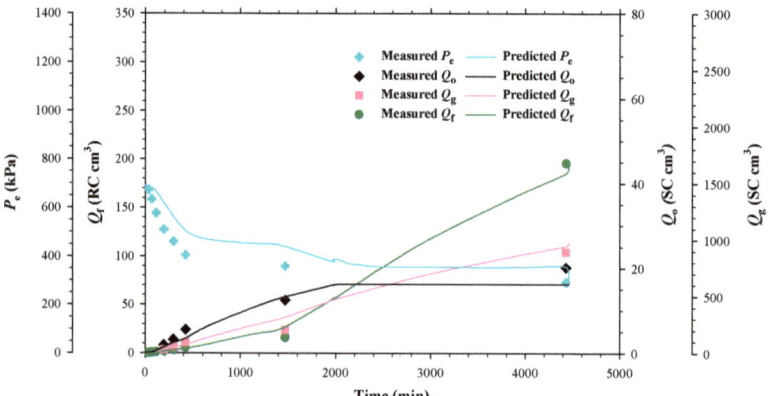

Figure 11. Measured and predicted pressures at the opposite end of the producer (P_e); cumulative total fluids production (Q_f) at the reservoir conditions; cumulative oil and gas productions (Q_o and Q_g) at the atmospheric pressure in Test #6 for the heavy oil–C_3H_8 system.

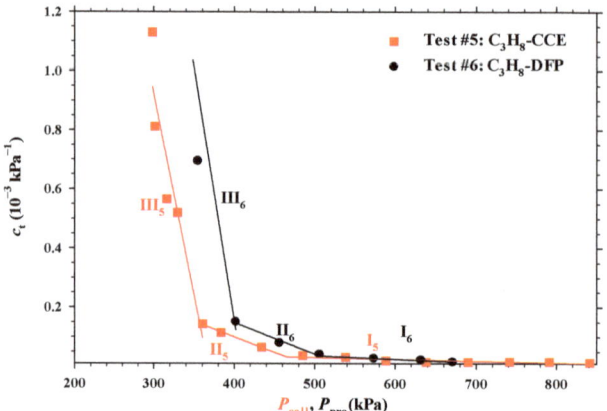

Figure 12. Measured total isothermal compressibilities (c_t) in Tests #5 and #6.

5. Conclusions

In this paper, the solvent exsolution and liberation processes of three different heavy oil–solvent systems in the porous media and bulk phases were studied experimentally and numerically. The following conclusions can be drawn from this study:

- The measured bubble-nucleation pressures (P_n) for the heavy oil–CO_2 system, heavy oil–CH_4 system and heavy oil–C_3H_8 system in the porous media were 0.24 MPa, 0.90 MPa and 0.02 MPa higher, respectively, than those in the bulk phases. This was because a lower supersaturation was needed for the bubble nucleation to occur in the porous media than that in the bulk phases.
- The measured total isothermal compressibility (c_t) vs. test pressure (P) data in the heavy oil–CH_4 system showed that the nucleation of CH_4 bubbles was found to be more instantaneous than that of CO_2 or C_3H_8 bubbles.
- Numerically, the obtained reaction frequency factors (rff_e) for the gas exsolution were all higher in the tests with the porous media than those with the bulk phases for the

three heavy oil–solvent systems. The higher rff_e indicated that the dissolved gas was more readily to be exsolved from the heavy oil.
- The reaction frequency factors (rff_l) for the gas liberation in the heavy oil–C_3H_8 system with the porous media was found to be lower than that for the heavy oil–CO_2 system, suggesting that the former system had more stable foamy oil than that in the latter system.
- The high rff_e but low rff_l in the heavy oil–CH_4 system with the porous media showed that CH_4 was an excellent solvent for inducing and maintaining foamy oil in the porous media.

Author Contributions: Conceptualization, W.Z. and Y.G.; Methodology, W.Z. and Y.G.; Validation, W.Z.; Formal analysis, W.Z.; Investigation, W.Z.; Data curation, W.Z.; Writing—original draft, W.Z.; Writing—review & editing, W.Z. and Y.G.; Supervision, Y.G.; Project administration, Y.G.; Funding acquisition, Y.G. All authors have read and agreed to the published version of the manuscript.

Funding: This research was funded by the Petroleum Technology Research Centre (PTRC) [Fund No.: HO-UR-02-2022]; the Nature Sciences and Engineering Research Council (NSERC) of Canada [Discovery Grant No.: RGPIN-2019-05564]; and the Mitacs [Grant No.: IT12361-2020].

Data Availability Statement: The original contributions presented in the study are included in the article, further inquiries can be directed to the corresponding author.

Acknowledgments: The authors acknowledge an innovation fund from the Petroleum Technology Research Centre (PTRC), a discovery grant from the Nature Sciences and Engineering Research Council (NSERC) of Canada and a research grant from Mitacs to Yongan Gu. In addition, the authors also thank the research group members for their technical support and discussions.

Conflicts of Interest: The authors declare no conflict of interest.

References

1. Maini, B.B. Foamy oil flow in heavy oil production. *J. Can. Pet. Technol.* **1996**, *35*, 21–24. [CrossRef]
2. Sheng, J.; Maini, B.; Hayes, R.; Tortike, W. Experimental study of foamy oil stability. *J. Can. Pet. Technol.* **1997**, *36*, 31–37. [CrossRef]
3. Wang, J.; Yuan, Y.; Zhang, L.; Wang, R. The influence of viscosity on stability of foamy oil in the process of heavy oil solution gas drive. *J. Pet. Sci. Eng.* **2009**, *66*, 69–74. [CrossRef]
4. Sheng, J.J.; Hayes, R.E.; Maini, B.B.; Tortike, W.S. Modelling foamy oil flow in porous media. *Transp. Porous Media* **1999**, *35*, 227–258. [CrossRef]
5. Wu, M.; Lu, X.; Yang, J.; Lin, Z.; Zeng, F. Experimental analysis of optimal viscosity for optimizing foamy oil behavior in the porous media. *Fuel* **2020**, *262*, 116602. [CrossRef]
6. Li, S.; Hu, Z.; Lu, C.; Wu, M.; Zhang, K.; Zheng, W. Microscopic visualization of greenhouse-gases induced foamy emulsions in re-covering unconventional petroleum fluids with viscosity additives. *J. Chem. Eng.* **2021**, *411*, 128411. [CrossRef]
7. Shi, W.; Ma, Y.; Tao, L.; Zhang, N.; Ma, C.; Bai, J.; Xu, Z.; Zhu, Q.; Zhong, Y. Study of the enhanced oil recovery mechanism and remaining oil state of heavy oil after viscosity-reducer-assisted CO_2 Foam flooding: 2D microvisualization experimental case. *Energy Fuels* **2023**, *32*, 18620–18631. [CrossRef]
8. Sun, X.; Cai, L.; Zhang, Y.; Li, T.; Song, Z.; Teng, Z. Comprehensive investigation of non-equilibrium properties of foamy oil induced by different types of gases. *Fuel* **2022**, *316*, 123296. [CrossRef]
9. Zhou, X.; Zeng, F.; Zhang, L.; Wang, H. Foamy oil flow in heavy oil–solvent systems tested by pressure depletion in a sandpack. *Fuel* **2016**, *171*, 210–223. [CrossRef]
10. Li, S.; Li, Z.; Wang, Z. Experimental study on the performance of foamy oil flow under different solution gas–oil ratios. *RSC Adv.* **2015**, *5*, 66797–66806. [CrossRef]
11. Zhou, X.; Yuan, Q.; Zeng, F.; Zhang, L.; Jiang, S. Experimental study on foamy oil behavior using a heavy oil–methane system in the bulk phase. *J. Pet. Sci. Eng.* **2017**, *158*, 309–321. [CrossRef]
12. Tang, G.-Q.; Firoozabadi, A. Gas- and liquid-phase relative permeabilities for cold production from heavy-oil reservoirs. *SPE Reserv. Evaluation Eng.* **2003**, *6*, 70–80. [CrossRef]
13. Modaresghazani, J.; Moore, R.; Mehta, S.; Anderson, M.; Badamchi-Zadeh, A. A novel method (CCE&C) to study transient phase behaviour in heavy oil and ethane. *Fuel* **2019**, *257*, 115946. [CrossRef]
14. Yao, J.; Zou, W.; Gu, Y. Solvent effects on the measured bubble-point pressures and pseudo bubble-point pressures of different heavy crude oil–solvent systems. *Petroleum* **2022**, *8*, 577–586. [CrossRef]
15. Dong, X.; Xi, Z.; Zadeh, A.B.; Jia, N.; Gates, I.D. A Novel Experimental Method CCEC and Modelling of Methane Dissolution and Exsolution in Heavy Oil. In Proceedings of the SPE 199942, Presented at the Canada Heavy Oil Conference, Calgary, AB, Canada, 29 September–2 October 2020.

16. Shi, Y.; Zhao, W.; Li, S.; Yang, D. Quantification of gas exsolution and preferential diffusion for alkane solvent(s)–CO2–heavy oil systems under nonequilibrium conditions. *J. Pet. Sci. Eng.* **2021**, *208*, 109283. [CrossRef]
17. Bauget, F.; Egermann, P.; Lenormand, R. A New model to obtain representative field relative permeability for reservoirs produced under solution-gas drive. *SPE Reserv. Evaluation Eng.* **2005**, *8*, 348–356. [CrossRef]
18. Uddin, M. Numerical studies of gas exsolution in a live heavy oil reservoir. In Proceedings of the SPE 97739, Presented at the International Thermal Operations and Heavy Oil Symposium, Calgary, AB, Canada, 1–3 November 2005.
19. Ivory, J.; Chang, J.; Coates, R.; Forshner, K. Investigation of cyclic solvent injection process for heavy oil recovery. *J. Can. Pet. Technol.* **2010**, *49*, 22–33. [CrossRef]
20. Zhou, X.; Zeng, F.; Zhang, L.; Jiang, Q.; Yuan, Q.; Wang, J.; Zhu, G.; Huang, X. Experimental and mathematical modeling studies on foamy oil stability using a heavy oil–CO_2 system under reservoir conditions. *Fuel* **2020**, *264*, 116771. [CrossRef]
21. Shahvali, M.; Pooladi-Darvish, M. Dynamic modelling of solution-gas drive in heavy oils. *J. Can. Pet. Technol.* **2009**, *48*, 39–46. [CrossRef]
22. Sun, X.; Zhang, Y.; Wang, S.; Song, Z.; Li, P.; Wang, C. Experimental study and new three-dimensional kinetic modeling of foamy solution-gas drive processes. *Sci. Rep.* **2018**, *8*, 4369. [CrossRef]
23. Arora, P.; Kovscek, A.R. A mechanistic modeling and experimental study of solution gas drive. *Transp. Porous Media* **2003**, *51*, 237–265. [CrossRef]
24. Ma, H.; Yu, G.; She, Y.; Gu, Y. A new hybrid production optimization algorithm for the combined CO_2-cyclic solvent injection (CO_2-CSI) and water/gas flooding in the post-CHOPS reservoirs. *J. Pet. Sci. Eng.* **2018**, *170*, 267–279. [CrossRef]
25. Li, Q.; Wang, Y.; Owusu, A.B. A modified Ester-branched thickener for rheology and wettability during CO_2 fracturing for im-proved fracturing property. *Environ. Sci. Pollut. Res.* **2019**, *26*, 20787–20797. [CrossRef] [PubMed]
26. Li, Q.; Liu, J.; Wang, S.; Guo, Y.; Han, X.; Li, Q.; Cheng, Y.; Dong, Z.; Li, X.; Zhang, X. Numerical insights into factors affecting collapse behavior of horizontal wellbore in clayey silt hydrate-bearing sediments and the accompanying control strategy. *Ocean Eng.* **2024**, *297*, 117029. [CrossRef]
27. Coombe, D.; Maini, B. Modeling Foamy Oil Flow. In *Workshop on Foamy Oil Flow*; Petroleum Recovery Institute: Calgary, AB, Canada, 1994.
28. Bayon, Y.M.; Cordelier, P.R.; Nectoux, A. A new methodology to match heavy-oil long-core primary depletion experiments. In Proceedings of the SPE 75133, Presented at the Improved Oil Recovery Conference, Tulsa, OK, USA, 13–17 April 2002.
29. Ma, H.; Huang, D.; Yu, G.; She, Y.; Gu, Y. Combined cyclic solvent injection and waterflooding in the post-cold heavy oil production with sand reservoirs. *Energy Fuels* **2017**, *31*, 418–428. [CrossRef]
30. Yao, J.; Zou, W.; Gu, Y. Experimental and theoretical studies of solvent bubble nucleation and liberation processes in different heavy crude oil–solvent systems. *J. Pet. Sci. Eng.* **2022**, *217*, 110949. [CrossRef]
31. Oskouei, S.J.P.; Zadeh, A.B.; Gates, I.D. A new kinetic model for non-equilibrium dissolved gas ex-solution from static heavy oil. *Fuel* **2017**, *204*, 12–22. [CrossRef]
32. Brooks, R.H.; Corey, A.T. Hydraulic properties of porous media and their relation to drainage design. *T ASAE* **1964**, *7*, 26–28.
33. Computer Modelling Group (CMG) Ltd. *STARS Users' Guide*; Computer Modelling Group (CMG) Ltd.: Calgary, AB, Canada, 2016.
34. Chen, T.; Leung, J.Y.; Bryan, J.L.; Kantzas, A. Analysis of non-equilibrium foamy oil flow in cyclic solvent injection processes. *J. Pet. Sci. Eng.* **2020**, *195*, 107857. [CrossRef]
35. Wang, H.; Torabi, F.; Zeng, F. Investigation of non-equilibrium solvent exsolution dynamics and bubble formation and growth of a CO_2/C_3H_8/heavy-oil system by micro-optical visualizations: Experimental and continuum-scale numerical studies. *Fuel* **2023**, *332*, 126188. [CrossRef]
36. Albartamani, N.S.; Ali, S.F.; Lepski, B. Investigation of foamy oil phenomena in heavy oil reservoirs. In Proceedings of the SPE 54084, Presented at the International Thermal Operations and Heavy Oil Symposium, Bakersfield, CA, USA, 17–19 March 1999.
37. Sahni, A.; Gadelle, F.; Kumar, M.; Tomutsa, L.; Kovscek, A.R. Experiments and Analysis of Heavy-Oil Solution-Gas Drive. *SPE Reserv. Eval. Eng.* **2004**, *7*, 217–229. [CrossRef]
38. Wang, H.; Zeng, F.; Torabi, F.; Xiao, H. Experimental and numerical studies of non-equilibrium solvent exsolution behavior and foamy oil stability under quiescent and convective conditions in a visualized porous media. *Fuel* **2021**, *291*, 120146. [CrossRef]
39. Bryan, J.; Butron, J.; Nickel, E.; Kantzas, A. Measurement of Non-Equilibrium Solvent Release from Heavy Oil during Pressure Depletion. In Proceedings of the SPE Canada Heavy Oil Conference, Calgary, AB, Canada, 13–14 March 2018.
40. Alshmakhy, A.; Maini, B.B. Foamy-oil-viscosity measurement. *J. Can. Pet. Technol.* **2012**, *51*, 60–65. [CrossRef]
41. Yang, C.; Gu, Y. Diffusion coefficients and oil swelling factors of carbon dioxide, methane, ethane, propane, and their mixtures in heavy oil. *Fluid Phase Equilibria* **2006**, *243*, 64–73. [CrossRef]

Disclaimer/Publisher's Note: The statements, opinions and data contained in all publications are solely those of the individual author(s) and contributor(s) and not of MDPI and/or the editor(s). MDPI and/or the editor(s) disclaim responsibility for any injury to people or property resulting from any ideas, methods, instructions or products referred to in the content.

Article

Hydrocarbon Accumulation Process and Mode in Proterozoic Reservoir of Western Depression in Liaohe Basin, Northeast China: A Case Study of the Shuguang Oil Reservoir

Guangjie Zhao [1,2], Fujie Jiang [1,2,*], Qiang Zhang [3,*], Hong Pang [1], Shipeng Zhang [3], Xingzhou Liu [4] and Di Chen [1,2]

1. College of Geoscience, China University of Petroleum (Beijing), Beijing 102249, China; zhaoguangjie23@126.com (G.Z.)
2. State Key Laboratory of Petroleum Resource and Prospecting, China University of Petroleum (Beijing), Beijing 102249, China
3. No.7 Exploration Institute of Geology and Mineral Resources in Shandong Province, Linyi 276000, China
4. PetroChina Liaohe Oilfield Exploration and Development Research Institute, Panjin 124000, China
* Correspondence: jfjhtb@163.com (F.J.); sddkqydds001@163.com (Q.Z.)

Abstract: The Shuguang area has great oil and gas potential in the Proterozoic and it is a major exploration target in the Western Depression. However, controlling factors and a reservoir-forming model of the Shuguang reservoir need further development. The characteristics of the reservoir formation in this area were discussed by means of a geochemical technique, and the controlling factors of the oil reservoir were summarized. The oil generation intensity of Es4 source rock was 25×10^6–500×10^6 t/km^2, indicating that the source rocks could provide enough oil for the reservoir. The physical property of the quartz sandstone reservoir was improved by fractures and faults, which provided a good condition for the oil reservoir. Two periods of oil charging existed in the reservoir, with peaks of 38 Ma and 28 Ma, respectively. A continuous discharge of oil is favorable for oil accumulation. Oil could migrate through faults and fractures. In addition, the conditions of source–reservoir–cap assemblage in the Shuguang area well preserved the oil reservoir. The lower part of the Shuguang reservoir was source rock, the upper part was reservoir, and it was a structure-lithologic oil reservoir. These results are crucial for further oil exploration.

Keywords: accumulation process; controlling factors; accumulation mode; Shuguang oil reservoir; Liaohe Basin

Citation: Zhao, G.; Jiang, F.; Zhang, Q.; Pang, H.; Zhang, S.; Liu, X.; Chen, D. Hydrocarbon Accumulation Process and Mode in Proterozoic Reservoir of Western Depression in Liaohe Basin, Northeast China: A Case Study of the Shuguang Oil Reservoir. *Energies* **2024**, *17*, 2583. https://doi.org/10.3390/en17112583

Academic Editor: Reza Rezaee

Received: 29 February 2024
Revised: 27 April 2024
Accepted: 28 April 2024
Published: 27 May 2024

Copyright: © 2024 by the authors. Licensee MDPI, Basel, Switzerland. This article is an open access article distributed under the terms and conditions of the Creative Commons Attribution (CC BY) license (https://creativecommons.org/licenses/by/4.0/).

1. Introduction

The revolutionary success of tight oil in the USA has reversed the downward trend of oil production in North America at one stroke [1,2] and changed the world's traditional energy pattern to some extent. As a result, through active exploration, tight oil areas with large-scale reserves have been explored in several basins in China, such as the eastern Bohai Bay Basin [3,4], Ordos Basin [5,6], Sichuan Basin [7], and western Junggar Basin [8]. Therefore, tight oil has become a critical field.

Tight oil refers to oil in tight reservoirs with a low overburden matrix permeability (less than 0.1 mD) and low porosity (less than 10%) [9]. Previous researchers focused on porosity [10] and permeability [11], diagenesis [12,13], fractures [14], migration [15], accumulation mechanism [16], and charging period [17]. However, research about the accumulation process and mode of tight reservoirs is not sufficient, and the comprehensive understanding of tight reservoirs needs further exploration. And, it has guiding significance for oil exploration.

The oil and gas in the Liaohe Basin have attracted extensive attention [18–20]. The Shuguang oilfield is in the north of the Western Depression of the Liaohe Basin. Previous studies mainly focused on reservoirs [21], favorable zone prediction [22], dynamic characteristics of accumulation [23], conditions for accumulation [24], evaluation of crude oil [25], and origin of crude oil [26]. However, the controlling factors of the Shuguang oilfield have not been comprehensively researched, which seriously hindered the pace of oil exploration in the Western Depression in the past two decades. Therefore, it is significant to reveal the main controlling factors of the Shuguang oilfield and point out its accumulation mode. The objectives of the study were to point out the controlling factors and accumulation mode of the Shuguang reservoir. The result has reference significance for the exploration of other similar oil reservoirs.

2. Geological Setting

The Liaohe Basin is a rift basin of Liaoning Province in the northeast of China (Figure 1a). Panjin city is in the southwest of the basin. The basin is composed of the Eastern Depression, Damintun Depression, and Western Depression; the Western Depression was mainly studied for this study. To the northwest of the basin are the Yanshanian Mountains and to the southeast are the Jiaoliao Mountains (Figure 1b). The east–west span of the basin is about 65 km, and the north–south span is about 470 km. The basin was formed in three rifting cycles starting from the early Jurassic [27]. The second rift cycle of the Cretaceous increased the kinds of volcanic rocks and sediments. The present basin resulted from the third rifting cycle caused by subduction [28].

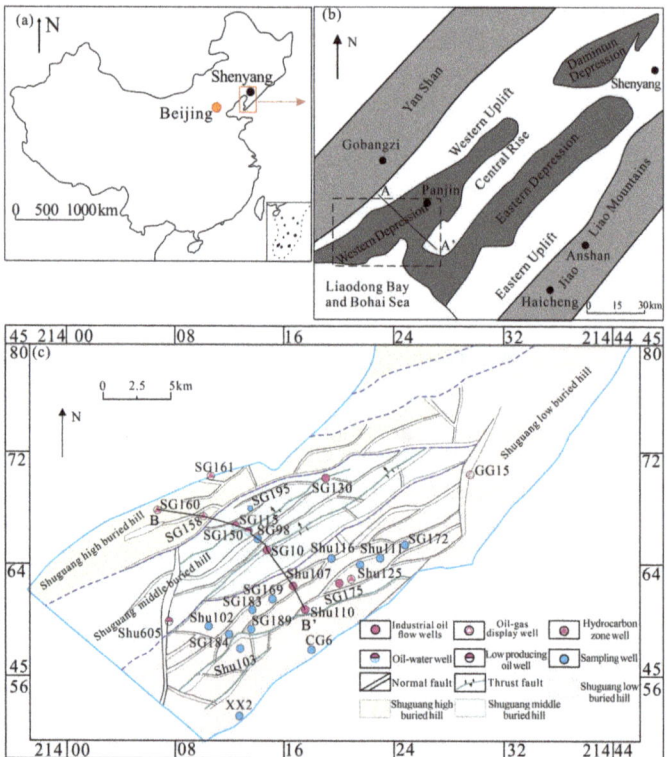

Figure 1. Location and sampling sites of the Shuguang oilfield. (**a**) Location map of the Liaohe Basin. (**b**) Selected profiles and division of main structural units. (**c**) Faults and wells in the Western Depression [19]. The sample wells are marked. A-A' line shows the profile position of Figure 3. B'-B line shows the profile position of Figure 14.

The Shuguang ancient buried hill belt is located in the western slope of the Western Depression, which is controlled by NE-trending faults. Due to the influence of NE-trending and near-EW-trending faults, the whole buried hill belt is divided into multiple buried hill fault blocks, and the long axis direction of each buried hill is mainly eastward. The strata in this area are in turn the Neozoic Guantao Formation, Dongying Formation, Shahejie Formation, Fangshenpao Formation, Mesozoic, Paleozoic, Proterozoic (Pt), and Archaean (Ar) from top to bottom (Figure 2). The main reservoirs in this area are the Pt and Archean reservoirs in Qianshan (Figure 1c). In this paper, the main target layers are the buried hill strata of Paleozoic, Pt, and Archaean.

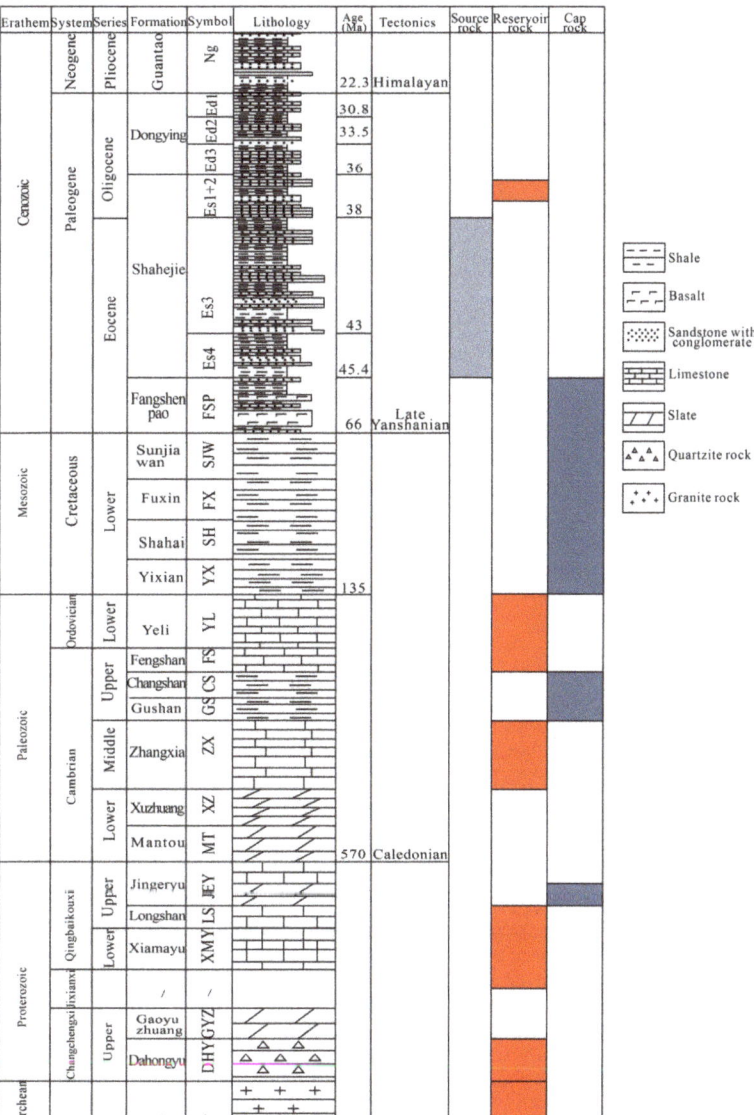

Figure 2. Generalized geological histogram indicating the structural events and source–reservoir–cap combination of the Shuguang area (modified from reference [19]).

The Shuguang oilfield is in the middle and northern section of the Western Depression, with several oil-generating depressions around it, including the Chenjia and Panshan sags in the east and the Qingshui Sag in the south, which are very rich in hydrocarbon resources. Normal faults and reverse faults are relatively developed, providing a good channel for regional oil migration (Figure 3). The Pt is the main reservoir, while some reserves also exist in the Paleozoic (Pz) and Archean. The mudstone of the Mesozoic and Fangshenpao Formation, the mudstone of the Paleozoic Cambrian, and the Slate of the Proterozoic Qingbaikou are good cap beds.

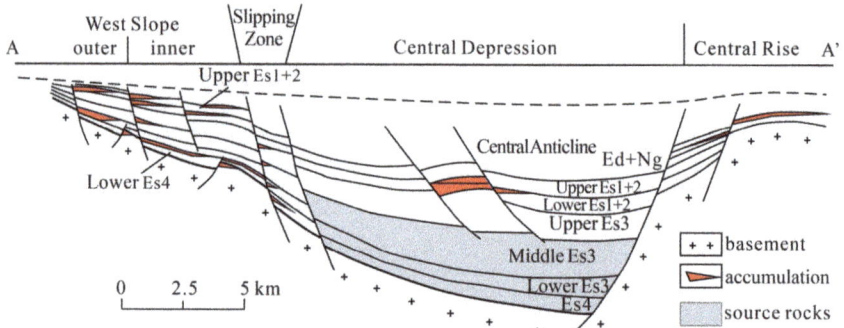

Figure 3. Profile map (geological section A-A' is from Figure 1b) indicating a common mode of accumulations in different tectonic zones in the Liaohe Basin (modified from reference [19]).

3. Samples and Methods

3.1. Samples

To characterize the Shuguang oil reservoir, samples were collected (Table 1). The geochemical characteristics of source rock, reservoir quality, and charging period of the samples were analyzed. The data of crude oil were provided by the Research Institute of Exploration and Development of Liaohe Oilfield.

Table 1. The data of the samples from Shuguang area.

Well	Layer	Depth (m)	Lithology	Rock Types
SG183	Es4	2911		
Shu103	Es4	3102		
SG184	Es4	3030		
Shu111	Es4	3276	Mudstone	Source rocks
SG169	Es4	2910		
Shu102	Es4	2301		
CG6	Es3	3181		
XX2	Es4	3710		
Shu103	Pt	3404.7	Dolomite limestone	
SG184	Pz	3180	Dolomite limestone	
SG189	Pt	4169	Quartzite	
SG183	Pz	3157	Dolomite limestone	
SG98	Pt	1750	Quartzite	Reservoir rocks
SG169	Pt	3398	Quartzite	
SG195	Pz	2255	Quartz sandstone	
Shu116	Pt	4005.6	Dolomite limestone	
Shu125	Pt	3788	Limestone	
SG172	Pz	4138	Calcite dolomite	

3.2. Methods

3.2.1. Geochemical Analyses

TOC (total organic carbon) and rock pyrolysis were measured by a Rock-Eval 6 pyrolysis analyzer. Pyrolysis includes two parameters: $S_1 + S_2$ (potential hydrocarbon generation from rock pyrolysis, mg/g) and Tmax (temperature of maximum rate of hydrocarbon generation from source rock, °C). Ro is the reflected light test by a Leica DM4500P polarizing microscope under the condition of sample fragmentation and oil immersion. The chloroform asphalt "A" was determined by a traditional fast Soxhlet extraction method.

3.2.2. Reservoir Characteristics

The blue dyed resin or liquid glue was poured into the pore space of the rock under a vacuum, so that the resin or liquid glue was consolidated, and then the rock was ground into thin slices. SEM images were produced at the China University of Petroleum (Beijing). The diameter of samples was 0.5~3.0 cm and the thickness were 0.2~1.0 cm. The observation instrument was a Zeiss SUPRA 55 Sapphire SEM. A sample of quartz sandstone was cut into sheets and then observed through an instrument. The helium expansion method was used to measure the porosity and permeability using an ultrapore–200A porometer. The diameter of the experimental reservoir core sample for porosity and permeability was 2.5 cm. The features of fractures were depicted by core description.

3.2.3. Fluid Inclusion

Petrographic observation was made by a NIKON-LV100 double-channel fluorescence, reflected light, and transmission light microscope. Micro thermometry was performed on a 300 μm thick, double-sided polished sheet on a LinkamTHMSG600 cooling and hot station. In the process of temperature measurement, the temperature rise rate of the hot and cold stations was controlled to within 0.1~5.0 °C/min and the correction test error was ±0.1 °C.

Moreover, BasinMod® 2012 1D software was utilized to construct burial history and thermal history by collecting data such as the strata thickness of single well, lithology, erosion thickness, Ro, geothermal gradient, and absolute age.

4. Results

4.1. Geochemical Features of Oil

The fluid physical properties reflect the preservation conditions and migration direction of crude oil to some extent. In this article, the oil test data of eight industrial oil flow wells in the Shuguang area in the Proterozoic and Archean oil groups were sorted out and analyzed, and the density of crude oil at 20 °C and the viscosity of crude oil at 50 °C were statistically analyzed so as to have a better understanding of the properties of Shuguang crude oil. At 20 °C, the density of crude oil basically ranged from 0.83 g/cm^3 to 0.97 g/cm^3 and light oil was generally less than 0.90 g/cm^3, while heavy oil was more than 0.90 g/cm^3. As can be seen from the table, the Shuguang area was dominated by light oil and the oil properties of the wells at the structural high point were obviously different from those of other wells, mainly due to the distribution of heavy crude oil in the top part of the early structure, which may be caused by the destruction of the crude oil formed in the early stage. Viscosity values ranged from 4.55 to 1256 mPa·s (Table 2).

Table 2. Data of crude oil in Shuguang oilfield.

Area	Well	Depth (m)	Density/g·cm^{-3} (°@20 °C)	Viscosity/mPa·s (°@50 °C)
Shuguang oilfield	SG183	3185.0	0.87	57.6
	Shu103	3392.5	0.84	8.73
	Shu111	3755.42	0.83	7.33
	SG98	1633.3	0.93	438
	Shu116	4023	0.88	31.2
	Shu125	3358.71	0.86	12.6
	Shu110	3734.48	0.84	10.88
	Shu107	3491.89	0.88	33.99
	Shu112	2824.03	0.87	28.6
	Shu123	3834.49	0.94	4.55
	SG175	3993.29	0.83	9.12
	SG100	1604.5	0.93	496.6
	SG103	1933.50	0.85	19.91

Table 2. Cont.

Area	Well	Depth (m)	Density/g·cm^{-3} (°@20 °C)	Viscosity/mPa·s (°@50 °C)
Shuguang oilfield	SG104	1944.8	0.87	87.07
	SG105	1917.50	0.86	23.6
	SG106	1885.5	0.86	18.2
	SG11	1791.83	0.85	15.2
	SG112	1555.5	0.93	441
	SG26	1652.77	0.93	1256
	SG32	1588.8	0.93	449.26
	SG158	1134.66	0.97	7953

Normal-density oil and heavy oil were mainly found in the Shuguang oilfield. Most of the normal-density oil was in the southeast of the Shuguang oilfield. The thermal cracking and migration fractionation were the critical factors controlling the physical properties of crude oil. Heavy oil was concentrated in the northwest of the Shuguang oilfield.

4.2. Condition of Source Rocks

Based on the research result, the source rocks in the Western Depression were thick and mainly in the Es4 Formation (Figure 4) [29]. Therefore, this paper focuses on evaluating the oil generation intensity of the Es4 Formation.

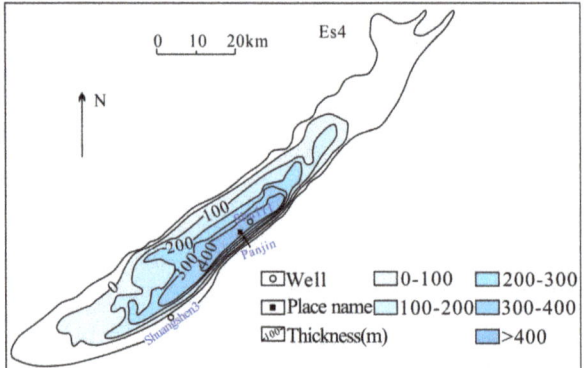

Figure 4. Thickness map of source rock of the Es4 Formation in Liaohe Basin.

The main indexes to evaluate the abundance of source rocks were TOC, PG, and chloroform asphalt "A" [30]. TOC content of mudstone ranged from 0.43% to 21.5% (average 4.42%). The PG of mudstone ($S_1 + S_2$, i.e., gas production) ranged from 0.49 to 62.31 mg HC/rock (average 19.4). According to the evaluation reference [31], the potential generation of mudstone was classified into non, poor, fair, good, and excellent grades. The mean value of chloroform asphalt "A" was 0.76%. The main types of the Shuguang reservoir were of a fair–good grade (Table 3, Figure 5a).

Table 3. The pyrolysis data of source rocks for the Shuguang area.

Area	Rock Stratum	TOC (%)	PG (mg/g)	"A" (%)	Tmax (°C)
Shuguang area	Es4	0.43–21.5/ 4.42 (25)	0.49–62.31/ 19.4 (16)	0.25–1.34/ 0.76 (19)	420–458

Note: The number before '/' is the data range; the number after '/' is the mean value; the number in '()' is the quantity of samples. TOC: total organic carbon; PG: potential generation; "A": chloroform bitumen "A"; Tmax: pyrolysis peak temperature.

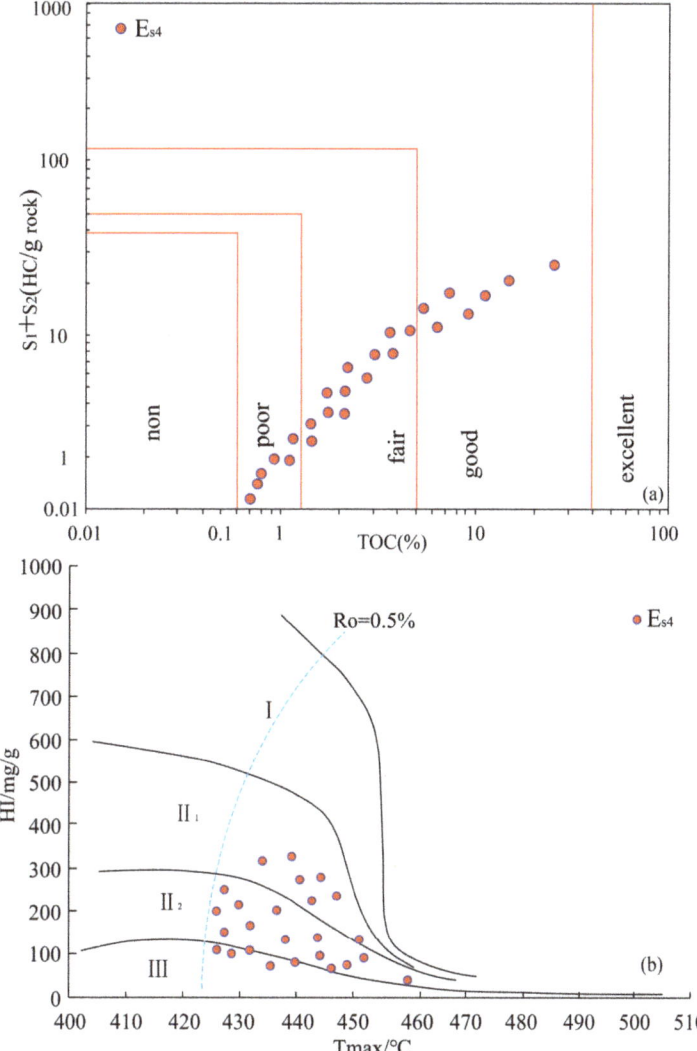

Figure 5. Identification chart of generation potential (**a**) and organic matter types (**b**) of source rocks.

The oil index (HI) of mudstone ranged from 3.20 to 290.00 mg HC/g TOC. Most of the Tmax values of the source rocks ranged from 420 °C to 448 °C (Figure 5b). As can be seen from the relationship between the two parameters in Figure 5b, the main types of kerogens were II_1–II_2 [31]. The Ro values of the samples ranged from 0.62% to 1.73%, and the main range was between 0.60% and 0.90%, indicating that source rocks had high hydrocarbon generation potential (Table 4).

When organic matter entered the oil window, it produced a lot of oil. In conclusion, when the Ro value was 0.60%~0.90% (within the oil generation window), the source rocks could produce more oil. In a word, the source rocks in the Shuguang area had the conditions for forming industrial gas reservoirs [19].

Table 4. The vitrinite reflectance test data of the Shuguang oilfield.

Well	Layer	Depth (m)	Lithology	Ro (%)	Reference
SG183	Es4	2911	Mudstone	0.82	This study
Shu103	Es4	3102	Mudstone	1.73	PLOEDRI
SG184	Es4	3030	Mudstone	0.84	This study
Shu111	Es4	3276	Mudstone	1.04	This study
SG169	Es4	2910	Mudstone	0.78	This study
Shu102	Es4	2301	Mudstone	0.85	This study
XX2	Es4	3710	Mudstone	1.58	This study

Note: PLOEDRI = PetroChina Liaohe Oilfield Exploration and Development Research Institute; Ro: vitrinite reflectance of source rock.

4.3. Characteristics of Reservoir

The thin section test (Figure 6) showed that the Proterozoic lithology was mainly quartz sandstone. Rock fragment is a sedimentary rock formed by the weathering, migration, and deposition of rock debris. The particles of debris were from particulate to coarse medium; the sorting was generally good and the roundness was mainly subroundness. The main sandstones were quartz arenite (Figure 6) [32], which differed from the Middle Jurassic Ravenscar sandstones of North America.

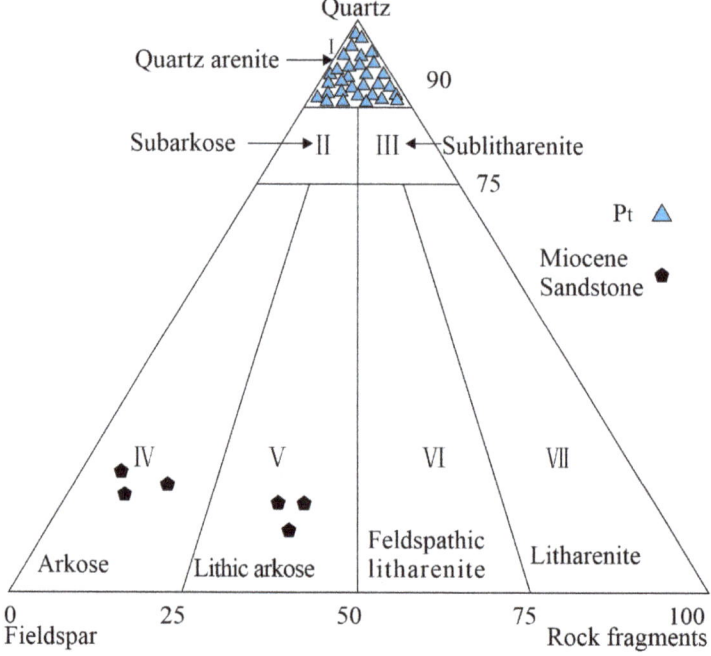

Figure 6. The Pt quartz sandstone compositional data in the Shuguang oilfield (the Miocene data are cited from Okunuwadje [33]).

Figure 7 shows that intergranular pores, intergranular dissolution pores, and fractures formed the major space of the Proterozoic reservoir. However, primary pores were rare. Unstable rock dissolved, leading to the formation of intragranular pores (Figure 7a). The different shapes of the intergranular pores were caused by the dissolution of calcite (Figure 7b,c), and some of the intergranular pores were due to the formation of quartz particles (Figure 7d). Part of the fractures existed in the core, showing the development of sandstone fractures (Figure 7e,f). To sum up, a comprehensive analysis was suitable to characterize the different pore structures of the reservoir.

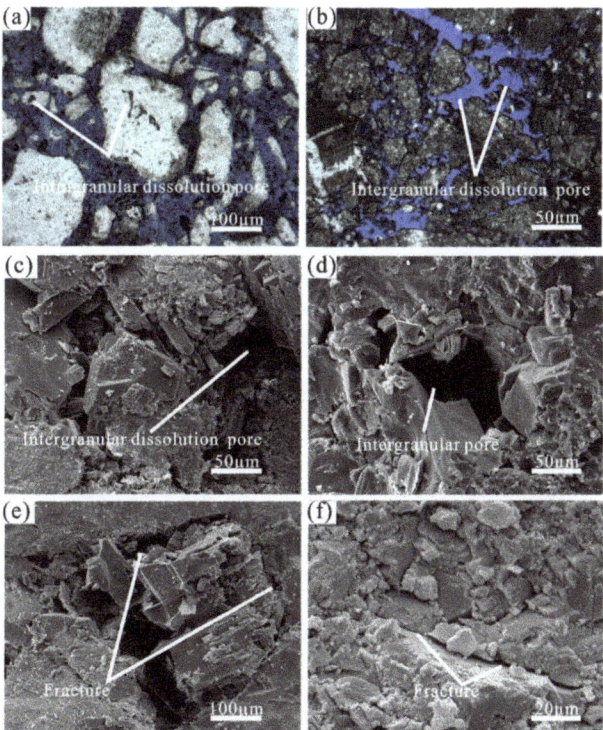

Figure 7. Main reservoir space types of sandstones in the Shuguang oilfield. Note: Intragranular dissolution pore (cast thin section), well Shugu175 (Pt), 3986.63 m (**a**); intergranular dissolved pore, (cast thin section), well Shugu175 (Pt), 3989.13 m (**b**); intergranular dissolved pore (SEM), well Shugu175 (Pt), 3987.83 m (**c**); intergranular pores (SEM), well Shugu169, (Pt) 3396.10 m (**d**); fracture (SEM), well Shugu172 (Pt), 3983.49 m (**e**); fracture (SEM), well Shugu173 (Pt), 4137.08 m (**f**).

Figure 8 indicates there was no obvious change trend between the porosity and permeability of Pt sandstone. More than 55% of Pt sandstones had porosity less than 10%. The porosity distribution ranged from 0.65% to 19.6%, and the 55% of the permeability was from 0.04 to 13.2 mD. To sum up, the Pt reservoir porosity was generally low. However, fractures and dissolution could make local physical properties better [20].

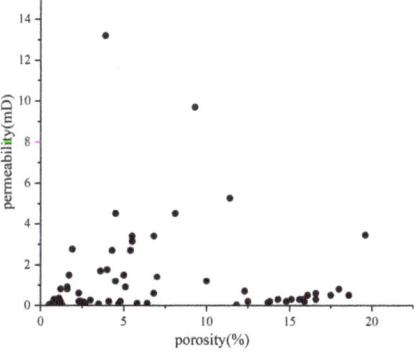

Figure 8. The relationship between porosity and permeability of the Pt sandstone reservoir from the Shuguang oilfield (N = 102).

4.4. Oil Charging Periods

The homogenization temperature (HT) of inclusions and the burial history could determine the hydrocarbon charging period of the reservoir [34]. Fluid inclusions developed well in the Pt reservoir, which were divided into oil and brine inclusions.

The fluid inclusions in this area could be divided into two phases according to the formation period and inclusions' characteristics of the main minerals.

(1) The first phase of inclusions: oil inclusions (Figure 9a,b) and the host minerals were mainly quartz, followed by calcite and feldspar. The size of the inclusions ranged from 1 to 70 μm, the HT of the brine inclusions were symbiotic, and the organic inclusions ranged from 100 to 110 °C. Under a polarizing microscope, it was gray-black, black, and brown-black, and fluorescence showed a dark-brown fluorescence. It may be formed mainly in the low-maturity stage and mainly in heavy oil.

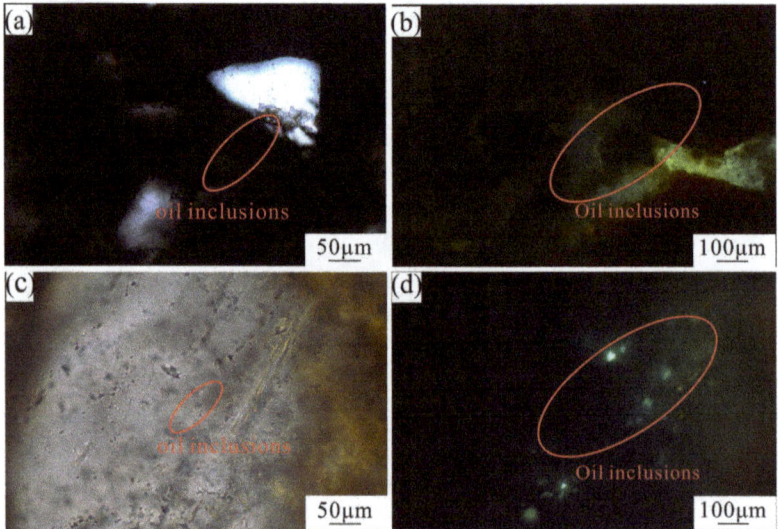

Figure 9. Photomicrographs of inclusions from for the Proterozoic strata. Note: well Shugu169, 3398.1 m, Pt, quartz sandstone, oil inclusions under transmitted light (**a**); well Shugu169, 3398.1 m, Pt, quartz sandstone, oil inclusions are yellow-green under UV light (**b**); well Shugu172, 4138.3 m, Pt, quartz sandstone, oil inclusions under transmitted light (**c**); well Shugu172, 4138.3 m, Pt, quartz sandstone, oil inclusions are yellow-green and blue-green under UV light (**d**).

(2) The second phase of inclusions: mainly oil inclusions (Figure 9c,d). The host minerals were mainly quartz followed by calcite. The size of the inclusions was 1~9 μm and they were grayish-brown, light grayish-brown, brown, etc. The fluorescence of blue-white, dark blue-white, or dark blue was displayed under UV excitation. The HT of the brine inclusions was 125~140 °C. They may be formed in the mature stage and mainly in mature oil [35].

Figures 10 and 11 show the oil charging periods were 38 Ma and 28 Ma, respectively, corresponding to the result of a previous study [36].

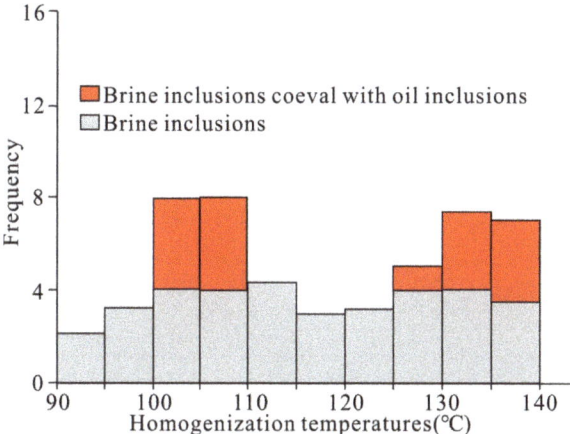

Figure 10. Homogenization temperature of oil inclusions in well Shugu169 in Shuguang area.

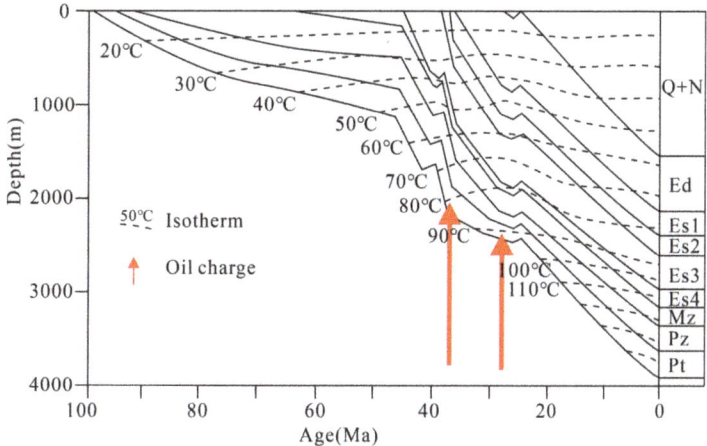

Figure 11. The timing of hydrocarbon filling by using homogenization temperatures of fluid inclusions and thermal history modeling for well Shugu169 in Shuguang area.

5. Discussion

5.1. Controlling Factors of Oil Accumulation

5.1.1. High-Quality Source Rocks in Organic Matter

Source rocks have a critical influence on the location of oil accumulation. Due to a large unconformity surface and fractures, oil accumulated near traps after hydrocarbon expulsion from source rocks. The analysis of Section 4.2 indicates that high-quality source rocks exist in Es4 with a thickness of 100–400 m (Figure 4). According to Section 4.2, the quality of source rocks was good; a previous study showed that the Es4 source rocks were good and the oil generation intensity was high [37]. The oil production intensity in the Shuguang area is about 25×10^6–500×10^6 t/km^2 (Figure 12). This result indicates that the source rocks can provide enough oil for a large accumulation of oil.

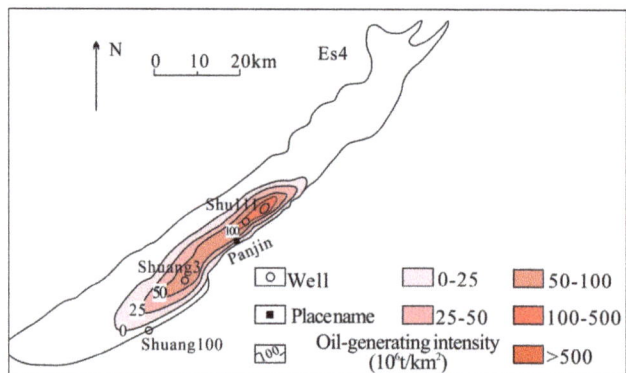

Figure 12. Accumulating oil generation intensity of Es4 source rocks from the Shuguang area (modified according to the data from PLOEDRI).

5.1.2. Oil Charging Characteristics

According to the analysis results in Section 4.4, oil charging periods were 38 Ma and 28 Ma, respectively (Figure 11). Tectonic activity is the key factor influencing the accumulation process of the Shuguang oil reservoir. On this basis, it can be concluded that there were two oil charging periods in the Shuguang reservoir.

From the Yanshanian movement period to the Himalayan orogeny period, the strata deformation was strong and developed from northeast to southwest. The tectonic deformation of this time had a dual function. First, it matched the major oil expulsion stage and caused late accumulation. Second, it disrupted and adjusted an early reservoir.

In general, the Pt reservoirs had different accumulation periods (Figure 13).

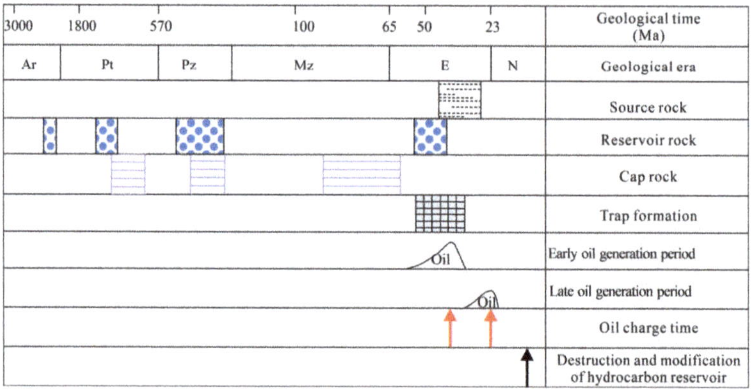

Figure 13. Comparison diagram of oil accumulation events in the Shuguang oil reservoir, including the reservoir-forming element and the oil charging period.

(1) Es stage (45–36 Ma): In the late Yanshanian movement, tectonic traps formed in the Pt strata, and the faults were still active at this time, with transport characteristics. The Es4 source rocks were less mature, and the heavy oil and low-maturity oil migrated to the middle and high buried hills along the active faults and accumulated in the traps of the Pt strata. The peak period of oil charging was 38 million years ago.

(2) Ed period (35–23 Ma): This period was the key period of reservoir formation. The traps were basically established. The Es source rocks gradually developed and matured. A large amount of mature oil was generated and then moved along the faults to the traps of the Pt strata in the low buried hill to form reservoirs. The peak of oil charging

was 28 million years ago. It is worth mentioning that the late Himalayan movement may have damaged and adjusted the reservoir at that time.

In conclusion, oil accumulation in the Shuguang oil reservoir is a continuous process. However, a previous scholar considered that three charge periods happened in the oil reservoir [38], which does not correspond to this study. It may have been caused by the difference in the selected inclusions.

5.1.3. Fracture Influencing the Distribution of Oil Reservoir

The pore structure of a reservoir affects the formation mechanism of that reservoir [39]. The major migration of reservoir formation is the pore, fracture, and the fault. The permeability of quartz sandstone in the Shuguang oilfield is low. Therefore, improving permeability through fractures is key to local fluid migration [40]. The linear fracture density (LFD) was used to study the fractures in the Pt reservoir [41]. Take the Dibei oil and gas reservoir in the Kuqa Depression as an example. Industrial oil flow from well YN2 was found in the J_1a Formation. The LFD of the strata is 4.18/m [42]. In contrast, the J_1y unit has no industrial oil and gas, and the LFD is 1.56/m. The result shows that LFD is critical to improve oil and gas production. Due to the structural extrusion in the Shuguang area, the LFD of the formation improved the porosity and permeability of the reservoir and controlled the distribution of the reservoir. Core and thin section results showed that the number of formation fractures was large in the Shuguang area (Figure 7e,f) and provide a good condition for oil accumulation. The analysis showed that fractures affect the distribution of oil, which corresponds to the result of the Dibei reservoir [39].

5.1.4. Preservation Condition

Good conditions for the preservation of a cap layer are conducive to the formation of large oil and gas reservoirs [43,44]. The combination of source, reservoir, cap, and fault in the Shuguang area is conducive to oil preservation. The main source rock is Es4 mudstone. Pt quartz sandstone is rich in fractures and is a good reservoir. The sealing ability of the overlaying Fangshenpao Formation and the Mesozoic cap layer is better, which can prevent vertical oil loss [45]. Oil migrates along faults and fractures and accumulates in nearby reservoirs.

5.2. Accumulation Mode

Oil generation intensity is high in the Shuguang area, indicating it can produce a large amount of an oil resource (Figure 12). Quartz sandstone is favorable for the formation of an oil reservoir. From the perspective of longitudinal sedimentary assemblage, the reservoir-forming mode of the Shuguang oil reservoir is of a lower generation and upper reservoir. Due to the differences in the source rocks, the oil accumulation mode differs from some scholars' conclusion [46].

Based on the above analysis, the Shuguang reservoir was confirmed as a structure-lithologic mode (Figure 14) with the following reservoir-forming characteristics:

(1) In the early stage, the source rock produced oil and migrated upward along the fault connecting the source rock to the reservoir. Some oil migrated laterally to the reservoir traps in the middle buried hill and high submerged hill. This situation applies to heavy oil and low-maturity oil charging in the early period.
(2) In the late period, mature oil from the source rocks moved along faults and fractures to the reservoirs in the low buried hill and accumulated under the cover of the cap.

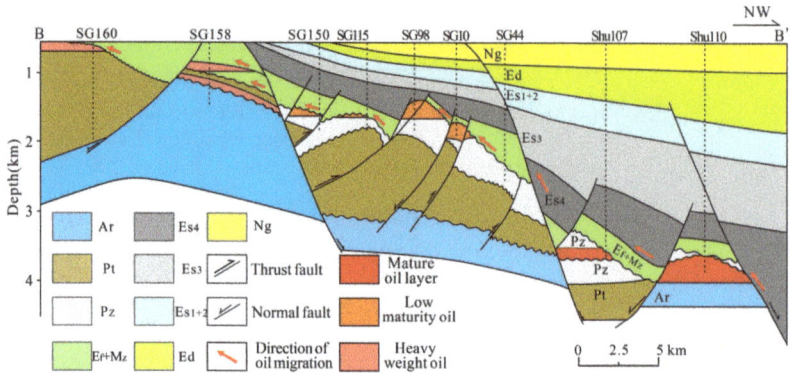

Figure 14. Oil accumulation mode in Shuguang area (the profile position is shown in Figure 1c, B-B').

5.3. Implications for Development

Reservoir physical properties control reservoir production. Oil migrates upward and along fractures and faults. If upper rock blocks the upward path, oil migrates laterally and gathers in the trap. A structural trap and stratigraphic trap are key to oil accumulation. A large number of thrust faults play a key role in oil migration and oil reservoir [14,47].

The fractures in the Shuguang reservoir mainly developed due to a strong structural extrusion. Fractures have a certain influence on the reservoir. If the fracture occurs at the top of the oil reservoir, it can reduce the sealing capacity of the cap layer, resulting in oil leakage. On the contrary, if fractures develop within the reservoir, the sandstone physical properties are optimized and the reservoir is adjusted, which is favorable for the formation of a "sweet spot" (a relatively oil- and gas-rich region) [39]. Well Shugu 169 has an LFD of 4.12/m with the highest daily energy, while wells Shugu175 and Shugu183 have a lower LFD and lower production. The reason for this phenomenon is that the development of fractures optimizes the physical properties of the reservoir. Such fractures can increase commercial production of oil and gas reservoirs in the United States [48] and China [14].

In addition, the Shuguang area is a potential area with a lot of oil. This point should be the critical focus of future work: by revealing the coupling relationship of key control factors, a new "sweet spot" of oil resources in the Shuguang area should be further determined.

6. Conclusions

(1) The source rocks in the Shuguang area were mainly developed in the Es4 Formation and have a strong oil generation intensity, which produces a lot of oil for the oil reservoir.

(2) During the Es stage (45–36 Ma), oil migrated along the active faults and accumulated in the tectonic traps of the Pt strata. During the Ed period (35~23 Ma), a lot of mature oil moved along the faults to the traps of the Pt strata to form reservoirs. The late Himalayan movement might have adjusted the reservoir at that time. The faults and fractures provide a good path for oil migration.

(3) The relationship between the source–reservoir–cap assemblage and trap and the fracture in the Shuguang area is favorable for oil accumulation. It is a structure-lithologic oil reservoir with lower generation and an upper reservoir. The "sweet spot" associated with fracture is the critical focus of future work.

Author Contributions: Conceptualization, G.Z. and F.J.; methodology, Q.Z.; software, H.P.; validation, S.Z., Q.Z., and X.L.; formal analysis, D.C.; investigation, X.L.; resources, D.C.; data curation, S.Z.; writing—original draft preparation, G.Z.; writing—review and editing, F.J.; supervision, H.P.; project administration, F.J.; funding acquisition, D.C. All authors have read and agreed to the published version of the manuscript.

Funding: This research was financially funded by the Science Foundation of China University of Petroleum, Beijing (2462022XKBH005), and China Postdoctoral Science Foundation (2022M723487).

Data Availability Statement: The original contributions presented in the study are included in the article, further inquiries can be directed to the corresponding author.

Acknowledgments: We appreciate Yanmin Guo and Zhi Tian for their help in collecting the data. Their insightful suggestions were also helpful for the manuscript.

Conflicts of Interest: The authors declare no conflicts of interest.

References

1. Lin, S.; Zou, C.; Yuan, X.; Yang, Z. Status quo and implications of tight oil development in the United States. *Lithol. Res.* **2011**, *23*, 25–30.
2. Zhang, J.; Bi, H.; Xu, H.; Zhao, J.; Yu, T.; Zhao, D.; Geng, Y. New development and reference significance of tight oil exploration and development abroad. *Acta Petrol. Sin.* **2015**, *36*, 127–137.
3. Meng, Q.; Bai, X.; Liang, J. Characteristics and exploration strategies of tight oil in Fuyu Reservoir in northern Songliao Basin. *Pet. Geol. Oilfield Dev. Daqing* **2014**, *33*, 23–29.
4. Pu, X.; Zhou, L.; Han, W. Gravity flow sedimentation and tight oil exploration in lower first member of Shahejie Formation in slope area of Qikou Sag, Bohai Bay Basin. *Pet. Explor. Dev.* **2014**, *41*, 138–149. [CrossRef]
5. Wu, W.; Deng, J.; Zhao, J. Accumulation conditions and models of tight oil reservoirs in Chang-7 of Huaqing area, the Ordos Basin. *Oil Gas Geol.* **2016**, *37*, 874–881.
6. Zhang, F.; Chen, Z.; Zhao, Z.; Gao, J.; Fu, L.; Li, C.; Zhang, L. Hydrocarbon Accumulation Periods in the Upper Paleozoic Strata of the Western Ordos Basin, China, Based on Fluid Inclusions and Basin Modeling. *ACS Omega* **2023**, *8*, 20536–20549. [CrossRef] [PubMed]
7. Chen, S.; Zhang, H.; Lu, J. Controlling factors of Jurassic Da'anzhai tight oil accumulation and high production in central Sichuan Basin, China. *Pet. Explor. Dev.* **2015**, *42*, 206–214. [CrossRef]
8. Su, Y.; Zha, M.; Qu, J. Simulations on oil accumulation processes and controlling factors in tight reservoirs of Lucaogou Formation of Jimsar Sag. *J. China Univ. Pet. (Ed. Nat.Sci.)* **2019**, *43*, 11–22.
9. Zou, C.; Zhang, G.; Yang, Z.; Tao, S.; Hou, L.; Zhu, R.; Yuan, X.; Ran, Q.; Li, D.; Wang, Z. Concepts, characteristics, potential and technology of unconventional hydrocarbons: On unconventional petroleum geology. *Pet. Explor. Dev.* **2013**, *40*, 413–428. [CrossRef]
10. Zhou, Y.; Ji, Y.; Xu, L.; Che, S. Controls on reservoir heterogeneity of tight sand oil reservoirs in Upper Triassic Yanchang Formation in Longdong Area, southwest Ordos Basin, China: Implications for reservoir quality prediction and oil accumulation. *Mar. Pet. Geol.* **2016**, *78*, 110–135. [CrossRef]
11. Bai, Y.; Zhao, J.; Wu, W. Methods to determine the upper limits of petrophysical properties in tight oil reservoirs: Examples from the Ordos and Songliao Basins. *J. Pet. Sci. Eng.* **2021**, *196*, 107983. [CrossRef]
12. Er, C.; Zhao, J.; Li, Y.; Si, S. Relationship between tight reservoir diagenesis and hydrocarbon accumulation: An example from the early Cretaceous Fuyu reservoir in the Daqing oil field, Songliao Basin, China. *J. Pet. Sci. Eng.* **2021**, *208*, 109422. [CrossRef]
13. Wang, M.; Yang, Z.; Shui, C.; Yu, Z.; Wang, Z.; Cheng, Y. Diagenesis and its influence on reservoir quality and oil-water relative permeability: A case study in the Yanchang Formation Chang 8 tight sandstone oil reservoir, Ordos Basin, China. *Open Geosci.* **2019**, *11*, 37–47. [CrossRef]
14. Lyu, W.; Zeng, L.; Zhang, B.; Miao, F. Influence of natural fractures on gas accumulation in the Upper Triassic tight gas sandstones in the northwestern Sichuan Basin, China. *Mar. Pet. Geol.* **2017**, *83*, 60–72. [CrossRef]
15. Jia, J.; Yin, W.; Wang, G.; Ma, L. Migration and accumulation of crude oil in Upper Triassic tight sand reservoirs on the southwest margin of Ordos Basin, Central China: A case study of the Honghe Oilfield. *Geol. J.* **2018**, *53*, 2280–2300. [CrossRef]
16. Ew, B.; Zw, C.; Xp, B.; Zz, D.; Zw, D.; Zw, B.; Yl, B.; Yue, F.B.; Zz, E. Key factors controlling hydrocarbon enrichment in a deep petroleum system in a terrestrial rift basin—A case study of the uppermost member of the upper Paleogene Shahejie Formation, Nanpu Sag, Bohai Bay Basin, NE China. *Mar. Pet. Geol.* **2019**, *107*, 572–590.
17. Liu, Y.; Ye, J.; Zong, J.; Wang, D. Analysis of forces during tight oil charging and implications for the oiliness of the tight reservoir: A case study of the third member of the Palaeogene Shahejie Formation, Qibei slope, Qikou sag. *Mar. Pet. Geol.* **2022**, *144*, 105819. [CrossRef]
18. Wei, W.; Zhu, X.; Meng, Y.; Xiao, L.; Xue, M.; Wang, J. Porosity model and its application in tight gas sandstone reservoir in the southern part of West Depression, Liaohe Basin, China. *J. Pet. Sci. Eng.* **2016**, *141*, 24–37. [CrossRef]
19. Hu, L.; Fuhrmann, A.; Poelchau, H.; Horsfield, B.; Zhang, Z.; Wu, T. Numerical simulation of petroleum generation and migration in the Qingshui sag, western depression of the Liaohe basin, northeast China. *AAPG Bull.* **2005**, *89*, 1629–1649. [CrossRef]
20. Liu, G.; Zeng, L.; Li, H.; Ostadhassan, M.; Rabiei, M. Natural fractures in metamorphic basement reservoirs in the Liaohe Basin, China. *Mar. Pet. Geol.* **2020**, *119*, 104479. [CrossRef]
21. Meng, K. Study on Distribution Characteristics of Inner Stratigraphic Framework and Reservoir Interval in Shuguang Buried Hill, Liaohe Basin. Master's Thesis, Northeast Petroleum University, Daqing, China, 2018. (In Chinese with English Abstract)

22. Yang, S. Optimum selection of Shuguang Buried Hill Belt in Western Liaohe Depression. Master's Thesis, Northeast Petroleum University, Daqing, China, 2016. (In Chinese with English Abstract).
23. Zhao, H.; Liu, X.; Meng, W.; Chen, Z.; Han, H. Subtle oil and gas reservoirs and their accumulation dynamics in Shuguang-Leijia area. *J. Jilin Univ. (Earth Sci.Ed.)* **2011**, *41*, 21–28.
24. Wang, G. Analysis of special reservoir distribution and hydrocarbon accumulation conditions in Shuguang-Lejia area. *Spec. Oil Gas Reserv.* **2009**, *16*, 49–52.
25. Jin, Z.; Liao, K. A Comprehensive Evaluation of High Viscous Crude Oil from Shuguang No. 1 Zone of Liaohe Oil Field. *Pet. Sci. Technol.* **2003**, *21*, 1077–1088. [CrossRef]
26. Xiong, Y.; Geng, A.; Wang, C.; Sheng, G.; Fu, J. The origin of crude oils from the Shuguang-Huanxiling Buried Hills in the Liaohe Basin, China: Evidence from chemical and isotopic compositions. *Appl. Geochem.* **2003**, *18*, 445–456. [CrossRef]
27. Li, D. Geological Structure and Hydrocarbon Occurrence of the Bohai Gulf Oil and Gas Basin (China). *Mar. Geol. Quat. Geol.* **1981**, *1*, 3–20.
28. Chen, Z.; Yan, H.; Li, J.; Zhang, G.; Zhang, Z.; Liu, B. Relationship between Tertiary volcanic rocks and hydrocarbons in the Liaohe basin People's Republic of China. *AAPG Bull.* **1999**, *83*, 1004–1014.
29. Ge, W. Study on Tight Oil Accumulation Characteristics of the Fourth Member of Shahejie Formation in Western Liaohe Depression. Master's Thesis, Southwest Petroleum University, Chengdu, China, 2018. (In Chinese with English Abstract).
30. Kotarba, M.J.; Bilkiewicz, E.; Wicaw, D.; Radkovets, N.Y.; Romanowski, T. Origin and migration of oil and natural gas in the central part of the Ukrainian outer Carpathians: Geochemical and geological approach. *AAPG Bull.* **2020**, *104*, 1323–1356. [CrossRef]
31. Chen, J.; Zhao, C.; He, Z. Evaluation criteria for hydrocarbon generation potential of organic matter in coal measures. *Pet. Explor. Dev.* **1997**, *24*, 1–5.
32. Folk, R.L. *Petrology of Sedimentary Rocks*; Galehouse, J.S., Ed.; Hemphill Publishing Company: Austin, TX, USA, 1971; 184p.
33. Okunuwadje, S.E.; Bowden, S.A.; Macdonald, D.I.M. Diagenesis and reservoir quality in high-resolution sandstone sequences: An example from the Middle Jurassic Ravenscar sandstones, Yorkshire CoastUK. *Mar. Pet. Geol.* **2020**, *118*, 104426. [CrossRef]
34. Lisk, M.; Brincat, M.; Gartrell, A. An integrated evaluation of hydrocarbon charge and retention at the Griffin, Chinook, and Scindian oil and gas fields, Barrow Subbasin, North West Shelf, Australia. *AAPG Bull.* **2006**, *90*, 1359–1380.
35. Meng, Y. Study on hydrocarbon migration history and prediction of favorable accumulation area in southern section of Liaohe Western Sag. *Bull.Mineral. Petrol. Geochem.* **2009**, *28*, 12–18.
36. Xu, B.; Guo, H.; Lin, T.; Qi, J.; Yang, H. Hydrocarbon accumulation periods in the western sag of Liaohe Depression. *Pet. Geol. Recovery Effic.* **2010**, *17*, 12–14.
37. Liu, Y. The Research on Interior Characteristics of Shuguang Buried Hill of Liaohe Depression. Master's Thesis, Northeast Petroleum University, Daqing, China, 2015. (In Chinese with English Abstract).
38. Liu, X. Fluid charging period of lacustrine carbonate rocks in Leijia area, western Liaohe Depression. *Fault-Block Oil Gas Field* **2020**, *27*, 432–437.
39. Pang, X.; Peng, J.; Jiang, Z.; Yang, H.; Wang, P.; Jiang, F.; Wang, K. Hydrocarbon accumulation processes and mechanisms in Lower Jurassic tight sandstone reservoirs in the Kuqa subbasin, Tarim Basin, northwest China: A case study of the Dibei tight gas field. *AAPG Bull.* **2019**, *103*, 769–796. [CrossRef]
40. Pitman, J.K.; Sprunt, E.S. Origin and distribution of fractures in Tertiary and Cretaceous rocks, Piceance Basin, Colorado, and their relation to hydrocarbon occurrence. *AAPG Bull.* **1985**, *69*, 860–861.
41. Van Golf-Racht, T.D. *Fundamentals of Fractured Reservoir Engineering*; Elsevier Scientific Publishing Company: New York, NY, USA, 1982.
42. Zeng, L.; Zhou, T. Distribution of reservoir fractures in Kuqa Depression, Tarim Basin. *Nat. Gas Ind.* **2004**, *24*, 23–25.
43. Jiang, Y.; Liu, J.; Su, S.; Liu, J.; Hu, H. Discussion on preservation conditions of buried hill oil and gas reservoirs in Bohai Bay Basin. *J. China Univ. Pet. (Nat. Sci.Ed.)* **2022**, *46*, 1–11.
44. Guo, Z.; Ma, Y.; Liu, W.; Wang, L.; Tian, J.; Zeng, X.; Ma, F. Main factors controlling the formation of basement hydrocarbon reservoirs in the Qaidam Basin, western China. *J. Pet. Sci. Eng.* **2017**, *149*, 244–255. [CrossRef]
45. Yu, J. Main Controlling Factors and Objective Evaluation of Hydrocarbon Accumulation in Gaosheng Proterozoic Buried Hill. Master's Thesis, Northeast Petroleum University, Daqing, China, 2016.
46. Li, J.; Lin, C. Inversion anticline structure self-generated self-reservoir reservoir-forming model. *Acta Petrol. Sin.* **2006**, *27*, 34–37.
47. Shanley, K.W.; Cluff, R.M.; Robinson, J.W. Factors controlling prolific gas production from low-permeability sandstone reservoirs: Implications for resource assessment, prospect development, and risk analysis. *AAPG Bull.* **2004**, *88*, 1083–1121. [CrossRef]
48. Spencer, C.W. Review of characteristics of low-permeability gas reservoirs in western United States. *AAPG Bull.* **1989**, *73*, 613–629.

Disclaimer/Publisher's Note: The statements, opinions and data contained in all publications are solely those of the individual author(s) and contributor(s) and not of MDPI and/or the editor(s). MDPI and/or the editor(s) disclaim responsibility for any injury to people or property resulting from any ideas, methods, instructions or products referred to in the content.

Article

Occurrence and Potential for Coalbed Methane Extraction in the Depocenter Area of the Upper Silesian Coal Basin (Poland) in the Context of Selected Geological Factors

Sławomir Kędzior and Lesław Teper *

Institute of Earth Sciences, Faculty of Natural Sciences, University of Silesia in Katowice, Będzińska 60, 41-200 Sosnowiec, Poland; slawomir.kedzior@us.edu.pl
* Correspondence: leslaw.teper@us.edu.pl

Abstract: Coalbed methane (CBM) is the only unconventional gas in Poland with estimated recoverable resources. The prospects for developing deep CBM have been explored in recent years by drilling deep exploration wells within the depocenter of the Upper Silesian Coal Basin. The purpose of this study is to analyze the occurrence and potential for CBM extraction in this area of the basin, which can be considered prospective due to the confirmed presence of significant amounts of gas and thick coal seams at depths > 1500 m. The study examined the vertical and horizontal variability of the gas content in the studied area, the coal rank in the seams, thermal conditions, and coal reservoir parameters. The gas content in the seams, reaching more than 18 m^3/t $coal^{daf}$ at a depth of 2840 m, and indicative estimated gas resources of 9 billion m^3 were found. The high gas content is accompanied by positive thermal and coal rank anomalies. The permeability and methane saturation of the coal seams are low, and therefore, potential methane production may prove problematic. However, the development of CBM extraction technologies involving directional drilling with artificial fracturing may encourage gas production testing in the study area.

Keywords: methane; resources; bituminous coal; formation temperature; Upper Silesian Coal Basin; Poland

Citation: Kędzior, S.; Teper, L. Occurrence and Potential for Coalbed Methane Extraction in the Depocenter Area of the Upper Silesian Coal Basin (Poland) in the Context of Selected Geological Factors. *Energies* **2024**, *17*, 2592. https://doi.org/10.3390/en17112592

Academic Editors: Shu Tao, Wei Ju, Shida Chen, Zhengguang Zhang and Jiang Han

Received: 22 April 2024
Revised: 23 May 2024
Accepted: 26 May 2024
Published: 28 May 2024

Copyright: © 2024 by the authors. Licensee MDPI, Basel, Switzerland. This article is an open access article distributed under the terms and conditions of the Creative Commons Attribution (CC BY) license (https://creativecommons.org/licenses/by/4.0/).

1. Introduction

Coalbed methane (CBM), belonging to unconventional gas resources, is still an attractive source of energy due to the possibility of borehole extraction without the necessity to extract the coal itself. CBM drilling technology was initiated and expanded in the USA in the second half of the 20th century; however, it is currently most developed in Australia and China [1]. Global CBM resources are estimated at 113–184 trillion m^3 [2]. Countries with the largest CBM potential resources include Russia, the USA, Canada, Australia, Indonesia and Poland [2]. In the latter of the unconventional reservoirs, only coalbed methane (CBM) has defined and calculated recoverable resources and developed reserves of 106,362.35 and 10,564.32 million m^3, respectively, for the Upper Silesian Coal Basin (USCB) only [3]. The volume of recoverable resources of CBM is comparable to the volume of conventional gas resources in Poland; however, due to different methods of resource calculation and uneven levels of deposit assessment, the vast majority of these resources (>90%) belong to the lowest category of appraisal corresponding to possible resources [3].

The exploration and appraisal of CBM deposits in Poland has a long history [4–6]; however, to date, industrial production of this gas has not been initiated except for minor production of coal-mine methane and abandoned-mine methane. Among the numerous projects aimed at initiating industrial production of CBM, it is worth mentioning the Geometan project implemented in 2017–2019 to demethanation of coal seams with surface boreholes a few years before the start of coal mining and thus reduce the methane hazard. The following undertaking is the deep exploratory well Orzesze-1 with a depth of 3710 m carried out in 2019–2020 to check the possibility of unconventional gas deposits developing in the place of the deepest subsidence and the largest thickness of coal-bearing sediments (depocenter) in the USCB [7].

The extraction of so-called deep coalbed methane, defined as occurring at depths > 1500 m, is of interest due to, among other things, the expected significant gas reserves thicker as well as regularly occurring coal seams and has been practiced in the USA and China [8,9]. Deep CBM production tests performed in the Laramide Basin in the western United States have confirmed the feasibility of cost-effective gas production to depths of approximately 3000 m [9].

The results of gas content tests in coal seams carried out in the central part of the USCB with a maximum thickness of carboniferous strata proved positive for the amount of gas present in coal seams and are encouraging for the initiation of more advanced tests for its extraction [7].

This paper aims to analyze the variability of gas content in selected wells with depths > 1500 m in the depocenter area of the USCB in the context of the amount of gas present (estimated resources), the spatial distribution of the resources, and its production potential. The study will also address factors associated with gas-bearing reservoirs, such as formation temperature, coal rank, and selected geological features of the area (carboniferous lithology and fault tectonics).

2. Study Area
2.1. Location

The study area is situated in the central part of the USCB within the Main Syncline and includes the area of the depocenter of the basin, i.e., the maximum depression of the carboniferous coal-bearing series. The thickness of the carboniferous sequence exceeds 4000 m at this location. The boundaries of the study area were chosen to include deep wells (>1500 m deep) located in the hanging wall of the Bełk Fault, which ensures that the depocenter area is recognized for its gas-bearing capacity at the locus of accumulation of a significant amount of gas, generally located in the hanging wall of regional faults [10]. The defined sector has an area of 53.4 km^2 and includes, among others, the eastern part of the former Dębieńsko mine, the southern part of the Budryk mine, and the western part of the Bolesław Śmiały mine (Figure 1).

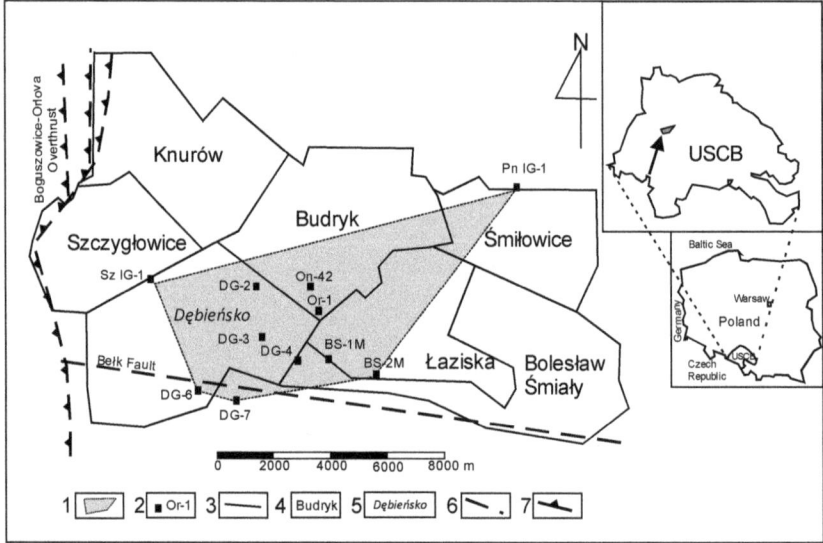

Figure 1. Location of the study area, 1—study area, 2—borehole, 3—the boundary of the coalfield, 4—the name of the operating minefield, 5—the name of the closed minefield, 6—fault, 7—overthrust.

2.2. Outline of Geological Structure

In the study area to the geological prospection/exploration depth, there are carboniferous formations belonging to the Mississippian and Pennsylvanian, covered by Triassic and Miocene strata in the form of isolated patches and quaternary sediments (Figure 2). The upper part of the Mississippian (Serpukhovian) and the entire Pennsylvanian (Bashkirian and Moskovian) are coal-bearing and developed in the form of molasses deposits consisting of alternating packages of sandstones, siltstones, and mudstones with numerous coal seams. The carboniferous coal-bearing formation has been subdivided into three series, differing in the proportion of sandstone layers and the number and thickness of coal seams [4,11]. The Paralic Series (Serpukhovian) contains distinctive horizons with marine faunas, indicative of periodic inundation by the sea, the Upper Silesian Sandstone Series (Bashkirian) contains thick sandstone packages with thick coal seams, while the Mudstone Series (Moscovian) contains fine clastic packages (siltstone and mudstone) and numerous but rather thin coal seams, sandstones are in the minority. The carboniferous roof is erosional. Overlying the eroded carboniferous surface lie, for most of the area, quaternary sediments up to 50 m in thick and isolated slabs of Triassic and Miocene deposits were found, with a total thickness of up to 100 m.

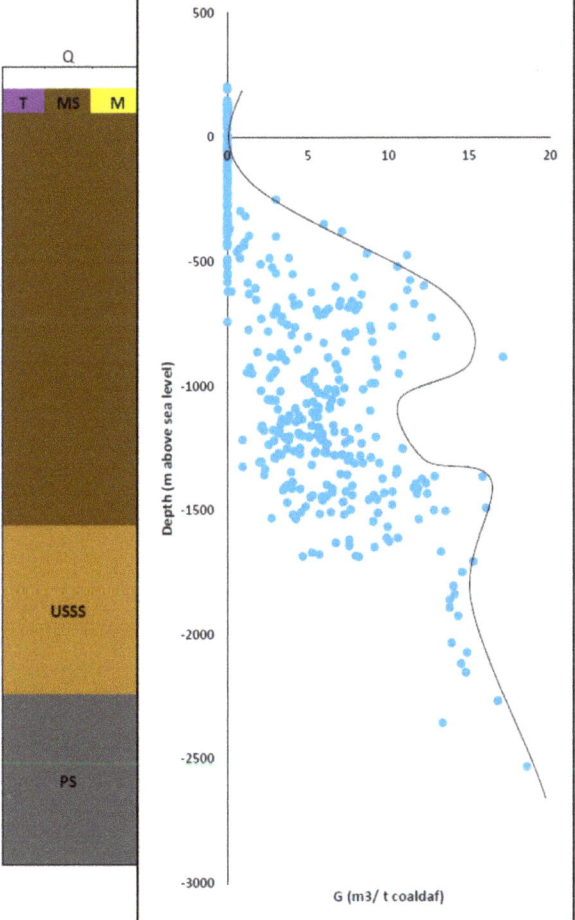

Figure 2. Vertical distribution of the gas content (G) with a simplified sketch of the stratigraphic column in the study area. Overburden: Q—Quaternary, M—Miocene, T—Triassic, Carboniferous series: MS—Mudstone Series, USSS—Upper Silesian Sandstone Series, PS—Paralic Series.

The study area, like the entire USCB, is tectonically involved. The strata generally run in an NE-SW direction and dip to the SE at an angle of several to 50°. To the west, N-S-oriented fold structures appear, which include the Knurów Anticline adjacent to the Orlova Thrust, which is regarded as the western boundary of the disjunctive part of the USCB. The fault network of this part of the basin is built of several systems with distinct geometry and origin [4,12]. Among them, the system with sub-latitudinal principal faults is of great importance [13]. One such dislocation is the WNW-ESE Bełk Fault forming the southwestern boundary of the study area, which is regional with throw up to several hundred meters to the south. In addition, in the study area, there are numerous faults with small throws dividing the coal-bearing series into blocks. Due to the borehole reconnaissance of the study area, these faults have not been definitively confirmed, but they have been found in neighboring mines.

The sub-latitudinal fault systems have been active in various tectonic regimes, beginning with the syn- and post-sedimentary periods in the Carboniferous and Permian, through the Triassic and Cretaceous to the Neogene [13]. Neotectonic activity is also recorded in these systems, which results in contemporary movements on the faults [14]. During periods of reactivation, the recurring left-lateral oblique-slip character of principal faults was noted [13]. Thanks to the horizontal component of such displacement, alternating segments with compressional (watertight) and tensile (water-bearing, conducive to fluid migration) conditions were episodically formed in the zones of these dislocations [13].

2.3. Methane Occurrence

Methane in the coal-bearing series is genetically related to coal since it is formed as a result of chemical alterations of organic matter caused by the coalification process. The main factor driving this process is the temperature of the rock mass and the pressure of the overburdened rock. The amount of generated methane depends on the thermal maturity of organic matter, i.e., the more mature the substance, theoretically, the more methane can be produced [15,16]. The coal-bearing series in the USCB reached its greatest depth at the turn of the Carboniferous and Permian, and at that time, the highest temperature affected the coal seams, causing the currently observed coal rank and generating significant amounts of methane estimated at trillions of m^3 [17]. However, the sorption capacity of coal is limited and dependent on pressure and temperature, so not all the gas produced remains in the coal, and the excess is expelled outside the deposit. Therefore, only an insignificant part of the generated gas is accumulated in coal seams today [17,18]. Another genetic type is methane produced by bacteria (microbial methane), but its presence was not detected in the study area.

There are two basic forms of methane occurrence in coal-bearing formations: sorbed methane, which is physically and chemically bound to the coal substance and located in coal micropores, and free methane, which fills macropores and fractures in the coal seam and is present in gangue rock (sandstone) and fault zones. Free methane and sorbed methane exist in mutual equilibrium expressed by gas pressure, an increase of which promotes gas adsorption by coal, while a decrease induced, for example, by geological processes or mining activities, causes gas desorption and migration toward other parts of the deposit or into the atmosphere [19].

The spatial distribution of the gas content in the USCB is related to its geological development. The long period of uplift of the basin lasting from the end of the Permian to the Paleogene caused erosion of the upper parts of the coal-bearing complex, changing hydrogeological conditions and, consequently, resulting in the escape of methane into the atmosphere. The free migration of gases was facilitated by the lack of a tight overburden of the coal-bearing series. Thus, the carboniferous rock mass was naturally degassed to depths varying from 500 to 1000 m, depending on local lithological and tectonic conditions [4,19].

The study area lies in the northern region of the USCB [4], where up to a depth of 500–600 m from the ground surface the carboniferous coal-bearing series has been naturally degassed in the geological past and the gas content in the coal seams is very low <2.5 m^3 CH$_4$/Mg coal^{daf}, deeper the amount of methane increases very rapidly until it reaches a methane-bearing maximum with a gas content of 10–18 m^3 CH$_4$/Mg coal^{daf} (Figure 2).

3. Methods

The study is based on archival data from 11 boreholes drilled to identify coal deposits and methane conditions over an area of more than 53 km^2. The depths of the boreholes range from 1612 to 3710 m, but gas-bearing tests were carried out to a depth of 2840 m in the Orzesze-1 borehole, the deepest in the USCB. The type and number of data obtained are shown in Table 1.

Table 1. Type and number of data from boreholes used in the study Th—coal seam thickness, G—gas content, M—moisture content, A—ash content, T—temperature, Ro—vitrinite reflectance, "-" no data.

Borehole	Symbol	Depth (m)	Th (%)	G (m^3/t)	M (%)	A (%)	T (°C)	Ro (%)
Bolesław Śmiały 1M	BS 1M	1836	34	34	34	34	6	3
Bolesław Śmiały 2M	BS 2M	1834	27	27	27	27	6	5
Dębieńsko-Głębokie 2	DG-2	1726	27	27	27	27	6	-
Dębieńsko-Głębokie 3	DG-3	1950	42	42	40	40	7	-
Dębieńsko-Głębokie 4	DG-4	2000	34	34	31	31	7	4
Dębieńsko-Głębokie 6	DG-6	2000	36	36	35	35	7	-
Dębieńsko-Głębokie 7	DG-7	2000	37	37	37	37	7	-
Orzesze-1	Or-1	3708	41	41	41	41	4	3
Ornontowice 42	On-42	1612	33	33	28	28	6	3
Szczygłowice IG-1	Sz IG-1	2000	45	45	44	44	6	5
Paniowy IG-1	Pn IG-1	1948	32	32	31	31	6	4
In total			388	388	375	375	68	27

Due to the depocenter area of the USCB and, therefore, the focus on deep gas, the wells were selected to be deeper than 1600 m and the recorded gas content high (>8 m^3/t coal^{daf}) and as little disturbed by mining activities as possible. These conditions are fulfilled in the area located between the Bełk Fault in the south, the decommissioned Dębieńsko mine in the west, and the working Budryk mine in the northwest. The eastern boundary is represented by the approximate border of the depocenter marked by the location of the Paniowy IG-1 borehole.

Two groups of parameters were analyzed. The first group consists of values related to the amount of gas in the deposit, i.e., gas content, thickness of coal seams, coal rank expressed by the vitrinite reflectance, and the temperature of the rock mass. They are important in the case of spatial changes in the quantity of gas in the rock mass and affect the amount of gas-in-place resources. The parameters from the second group determine the gas volume that can be extracted from the deposit and include gas saturation, permeability, and hydrogeological conditions [20]. The results from both groups of parameters indicate trends in gas content variability in the study area and indicate the possibilities of extracting CBM as a raw material.

The main parameter used in this study is the gas content in the coal seams, defined as the volume of gas contained in a mass unit of dry ash-free (pure) coal substance, expressed as m^3 of gas per 1 ton of pure coal substance. Seam gas content was determined using two methods. The KPG hermetic canisters method involves placing a 0.1 m piece of coal core in a hermetically sealed container and subjecting it to two-stage vacuum degassing at a pressure of approximately 7 mmHg [4]. The amount of gas obtained is measured and converted per unit mass of pure coal substance. Gas content was measured in all 11 boreholes using this method. In boreholes Bolesław Śmiały 1M, 2M, and Orzesze-1, the free degasification USBM method was additionally applied at atmospheric pressure [21]. This is based on cyclic (daily) measurements of gas release through a coal core placed in 0.3 or 0.6 m containers until, for 5 consecutive readings, the amount of gas falls to zero or the amount of gas released is less than 5–10 cm^3/day. Both methods usually give similar results.

In the study area, the spatial distribution of gas was analyzed both vertically in 11 wells and horizontally between wells. Vertical variability was analyzed by drawing up a summary graph of the variability of gas content in the 11 boreholes. Horizontal variability was analyzed by separating 4 levels: -970, -1220, -1470 and -1720 m a.s.l. For each level, the average gas content in individual wells was calculated by selecting data from an interval of 250 m above the level, i.e., for example, for the -970 m a.s.l., data were collected from an interval of -720 to -970 m a.s.l., and for the -1220 m a.s.l. from an interval of -970 to -1220 m a.s.l., etc. For the mean gas content values calculated in this way, horizontal variability maps, as well as a gas content cross-section were made. The maps and cross-sections were produced in the Surfer 12 computer program using the natural neighbor interpolation method.

The estimated gas resources were then calculated using the formula:

$$Q = Vc \times (G - Gr) \times b \times T \tag{1}$$

where Q—estimated gas resources, Vc—mass of coal in seams with thickness < 0.6 m within the surveyed area, G—average gas content for the horizon, Gr—average residual gas content, b—coal mass correction factor according to dry ash-free basis, T—temperature factor.

Resources were calculated for each of the levels mentioned above and then summed. The desorbing gas content was used, i.e., the total gas content minus the residual gas content, which is the amount of gas present at atmospheric pressure and, therefore, not desorbed. It was determined using the USBM method. Coal seams with a thickness of >0.6 m were considered for the calculation of resources.

The resources calculated in this way provide information on the amount of gas currently accumulated in seams (gas in place) but not on the amount of gas that can be recovered from the reservoir during drilling, i.e., the recoverable resources. To calculate the recoverable resources, it is necessary to establish the so-called recovery factor, i.e., the ratio of recoverable gas to the total amount of gas in the reservoir. This factor varies and depends on parameters related to the productivity of the field, i.e., the permeability of the reservoir, its gas saturation, waterlogging, etc. It is most often determined after the test phase of gas production. As the trial exploitation was not carried out in the study area, it is not possible to establish this coefficient precisely. Therefore, an attempt was made to roughly estimate gas production possibilities based on reservoir permeability.

The coal permeability test was performed in the laboratory of the Oil and Gas Institute in Krakow on cylinder-shaped samples with a diameter of 1 inch and a length of 3–4 cm. The coal samples came from the Orzesze-1 well [7]. Nitrogen was passed through these samples at a given pressure, and the amount of gas flowing per unit of time was measured. The measurement was repeated several times for different pressure values, and then the permeability was calculated using the Darcy equation. The permeability results are presented for two pressure variants at 500 psi load corresponding to the near-wellbore zone disturbed by drilling works and at 2900 psi corresponding to the undisturbed rock mass. A total of 40 measurements from each variant were used. The variation of permeability of each variant with depth is presented.

The gas saturation of the coal seams was calculated using 11 measurements of the coal sorption capacity from the Orzesze-1 borehole and data on the measured gas content of the seam according to the following formula

$$S = G/Gc \times 100 \, (\%) \tag{2}$$

where S—saturation of the seam with gas, G—measured gas content in the seam, and Gc—sorption capacity of the seam.

In addition, the results of gas production tests conducted in other areas of the USCB were used.

The high gas content in the studied area is accompanied by a thermal positive anomaly [22,23] and an increased coal rank of the seams expressed by a slightly higher vitrinite reflectance in comparison with neighboring areas [24]. Therefore, rock mass temperature data were measured in 11 boreholes, and vitrinite reflectance data were collected. The sources of data on rock temperature and vitrinite reflectance are literature data [22,24]. Measurements of vitrinite reflectance were performed in the laboratories of geological companies (e.g., Katowice Geological Enterprise and Polish Geological Institute, Warsaw, Poland). Temperature measurements are taken in boreholes after drilling is completed, and about 120 h have elapsed to eliminate disturbance to the natural temperature of the rock mass caused by drilling [22]. The variability of these parameters was presented as a map of temperature variability and the vitrinite reflectance at -1000 m a.s.l. The aim is to investigate how temperature and coalification degree have contributed to shaping the recent gas-bearing nature of the study area.

4. Results and Discussion

4.1. Gas Content

Table 2 and Figure 2 show the vertical distribution of gas content in 11 wells drilled in the study area. This distribution coincides with the USCB's northern pattern of the gas content proposed by Kotas [4]. The data presented show that up to a depth of about -250 m above sea level (500–600 m below ground level), the gas content is low or very low (<1 m^3/t coal^{daf}). This is due to the natural process of degassing of the rock mass in the geological past between the Permian and Miocene periods (several hundred million years) due to the steady uplift of the carboniferous rock mass and erosion of its upper parts (see Section 2). The lack of a tight carboniferous overburden allowed the gas to escape into the atmosphere.

Table 2. Minimum, maximum, average gas content and total thickness of coal seams in boreholes at individual levels.

Borehole	Level (m a.s.l.)		Methane Content (m^3/t coal^{daf})					Total Seams Thickness (m)
	From	To	Minimum	Maximum	Average	Standard Deviation	Data Number	
BS 1M	−720	−970	3.40	7.35	5.48	1.62	3	2.25
	−970	−1220	5.24	5.91	5.72	0.38	5	4.4
	−1220	−1470	3.70	10.56	7.45	1.84	15	15.15
BS 2M	−720	−970	0.02	7.34	2.85	2.13	7	5.75
	−970	−1220	2.29	4.26	3.28	0.64	6	4.6
	−1220	−1470	2.00	6.19	3.50	1.38	7	5.8
DG-2	−720	−970	5.41	9.22	7.08	1.43	4	4.3
	−970	−1220	3.73	9.18	6.43	1.73	12	16
	−1220	−1470	5.26	8.35	7.18	1.16	4	5.4
DG-3	−720	−970	7.94	12.96	10.45	2.51	2	1.8
	−970	−1220	3.30	9.22	5.47	1.82	6	8.7
	−1220	−1470	4.53	12.08	9.47	2.37	10	12.15
	−1470	−1720	9.05	16.05	11.72	2.66	6	8.2
DG-4	−720	−970	8.88	9.02	8.95	0.07	2	1.4
	−970	−1220	2.12	5.82	4.34	1.24	5	4.8
	−1220	−1470	5.46	10.87	7.76	1.94	10	14.4
DG-6	−720	−970	3.01	5.03	3.81	0.87	3	2.9
	−970	−1220	2.75	3.31	3.03	0.28	2	1.2
	−1220	−1470	2.97	7.19	4.37	1.65	4	5.2
	−1470	−1720	4.57	11.89	6.71	3.02	4	7.4

Table 2. Cont.

Borehole	Level (m a.s.l.)		Methane Content (m³/t coal^daf)					Total Seams Thickness (m)
	From	To	Minimum	Maximum	Average	Standard Deviation	Data Number	
DG-7	−720	−970	1.23	2.53	1.72	0.46	5	5.4
	−970	−1220	0.96	3.39	2.29	1.01	3	2.6
	−1220	−1470	3.72	6.18	5.24	0.96	5	4.2
	−1470	−1720	4.87	9.13	7.22	1.38	6	8.9
Or-1	−720	−970	9.91	12.41	10.98	1.05	3	2.35
	−970	−1220	4.61	7.36	5.66	0.84	9	12.27
	−1220	−1470	7.60	14.71	10.63	2.11	9	12.56
	−1470	−1720	9.94	14.32	12.19	1.79	3	5.32
	−1720	−1970	12.20	13.15	12.72	0.34	6	12.99
On-42	−720	−970	6.74	17.11	10.93	4.46	3	3.1
	−970	−1220	3.13	8.92	6.09	1.69	9	11.5
Sz IG-1	−720	−970	2.65	10.22	4.70	2.38	9	11.25
	−970	−1220	2.10	5.05	3.39	0.93	6	9.35
	−1220	−1470	0.96	4.01	2.90	1.16	4	14.7
	−1470	−1720	2.74	5.27	4.00	1.27	2	1.5
Pn IG-1	−720	−970	3.72	9.32	5.83	1.90	6	7.25
	−970	−1220	2.17	6.70	4.38	1.52	6	7.65
	−1220	−1470	2.12	6.13	4.18	1.52	6	14.65
	−1470	−1720	5.74	7.57	6.96	0.86	3	6.3

The depth of the roof of the occurrence of methane coal seams is determined by the gas-bearing surface of 4.5 m³/t coal^daf (Figure 3), which in the study area is located at a depth of −300 to −1250 m above sea level (a.s.l.) and shows a morphologically varied character constituting a kind of dome with a culmination in the area of the Orzesze 1 and Ornontowice 42 wells in the central part of the area. The location of the roof reaches a level of −300 to −450 m a.s.l. here, and from this point, the roof dips in all directions and reaches its minimum value in the area of the Dębieńsko-Głębokie 6 and 7 boreholes located near the Bełk Fault in the south of the study area (Figure 3). Below the depth of the surface of the roof of the methane seams, the gas content increases rapidly, reaching high values often >8 m³/t coal^daf. This is a high methane-bearing zone, the depth range of which has not been recognized. Data from the Orzesze-1 well show that this zone does not disappear at a depth of −2500 m a.s.l. (2800 m below ground level) (Figure 2), making it very extensive vertically and reaching >2250 m in thickness. Within it, two depth sub-zones of increased gas content are noticeable: the first at a depth of −750 m a.s.l. (about 1000 m below ground level) with a maximum value of about 17 m³/t coal^daf, and the second broader one in the depth interval from −1250 m a.s.l. (1500 m below ground level) to the reconnaissance limit at a depth of −2500 m a.s.l. (about 2800 m below ground level). Gas content here ranges from 12 to 18 m³/t coal^daf. The two zones are separated by an interval of reduced gas content up to 10 m³/t coal^daf.

The horizontal distribution of gas content is shown in Figures 4 and 5. It can be seen that the average gas content of the seams increases towards the center of the area and in the vicinity of the wells Dębieńsko-Głębokie 2 and 3, Ornontowice 42, Orzesze 1 and Bolesław Śmiały 1M it takes maximum values from 7 to 11 m³/t coal^daf at the level of −1470 m a.s.l. From the visualization of the data presented in the geological-gas cross-section (Figure 6) and maps (Figures 3–5), it can be seen that the distribution of gas content in the study area has a dome shape with a peak in the central part of the area, from which the values decrease in all directions.

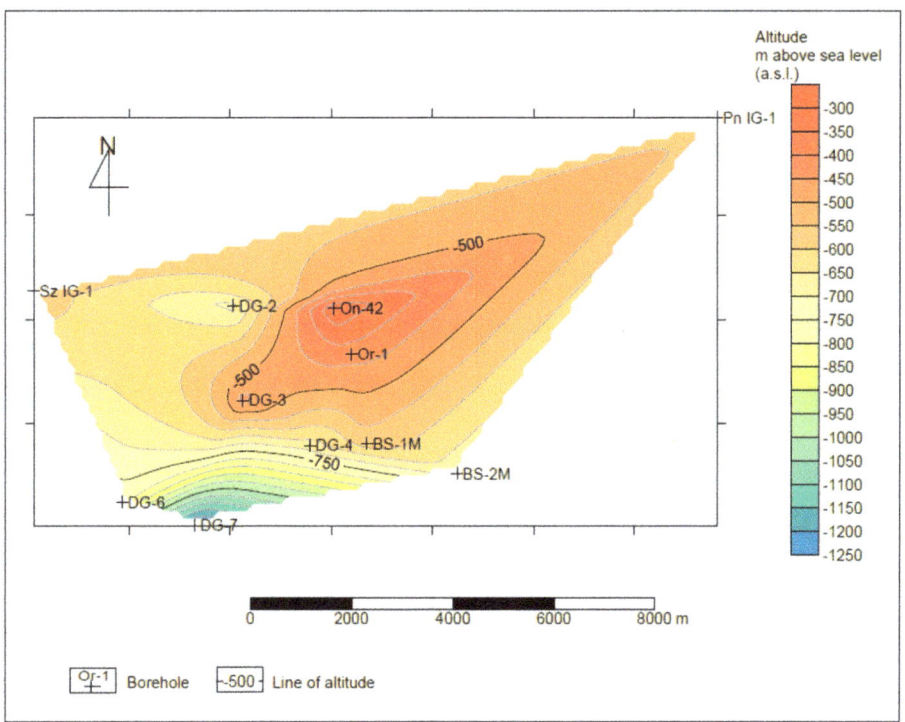

Figure 3. Roof surface of seams with gas content > 4.5 m3/t coal^{daf} (m above sea level).

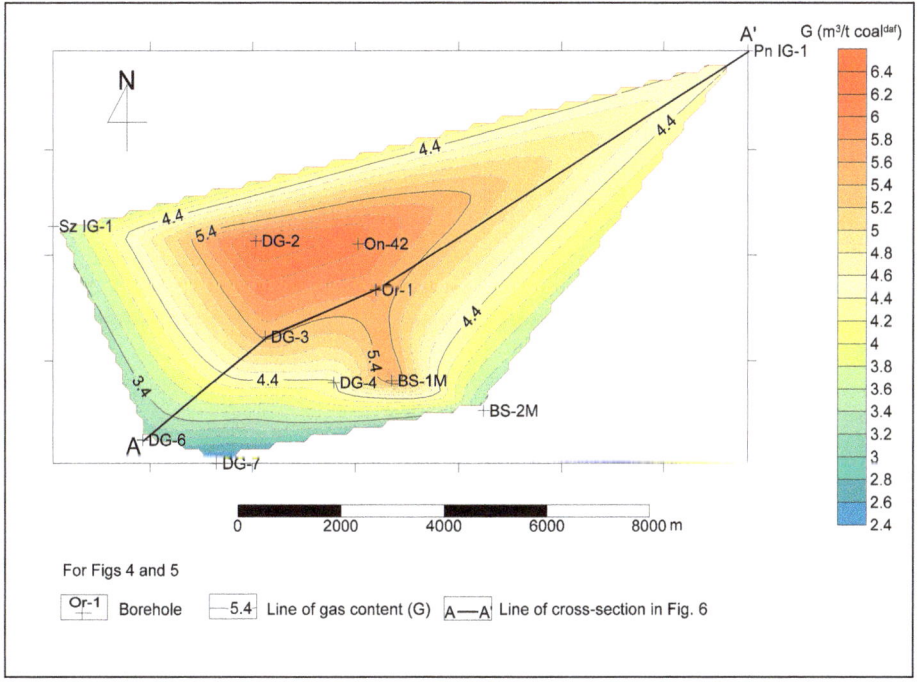

Figure 4. Gas content (G) at −1220 m above sea level.

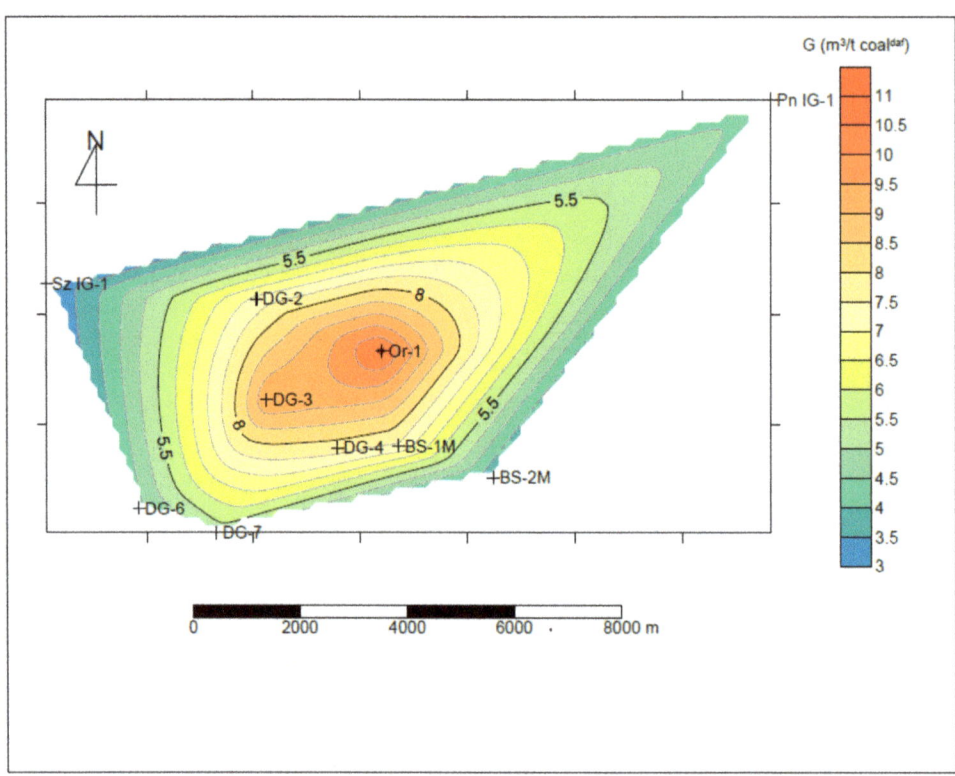

Figure 5. Gas content (G) at −1470 m above sea level.

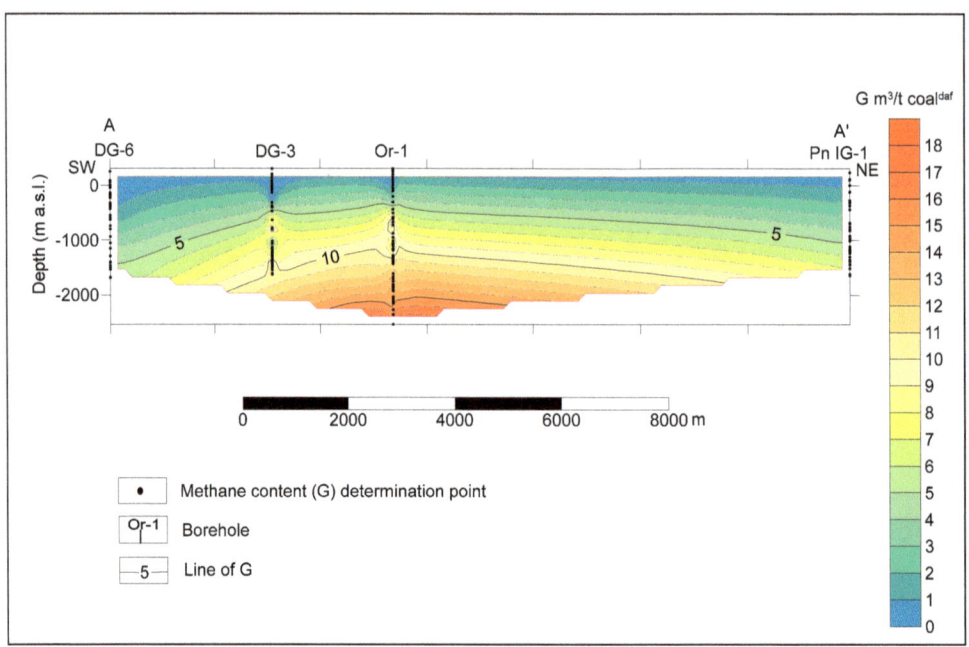

Figure 6. Gas cross-section through the study area.

The composition of the gas is characterized by the predominance of CH_4 (>90%). Of note is the occurrence of CO_2 at a level of about 15%, not previously recorded in the USCB at depths > 2300 m below ground level (Figure 7). The presence of CO_2 is accompanied by a decrease in methane content to about 80%, which may be due to the replacement of CH_4 by CO_2. Other gas components, namely higher hydrocarbons and N_2, are in the minority (up to 2–3%, Figure 7).

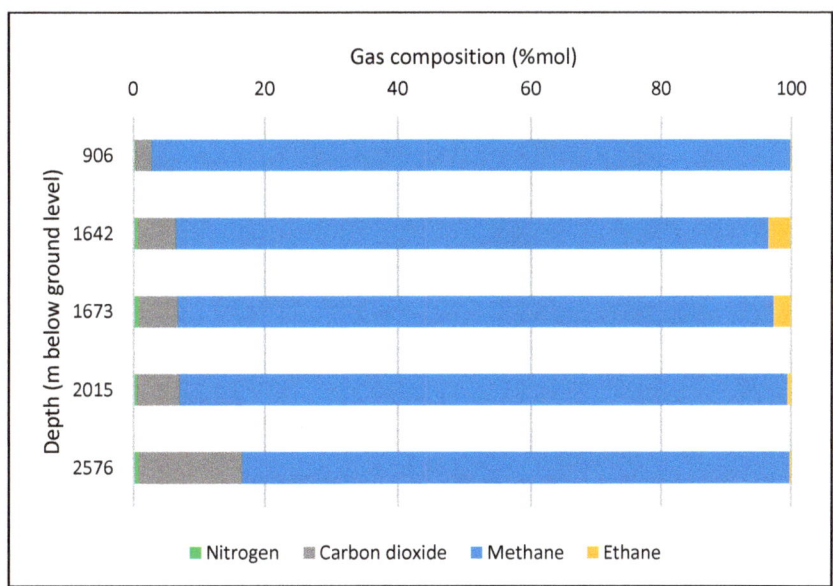

Figure 7. The molecular composition of the gas.

4.2. Coal Rank

Figure 8 shows the variation of coal rank of the seams determined using the average vitrinite reflectance (Ro) for a depth interval of −750 to −1000 m a.s.l. A clear upward trend toward the west is evident, which is consistent with the results of earlier studies [11,24]. The minimum value of Ro was recorded in the Paniowy IG-1 borehole in the eastern part of the area (<1%), while the maximum value was recorded in the Dębieńsko-Głębokie 8 borehole (1.34%) in the southwestern part. According to Kotas [11], the coal rank of the seams in the western and central parts of the USCB is zoned. There are alternating areas of high and low degrees of coalification at the same depth with a simultaneous depth-dependent decrease in the coalification field towards the east, as can be seen in Figure 8. The study area partly coincides with the latitudinal positive coal rank anomaly described by Kotas et al. [25], occurring in the Leszczyny, Orzesze, and Ornontowice areas.

Vertically, the coal rank shows an increase in depth according to Hilt's rule. Ro measured in boreholes BS-1M, BS-2M, and Orzesze 1 ranges from 0.75% at a depth of 450–880 m through 1.60% at a depth of 1700–1800 m to 2.90% at a depth of about 3000 m. This is manifested by the occurrence of coal rank ranging from high volatile bituminous coal in the carboniferous roof part through medium and low volatile bituminous coal to anthracite at depths > 2500 m [7].

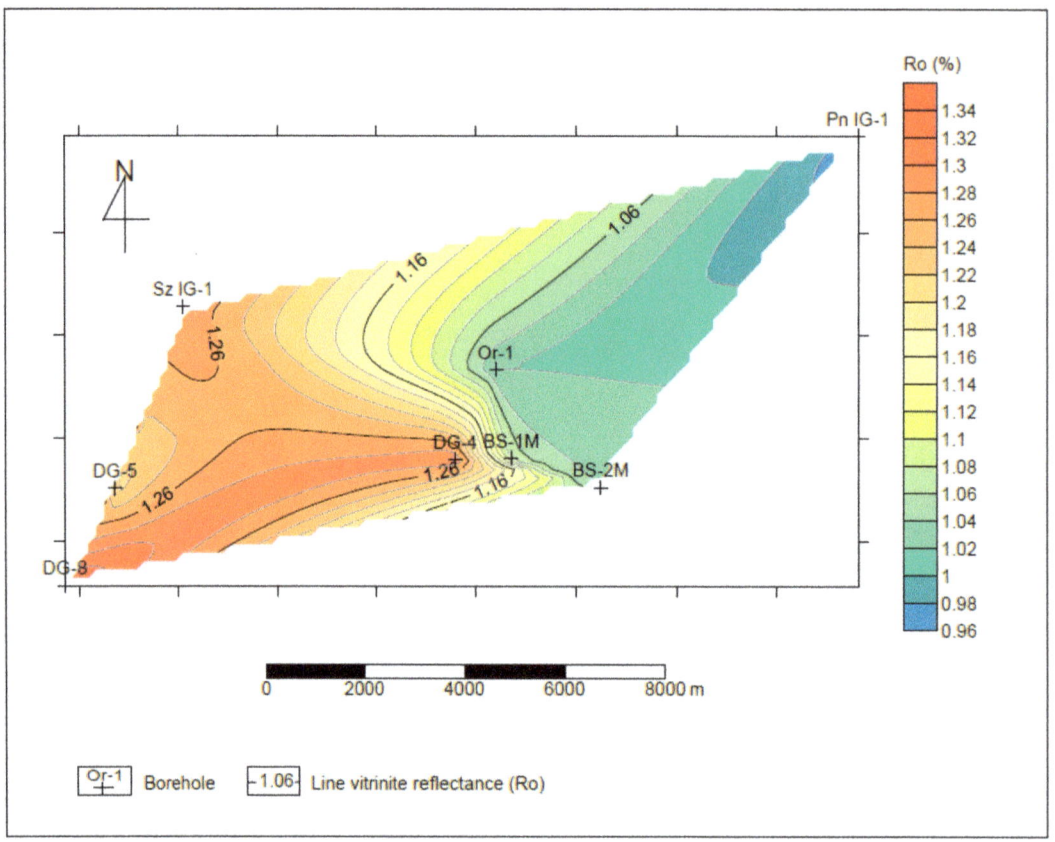

Figure 8. Vitrinite reflectance (Ro) of the deposits at the level of −1000 m above sea level.

4.3. Formation Temperature

In the study area, temperature was measured in all 11 boreholes. Its changes are observed both vertically and horizontally. Temperature values were shown to increase with depth from about a dozen degrees near the carboniferous roof to >100 °C at a depth of about 3000 m in the Orzesze-1 well. The geothermal gradient varies between the carboniferous stratigraphic series and in the study area is about 3.25 °C/100 m for the Mudstone Series, 4.00–4.50 °C/100 m for the Upper Silesian Sandstone Series, and about 3.50 °C/100 m for the Paralic Series [22]. The horizontal distribution of temperature in the study area shown in Figure 9 indicates a general increase in values to the northeast toward the Szczygłowice IG-1 borehole with a clear positive anomaly in the region of the Orzesze-1 borehole. This picture shows general agreement with the results presented by Karwasiecka [22]. A positive thermal anomaly in the Ornontowice and Orzesze area is evident at every documentation level shown in the Geothermal Atlas of the USCB [22]. Both this anomaly and others found in the USCB generally coincide with the position and direction of regional sub-latitudinal dislocations, of which the Bełk Fault is one in the study area. This localization of thermal anomalies indicates heat transport by these dislocations from great depths toward the surface. The source of heat could be the decay of radioactive elements (radiogenic heat) [26] and/or additional heating after the tectonic inversion of the basin in the Asturian phase [27]. Circulating hydrothermal solutions in the Mesozoic may have played a special role in this case, leading, for example, to the formation of giant zinc-lead ore deposits bound to the Triassic formations in the northern margin of the USCB [28]. This heating is unlikely to have led to a renewal of the coalification process [29] but may have remobilized methane accumulated in the coal seams. The two processes may overlap and are presumably linked to heat sources that are not very deep. Previous interpretations of the heat transport possibility in the area of large dislocations in the south of the USCB, in the Jastrzębie and Czechowice regions, have shown that the depth of these sources may be about 10 km from the ground surface, where crystalline rocks (granites and gneisses) occur that underlie the USCB sedimentary series and maybe the provider of radiogenic heat [22,30]. Analysis of the maps in Figure 4, Figure 5, Figure 8, and Figure 9 indicates some convergence of methane content in seams, high coal rank, and positive thermal anomaly. However, the trends of changes in vitrinite reflectance in Figure 8 and temperature in Figure 9 do not completely coincide. The high formation temperature at the level of −1000 m, culminating in the area of the Orzesze-1 well (53.5 °C), is accompanied by lower vitrinite reflectance (1.01%) compared to the western part of the area. This is because the vitrinite reflectance shows a regional USCB trend of increasing from east to west, which results from the rate of subsidence of the area of deposition of carboniferous sediments and, consequently, the course of the coalification process at the turn of the Carboniferous and Permian [29]. In turn, the thermal anomaly occurring in the research area (Figure 9) is the result of the subsequent heating of the rock mass, probably initiated in the Asturian phase and then continued in the Mesozoic [27,28]. Thus, it overlaps the previously existing coal field of the basin.

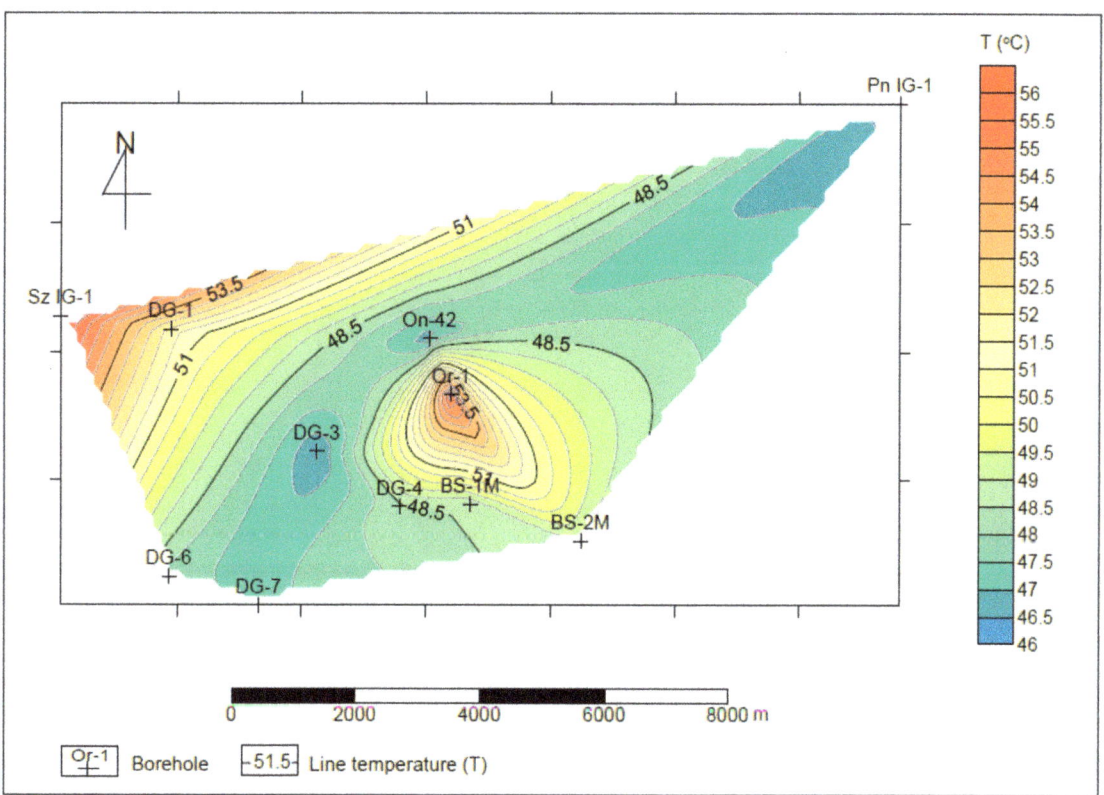

Figure 9. Formation temperature (T) distribution at the level of −1000 m above sea level.

The relationship between gas-bearing capacity and temperature was postulated by Tarnowski [10], who indicated the presence of magmatic intrusions around which the pressure of gas in the seams and gas-bearing capacity were low or zero, while at some distance from them, a sort of halo was formed with rapidly increasing gas pressure and high gas content.

The observed increase in gas content towards the central part of the study area (Figures 4 and 5) and, thus, the delineated dome (see Section 4.1) may indicate that gas migration occurred from the bottom up above the local hypothetical heat source. This may be pointed out by variations in the molecular gas composition manifested by the fact that higher hydrocarbons (C_2H_6 and C_3H_8) and CO_2 are found deeper than methane (Figure 7), the opposite of the gas composition resulting from the origin of its components [16]. On this basis, it can be assumed that the gas found in the so-called first subzone of increased gas content may be of migratory origin. Migration pathways are faults and accompanying fractures and permeable sandstones. The lack of information as to the fading of the deeper methane subzone at a depth of 2800 m makes it difficult to assess the mode of gas migration at this depth. Studies on the isotopic composition of carbon in methane indicate that up to a depth of about 1000 m, there is migration gas, and below that, indigenous gas occurs [31].

The coincidence of temperature anomalies with increased gas content is also observed in other areas of the USCB, especially in the south (Jastrzębie, Pszczyna, and Czechowice–Dziedzice regions) [22].

4.4. Gas Operating Conditions

The estimated gas-in-place resources are summarized in Table 3. They amount to 8.9 billion Nm^3 for the entire field, with the largest resources (3.3 billion Nm^3 of gas) recorded for the −1470 m a.s.l. This is due to the highest average thickness of coal seams at this level. The resource estimate presented does not mean the actual amount of gas to be extracted (see Section 3), as this can be calculated based on a recovery factor that depends on the reservoir parameters of the coal seams. These parameters include permeability of coal seams, hydrogeological conditions, and saturation of coal seams with gas.

Table 3. Summary of estimated gas resources at individual levels and in total in the study area. B—recalculation coefficient for the dry and ash-free basis of coal, G—gas content, Gr—residual gas content, M—moisture content, A—ash content, Q—methane resource.

Level (m a.s.l.)		Coal Mass (t)	B	G (m^3/t)	Gr (m^3/t)	M (%)	A (%)	Q (m^3)
From	To							
−720	−970	329,415,243.91	0.86	5.70	0.69	1.41	12.80	1,275,154,180.25
−790	−1220	573,079,043.18	0.86	5.10	0.95	1.25	13.03	1,837,238,965.39
−1220	−1470	767,209,827.76	0.87	6.96	1.40	1.21	12.16	3,329,197,046.04
−1470	−1720	458,504,627.04	0.85	8.46	1.53	1.36	14.11	2,421,325,988.99
In total								8,862,916,180.67

4.4.1. Permeability

This parameter is responsible for the migration of fluids (gas) in the rock mass. In the case of coal seams, it is determined by the presence of a complex system of fractures (cleat system), among which one can distinguish fractures of the primary system (face cleats) and perpendicular subordinate system (butt cleats). In addition, master cleats occur, i.e., fractures that cut not only the coal seam but also the surrounding rocks [32,33]. In the study area, permeability was measured only in the Orzesze-1 borehole on 40 coal samples, and its variation in depth is shown in Figure 10. Figure 10a illustrates permeability results at a seal of 500 psi, which corresponds to the conditions of the rock mass relaxed by drilling the borehole, while the results illustrated in Figure 10b correspond to a 2900 psi seal adequate to the conditions of an intact rock mass. In both cases, the permeability of the coal seams is low, ranging from 0.004 to 47.6 mD (average 3.97 mD) for a 500 psi load and from 0.0001 to 0.34 mD (average 0.05 mD) for a 2900 psi seal. Values >2 mD appear only in the set of results for 500 psi loading in 19 samples and a value exceeding 10 mD in only 2 cases for the same data set. The highest permeability occurs in the depth interval from about −1430 to −1600 m a.s.l. (1740–1900 m below ground level) (Figure 10), but in general, the permeability of coal seams is too low (<0.1 mD in the bulk of samples for the dataset at 2900 psi loading), considering the gas flows in a profitable amount. This is most likely due to the great depth of the survey (>800 m below the ground surface), at which significant rock mass stress (>800 m column of rocks) causes the tightening of fractures and thus reduces the permeability of coal seams. The petrographic structure of the coal substance and the tectonic involvement of the study area are also not negligible [5]. Low coal permeability (up to 0.1 mD) is also a feature of other regions of the USCB, and higher values appear rarely [5].

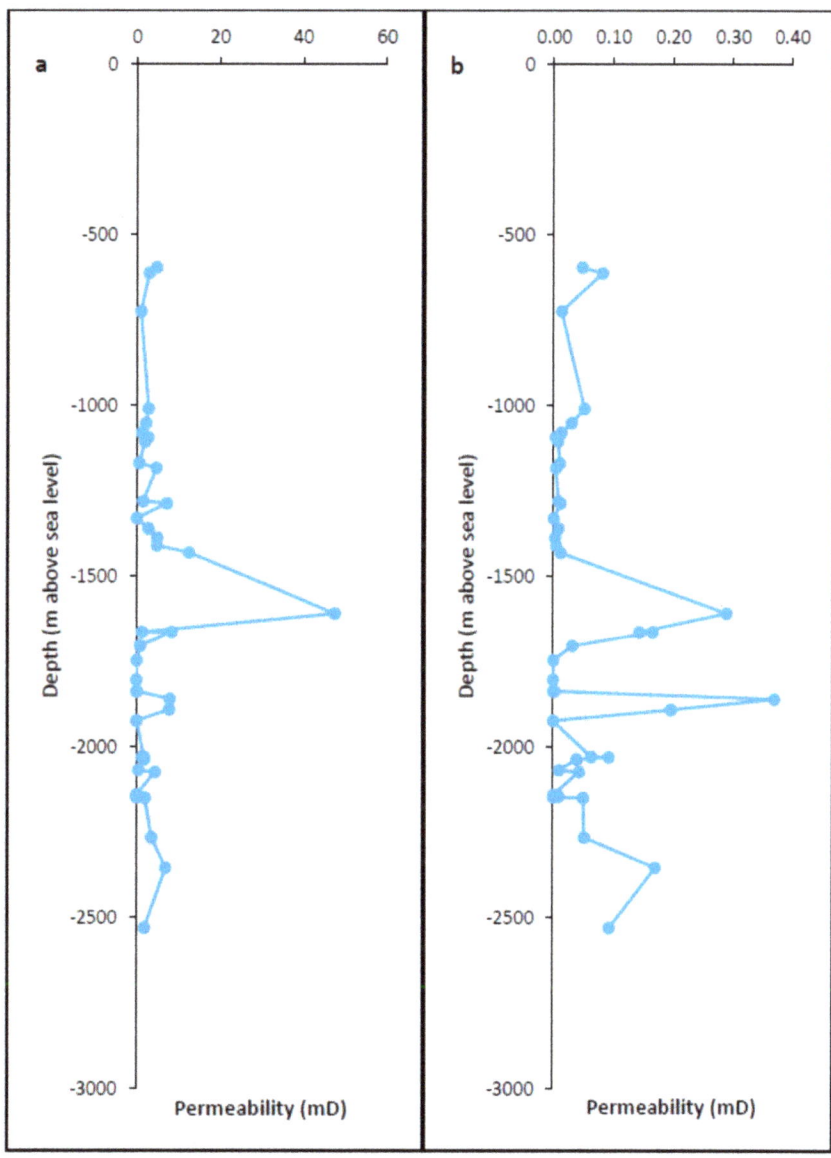

Figure 10. Variability of coal seams permeability in the Orzesze-1 well, (**a**)—for a 500 psi seal, (**b**)—for a 2900 psi seal.

4.4.2. Hydrogeological Conditions

The study area lies within the hydrogeologically open region of the USCB, where there is free media communication between the ground surface and the carboniferous complex. Quaternary and partly Triassic and Miocene formations occurring in the overburden are not a barrier to migrating meteoric waters. Aquifers occur in the quaternary and carboniferous. In the latter, sandstones, primarily of the Upper Silesian Sandstone Series, are water-bearing; however, in the Mudstone Series, the Orzesze strata also contain inserts of water-bearing sandstones several meters thick. The sandstones of the Upper Silesian Sandstone Series have the best filtration parameters, but at great depths (>1000 m), the filtration conditions worsen significantly, and the rock mass becomes practically impermeable. The filtration coefficient at this depth drops below 10^{-8} m/s. The water is also becoming increasingly saline (6 to more than 60 g/L), containing sulfates and chlorides.

From the perspective of methane exploitation, water pumping is necessary because of the need to lower the hydrostatic pressure to at least the critical desorption pressure, which enables the initiation of methane desorption from the coal seam and, thus, its exploitation. The amount of water pumped out varies and depends on the duration of operation and the watering of gas exploitation intervals. In the southern part of the USCB, an average of 1.1 to 2.7 thousand cubic meters of water was pumped out during the methane production tests, with 23–40 thousand cubic meters of gas captured during the 116–130 days of the test [5]. Due to the salinity of the pumped water, plans should be made for its effective disposal or injection back into the rock mass.

4.4.3. Sorption Capacity and Methane Saturation of the Seams

Figure 11 shows the sorption isotherm versus measured gas content for a sample from seam 420/1, at a depth of 2234 m in the Orzesze-1 well. The sorption capacity of coal at this depth is about 25 m^3 CH$_4$/t coal^{daf} with a measured gas content of 15 m^3/t coal^{daf}. It follows that the saturation of the sample with methane is only 60%. The situation is similar for other coal samples from this borehole, whose sorption capacity is in the range of 16–40 m^3 CH$_4$/t coal^{daf} and gas content is 8–18 m^3 CH$_4$/t coal^{daf}, which means saturation from 42 to 92%, on average 42–60% [7]. Similar results were obtained from the Bolesław Śmiały-1 well. This shows significant undersaturation of coal seams with gas under reservoir conditions. In contrast to other USCB regions, where methane saturation increases with depth and at >1000 m from the ground is >90% [34], here we are dealing with saturation much lower regardless of depth.

According to previous global studies [35,36], the saturation of coal seams with methane is linked to the geological evolution of coal basins, among other factors. When a coal seam reaches its final degree of coalification, the seam is mostly saturated with methane at 100%, and the excess gas is expelled outside. As a result of the subsequent uplift of the coal-bearing series and the lowering of the temperature of the rock mass, the seam increases its sorption capacity, and if there is no resumption of gas generation, the seam becomes undersaturated with methane. This phenomenon also occurred in the USCB [5], which was uplifted from the Permian to the Neogene, causing the temperature of the rock mass to decrease to the current values and, as a result, undersaturation of the seams with methane over a significant area.

Thermal events in the geological past described in Section 4.3, which can cause methane remobilization, probably contributed to the secondary degassing of coal seams and further increase of their undersaturation. The positive thermal anomaly in the study area may thus have played a role in the observed incomplete saturation of the seams with methane.

The phenomenon of undersaturation of seams with gas under high-pressure conditions at great depth (>1500 m) has an unfavorable effect on the extraction of methane from the coal seam through the production well due to the observed delay in gas flow after the critical desorption pressure is reached.

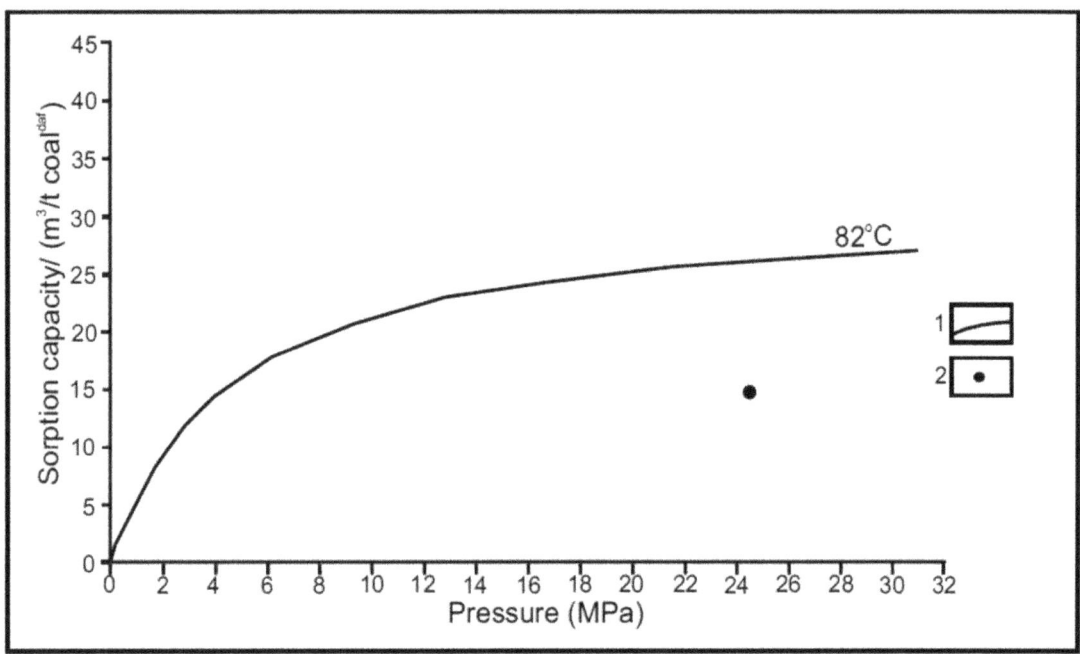

Figure 11. Gas sorption isotherm (1) and total gas content (2) of a coal sample from a depth of 2234 m (seam 420/1 in Orzesze-1 well).

4.4.4. Gas Production Prospects

Factors such as hydrostatic pressure, permeability, and saturation of seams with methane determine the success of methane well production. Hydrogeological conditions are also important. Methane extraction from deep coal seams in the depocenter area of the USCB can be particularly difficult due to the low permeability of coal seams. The results of studies and observations carried out during borehole methane production in the USA have shown that at high hydrostatic pressures at depth (>1000 m), a significant reduction is necessary to achieve the critical desorption pressure, so vast amounts of water must be pumped out [37]. The low permeability of coal seams can make this very difficult and, therefore, slow down the process of achieving the critical desorption pressure and thus delay the flow of gas into the well. The low saturation of the seams with gas can be an additional negative factor here.

However, the advantage of the deep interval for potential gas production is the high gas content in the seams, in most cases exceeding 4.5 m^3/t coal^{daf} at depths below -720 m a.s.l. (about 1000 m below ground level), and thick coal seams (>0.6 m). Among the thickest are seams of the Upper Silesian Sandstone Series (2.5 to 11.8 m thick), but they occur at depths > 2000 m in the study area. The considerable thickness of the coal seams and the high gas content have determined the estimated gas resources at 8.9 billion m^3. As already mentioned, this figure is only an indicative estimate and applies only to gas in place. The low permeability of the reservoir, which significantly determines the amount of gas production, leads us to conclude that the extraction rate will not be high. According to international experience, at a permeability of 1 mD, the utilization of gas-in-place resources is up to 25%, at 5 mD up to 47%, and at 25 mD up to 75%, respectively [20,38,39]. Taking the above into account and considering the low permeability of the reservoirs, only periodically increasing above 1 mD at a seal of 500 psi, the resource utilization rate in the study area will be low, below 50%.

Nevertheless, it should be remembered that the technology of exploitation, which has greatly modernized over the past two decades, is of great importance in the amount of gas extracted. A major innovation was the introduction of horizontal boreholes in the USA at the turn of the 21st century, which, covering a much larger area of coal seam drainage, contributed to a significant increase in gas extraction and the coverage of production in areas previously considered unpromising [40]. Artificial fracturing of coal seams also proved to be very helpful, leading to an increase in permeability and, therefore, improving the gas yield from the wells. Horizontal borehole technology combined with artificial fracturing was introduced in the USCB after 2010, following the failed tests of gas production with vertical boreholes providing access to multiple seams simultaneously, conducted in the basin in the 1990s [41]. The USCB's breakthrough was the use of a doublet of combined vertical and horizontal boreholes, with the vertical one acting as a production well. This ensured efficient dewatering of the deposit [41]. Tests conducted with this method in the southeastern part of the USCB (Międzyrzecze region) gave very good results, as daily gas production of 10,000 m^3 was achieved, and after pressure stabilization, about 5000 m^3. The trials were carried out at a depth of 1000 m in the Upper Silesian Sandstone Series by the Polish Oil and Gas Company and the Polish Geological Institute under the Geometan project. This shows that the use of new technologies can significantly increase the amount of gas extracted and significantly improve the recovery rate from a reservoir, even in problematic areas for gas production. Global experience shows that increased gas production from a well can also be achieved by combining hydraulic splitting technology with the use of directional drilling [42] or by applying coiled tubing fracturing technology implemented in several horizontal well clusters [43].

The combination of methods applied in the USCB and in other countries prompts us to consider the possibility of conducting CBM production tests in the depocenter area of the basin, which, if the results are positive, may contribute to expanding the gas resource base in Poland, given that the composition of the gas extracted from the virgin areas of the basin allows for a wide range of uses.

5. Conclusions

1. In the study area, the gas content of the coal seams is arranged in a zonal manner. Up to a depth of about 500–600 m from the ground, there is a naturally degassed zone, below which a vertically extensive high-methane zone is present with two sub-zones visible, an upper zone with gas content up to 17 m^3/t coal^{daf} and a lower zone with gas content up to 18.6 m^3/t coal^{daf} at a depth of 2840 m. The two zones are separated by an interval of reduced gas content.

2. Horizontally, the distribution of gas content is observed in the form of a dome with a maximum in the central part of the study area, from which gas volume decreases in all directions.

3. The dome of gas content coincides with the positive temperature anomaly and the coal rank of the seams. The transport of heat by faults from a deep source causing secondary migration of methane and its accumulation at some distance from the hypothetical heat source is not excluded.

4. Significant estimated methane resources of 8.9 billion m^3 are accompanied by not very favorable parameters related to reservoir productivity, i.e., low permeability of seams (in a significant part of the profile not exceeding 2 mD) and low saturation of seams with methane (on average 40–60%). The poor permeability is due to the considerable depth of the seams in the depocenter area of the basin. These parameters mean that the gas extraction rate from the reservoir may be low.

5. Leveraging new CBM production technologies used globally and in the USCB, for example, the use of hydraulic fracturing technology combined with horizontal boreholes in the form of well doublets can help improve reservoir productivity.

6. Given the significant estimated gas-in-place resources and the track record of deep gas production worldwide, it is worthwhile to conduct gas production tests in the study area, as the gas resource base in Poland could be expanded if successful.

Author Contributions: Conceptualization: S.K.; Methodology: S.K.; Investigation and sampling: S.K.; Formal analysis: S.K.; Writing—original draft: S.K.; Writing—review and editing: L.T.; Funding acquisition: L.T.; Supervision: L.T.; Data curation and Visualization: S.K. All authors have read and agreed to the published version of the manuscript.

Funding: This study was funded by the University of Silesia, Institute of Earth Sciences research program no WNP/INOZ/2020_ZB32.

Data Availability Statement: The data presented in this study are available on request from the corresponding author. The data are not publicly available due to legal reasons.

Acknowledgments: This study was undertaken in the framework of the University of Silesia in Katowice project: Pre-mining, mining, and post-mining areas–space of threats and opportunities (WNP/INOZ/2020_ZB32). Special thanks are due to the staff of the National Geology Archive of the Polish Geological Institute for their assistance and valuable advice. The reviewers are also thanked for their suggested corrections and appreciated comments.

Conflicts of Interest: The authors declare no conflict of interest.

References

1. Flores, R.M.; Moore, T.A. *Coal and Coalbed Gas: Future Directions and Opportunities*, 2nd ed.; Elsevier Inc.: Amsterdam, The Netherlands, 2024; 748p.
2. Mastalerz, M. Coal Bed Methane: Reserves, Production and Future Outlook (Chapter 7). In *Future Energy (Second Edition). Improved, Sustainable and Clean Options for Our Planet*; Elsevier: Amsterdam, The Netherlands, 2014; pp. 145–148.
3. *Balance of Resources of Mineral Deposits in Poland as of December 31, 2022*; Polish Geological Institute National Research Institute: Warszawa, Poland, 2023.
4. Kotas, A. (Ed.) *Coal-Bed Methane Potential of the Upper Silesian Coal Basin*; Prace PIG CXLII: Warszawa, Poland, 1994; ISSN 0866-9465.
5. Kędzior, S. *A Near-Roof Gas-Bearing Zone in Carboniferous Rocks of the Southern Part of the Upper Silesian Coal Basin—Occurrence, Coal Reservoir Parameters and Prospects for Methane Extraction*; Wydawnictwo Uniwersytetu Śląskiego: Katowice, Poland, 2012; ISBN 978-83-226-2093-9, ISSN 0208-6336.
6. Kwarciński, J.; Hadro, J. Coalbed methane in the Upper Silesian Coal Basin. *Przegląd Geol.* **2008**, *56*, 485–490.
7. Kędzior, S. *The Occurrence of Methane in the Deep Parts of the Carboniferous Formations in the Upper Silesian Coal Basin, Poland. Case Study of the Orzesze1 Deep Exploratory Well. Bulletin of the Geological Society Special Publication, No. 12, 76*; Geological Society of Greece: Athens, Greece, 2023; ISBN 978-618-86841-0-2, ISSN 2945-1426.
8. Tang, S.I.; Tang, D.Z.; Li, S.; Xu, H.; Tao, S.; Geng, Y.G.; Ma, L.; Zhu, X.G. Fracture system identification of coal reservoir and the productivity differences of CBM wells with different coal structures: A case in the Yanchuannan Block, Ordos Basin. *J. Pet. Sci. Eng.* **2018**, *161*, 175–189. [CrossRef]
9. Song, L.; Qin, Y.; Tang, D.; Shen, J.; Wang, J.; Chen, S. A comprehensive review of deep coalbed methane and recent developments in China. *Int. J. Coal Geol.* **2023**, *279*, 104369.
10. Tarnowski, J. *Geological Conditions of Methane Occurrence in the Upper Silesian Coal Basin (USCB)*. Zeszyty Naukowe Politechniki Śląskiej, z. 166; Dział Wydawnictw Politechniki Śląskiej: Gliwice, Poland, 1989.
11. Kotas, A. Upper Silesian Coal Basin. Geological Institute. In *Geology of Poland, Mineral Deposits*; Osika, R., Ed.; Publishing House Wydawnictwa Geologiczne: Warszawa, Poland, 1990; Volume VI, pp. 77–92. ISBN 83-220-0385-4.
12. Idziak, A.; Teper, L. Fractal Dimension of Faults Network in the Upper Silesian Coal Basin (Poland): Preliminary Studies. *Pure Appl. Geophys.* **1996**, *147*, 239–247. [CrossRef]
13. Teper, L.; Sagan, G. Geological History and Mining Seismicity in Upper Silesia (Poland). In *Mechanics of Joined and Faulted Rock II*; Rossmanith, H.P., Ed.; Balkema, Rotterdam-Brookfield: Rotterdam, The Netherlands, 1995; pp. 939–943.
14. Lewandowski, J. Neotectonic structures of the Upper Silesian region, southern Poland. *Stud. Quat.* **2007**, *24*, 21–28.
15. Jüntgen, H.; Karweil, J. Gasbildung und Gasspeicherung in Steinkohlenflozen. Part I and II. *Erdöl Kohle-Erdgas-Petrochem.* **1966**, *19*, 251–258, 339–344.
16. Hunt, J.M. *Petroleum Geochemistry and Geology*, 2nd ed.; W.H. Freeman and Co.: New York, NY, USA, 1995.
17. Kędzior, S. Distribution of methane contents and coal rank in the profiles of deep boreholes in the Upper Silesian Coal Basin, Poland. *Int. J. Coal Geol.* **2019**, *202*, 190–208. [CrossRef]
18. Słoczyński, T.; Drozd, A. Methane potential of the Upper Silesian Coal Basin carboniferous strata—4D petroleum system modeling results. *Nafta-Gaz* **2018**, *10*, 703–714.

19. Kędzior, S.; Teper, L. Coal Properties and Coalbed Methane Potential in the Southern Part of the Upper Silesian Coal Basin, Poland. *Energies* **2023**, *16*, 3219. [CrossRef]
20. Scott, A.R. Hydrogeologic factors affecting gas content distribution in coal beds. *Int. J. Coal Geol.* **2002**, *50*, 363–387. [CrossRef]
21. Diamond, W.P.; Inani, M.C.; Aul, G.N.; Thimons, E.D. Instruments, Techniques, Equipment. *USBM Bull.* **1980**, *687*, 79–83.
22. Karwasiecka, M. *Geothermal Atlas of the Upper Silesian Coal Basin*; Polish Geological Institute: Warszawa, Poland, 1996.
23. Karwasiecka, M. The geothermal field of the Upper Silesian Coal Basin. *Tech. Poszuk. Geologicznych. Geosynoptyka Geoterm.* **2001**, *5*, 41–49.
24. Jurczak-Drabek, A. *Petrographical Atlas of Coal Deposits Upper Silesian Coal Basin*; Polish Geological Institute: Warszawa, Poland, 1996.
25. Kotas, A.; Buła, Z.; Gądek, S.; Kwarciński, J.; Malicki, R. *Geological Atlas of the Upper Silesian Coal Basin, Part II Coal Quality Maps*; Wydawnictwa Geologiczne: Warszawa, Poland, 1983.
26. Leśniak, L.; Leśniak, A. Modelling of temperature distribution in the fault zone in the Czechowice region. *Sci. Bull. Stanisław Staszic Acad. Min. Metallurgy. Geol.* **1994**, *20*, 221–235.
27. Karwasiecka, M. The comparison of geothermal environment in the Upper Silesian Coal Basin and Lublin Coal Basin. *Geologia* **2008**, *34*, 335–357.
28. Sas-Gustkiewicz, M.; Dżułyński, S. On the origin of strata-bound Zn-Pb ores in the Upper Silesia, Poland. *Ann. Soc. Geol. Pol.* **1998**, *68*, 267–278.
29. Botor, D. Timing of coalification of the Upper Carboniferous sediments in the Upper Silesia Coal Basin on the basis of apatite fission track and helium dating. *Miner. Resour. Manag.* **2014**, *30*, 85–103. [CrossRef]
30. Marcak, H.; Leśniak, A. Interpretation of the thermal field of Upper Silesia. *Tech. Poszuk. Geologicznych. Geosynoptyka Geoterm.* **1989**, *5*, 47–56.
31. Kotarba, M.J. Composition and origin of gases in the Upper Silesian and Lublin Coal Basins, Poland. *Org. Geochem.* **2001**, *32*, 163–180. [CrossRef]
32. Laubach, S.E.; Marrett, R.A.; Olson, J.E.; Scott, A.R. Characteristics and origins of coal cleat. A review. *Int. J. Coal Geol.* **1998**, *35*, 175–207. [CrossRef]
33. Dawson, G.K.W.; Esterle, J.S. Controls on coal cleat spacing. *Int. J. Coal Geol.* **2010**, *82*, 213–218. [CrossRef]
34. Kędzior, S.; Dreger, M. Geological and Mining Factors Controlling the Current Methane Conditions in the Rydułtowy Coal Mine (Upper Silesian Coal Basin, Poland). *Energies* **2022**, *15*, 6364. [CrossRef]
35. Hildenbrand, A.; Krooss, B.M.; Busch, A.; Gaschnitz, R. Evolution of methane sorption capacity of coal seams as a function of burial history—A case study from Campine Basin, NE Belgium. *Int. J. Coal Geol.* **2006**, *66*, 179–203. [CrossRef]
36. Weniger, P.; Franců, J.; Hemza, P.; Krooss, B.M. Investigations on the methane and carbon dioxide sorption capacity of coals from the SW Upper Silesian Coal Basin, Czech Republic. *Int. J. Coal Geol.* **2012**, *93*, 23–39. [CrossRef]
37. Pashin, J.C. Variable gas saturation in coalbed methane reservoirs of the Black Warrior Basin: Implications for exploration and production. *Int. J. Coal Geol.* **2010**, *82*, 135–146. [CrossRef]
38. Nieć, M. Methane deposits in coal-bearing formations. In *Proceedings of Underground Exploitation School*; Polish Academy of Sciences: Kraków, Poland, 1993; pp. 281–301.
39. Moore, T. Coalbed methane: A review. *Int. J. Coal Geol.* **2012**, *101*, 36–81. [CrossRef]
40. Tao, S.; Tang, D.; Xu, H.; Gao, L.; Fang, Y. Factors controlling high-yield coalbed methane vertical wells in the Fanzhuang Block, Southern Qinshui Basin. *Int. J. Coal Geol.* **2014**, *134–135*, 38–45. [CrossRef]
41. Hadro, J.; Wójcik, I. Coalbed methane: Resources and recovery. *Przegląd Geol.* **2013**, *61*, 404–410.
42. Zhang, J.; Niu, Y.; Chen, J.; Guo, Y.; Guo, L. Research on Deep Coalbed Methane Localized Spotting and Efficient Permeability Enhancement Technology. *Appl. Sci.* **2022**, *12*, 11843. [CrossRef]
43. Huang, J.; Liu, S.; Tang, S.; Shi, S.; Wang, C. Study on the Coalbed Methane Development under High In Situ Stress, Large Buried Depth, and Low Permeability Reservoir in the Libi Block, Qinshui Basin, China. *Adv. Civ. Eng.* **2020**, *2020*, 6663496. [CrossRef]

Disclaimer/Publisher's Note: The statements, opinions and data contained in all publications are solely those of the individual author(s) and contributor(s) and not of MDPI and/or the editor(s). MDPI and/or the editor(s) disclaim responsibility for any injury to people or property resulting from any ideas, methods, instructions or products referred to in the content.

Article

Experimental Study of Forced Imbibition in Tight Reservoirs Based on Nuclear Magnetic Resonance under High-Pressure Conditions

Xiaoshan Li [1], Liu Yang [2,*], Dezhi Sun [3], Bingjian Ling [3] and Suling Wang [4]

1. Research Institute of Exploration and Development, Xinjiang Oilfield Company, PetroChina, Karamay 834000, China
2. State Key Laboratory for Tunnel Engineering, China University of Mining and Technology (Beijing), Beijing 100083, China
3. China Railway Tianjin Metro, Tianjin 300450, China
4. School of Mechanical Science and Engineering, Northeast Petroleum University, Daqing 163318, China
* Correspondence: shidayangliu@126.com

Abstract: This study utilizes nuclear magnetic resonance (NMR) techniques to monitor complex microstructures and fluid transport, systematically examining fluid distribution and migration during pressure imbibition. The results indicate that increased applied pressure primarily affects micropores and small pores during the initial imbibition stage, enhancing the overall imbibition rate and oil recovery. Higher capillary pressure in the pores strengthens the imbibition ability, with water initially displacing oil from smaller pores. Natural microfractures allow water to preferentially enter and displace oil, thereby reducing oil recovery from these pores. Additionally, clay minerals may induce fracture expansion, facilitating oil flow into the expanding space. This study provides new insights into fluid distribution and migration during pressure imbibition, offering implications for improved oil production in tight reservoirs.

Keywords: applied pressure; imbibition; nuclear magnetic resonance; tight oil reservoir

Citation: Li, X.; Yang, L.; Sun, D.; Ling, B.; Wang, S. Experimental Study of Forced Imbibition in Tight Reservoirs Based on Nuclear Magnetic Resonance under High-Pressure Conditions. *Energies* **2024**, *17*, 2993. https://doi.org/10.3390/en17122993

Academic Editors: Shu Tao, Wei Ju, Shida Chen, Zhengguang Zhang and Jiang Han

Received: 26 April 2024
Revised: 6 June 2024
Accepted: 11 June 2024
Published: 18 June 2024

Copyright: © 2024 by the authors. Licensee MDPI, Basel, Switzerland. This article is an open access article distributed under the terms and conditions of the Creative Commons Attribution (CC BY) license (https://creativecommons.org/licenses/by/4.0/).

1. Introduction

In the past decade, significant breakthroughs have been made in the development of unconventional oil and gas, mainly due to the development of hydraulic fracturing and horizontal drilling technologies [1,2]. Spontaneous imbibition is a process in which the wetting phase displaces the non-wetting phase in porous media under the action of capillary force. According to the movement direction of wetting phase and non-wetting phase, it can be divided into cocurrent and countercurrent spontaneous imbibition [3]. Spontaneous imbibition refers to the process in which water is automatically drawn into the core and drives out crude oil without pressure, while forced imbibition refers to the process in which water is forced into the core and drives out oil under pressure. The imbibition of conventional reservoirs has been studied extensively in recent years. A large number of studies have shown that imbibition in tight oil reservoirs is one of the important driving forces for crude oil recovery [4–6]. The spontaneous imbibition induced by capillary force is especially remarkable because the pore size is in micron or even nanometer range [7]. Production data from the field also show that shutting down the well for a while after fracturing allows the fracturing fluid to be fully absorbed in the reservoir, exerting the effect of imbibition suction to drive oil and improve recovery [8–10]. Studying multiphase flow is crucial for optimizing processes and ensuring safety across diverse scientific and engineering fields, including oil and gas extraction, chemical manufacturing, and environmental remediation. This understanding enables the efficient handling of complex interactions between different phases, leading to advancements in technology and improved operational outcomes.

In recent years, a large number of scholars have carried out experimental studies on imbibition. The traditional imbibition experiments are mainly mass and volume methods [11]. However, the experimental results of these two methods are easily influenced by artificial factors and cannot reflect the existence state of fluid in the core. The nuclear magnetic resonance (NMR) technique is a high-precision and high-efficiency detection tool. It has diverse applications across multiple disciplines, including medicine, chemistry, physics, and oil and gas development [12–14]. Li et al. [15] studied the effect of initial water saturation on spontaneous imbibition in tight gas reservoirs using NMR, which found that 90% of the residual gas is in pores with pore sizes greater than 1 μm. Yang et al. [16] combined physical simulation experiments combined with NMR, which found that the more hydrophilic the core, the higher the imbibition rate and recovery. Yang et al. [17] conducted spontaneous imbibition experiments of tight reservoirs and found through T_2 spectroscopy that pores correspond to larger capillary pressures, which have a more effective imbibition and oil displacement effect. Farahani [18] studied the kinetics and spatial characteristics of thermally induced methane hydrate formation in synthetic and natural sediment samples using magnetic resonance imaging. Xiaomin [19] pointed out that NMR techniques provide qualitative and quantitative observations of macro-fractures, microfractures and micro- and nanoscale fractures in reservoirs. Hassanpouryouzband [20] reviewed the different properties of natural gas hydrates and their formation and dissociation kinetics. Cui, Xiaojun [21] extended several methods for measuring permeability and diffusivity with consideration of gas adsorption.

Due to the presence of confining pressure in the reservoir, the driving force of imbibition is not only capillary force, but also the pressure applied to the fluid. Several scholars have conducted studies on pressurized imbibition. Xu et al. [22] conducted imbibition and backflow experiments on sandstone samples, which found that the water that seeped into the core was mainly retained in the nano-micropores and nano-mesopores. Compared with spontaneous imbibition, forced imbibition significantly increased the recovery of cores and increased the retention of fracturing fluid. Wang et al. [23] found that the recovery and imbibition rate of core samples were positively correlated with pressure by imbibition experiments under different applied pressures. Jiang et al. [24] used the tight sandstone in Ordos Basin as a research object, which found that the final recovery rate of core samples increased when the surrounding pressure increased.

Currently, a large number of experimental studies on imbibition have been conducted mainly under atmospheric pressure. Although research on forced imbibition has achieved certain results, there is still a lack of understanding of factors such as pore structure and mineral composition of rock cores, as well as the impact of external pressure on imbibition. In this paper, the cores of four formations were selected to test their porosity, permeability, mineral composition, and other information. The device for the forced imbibition experiment was also designed, and imbibition experiments under different pressures were conducted. Our experiments were conducted to study the distribution and transport pattern of fluid during the imbibition process, which use the NMR detection.

2. Experimental Materials and Methods
2.1. Rock Samples and Fluids

The samples of tight oil reservoirs are taken from Ordos Basin, Songliao Basin and Junggar Basin, which are rich in tight oil and gas reservoir resources. Since the samples are taken from different areas and at different stratigraphic depths, the influence of different stratigraphic fractures as well as clay minerals can be observed. Cylinders of 2.5 cm in diameter are drilled from larger rock outcrops and machined and cut by wireline cutters into cylinders of about 50 mm in length, with the ends machined flat. The cores were dried at 105 °C after removal of residual oil until the quality no longer changed. We used a Vernier scale and balance to accurately measure the length, diameter, and quality of core samples. During the experiment, the core was subjected to a certain pressure environment for imbibition by external pressurization. The basic information is listed in Table 1, including the source, size, and saturated oil mass.

Table 1. The physical properties of core samples.

Label	Formation	Pressure, MPa	Diameter, cm	Length, cm	Mass of Oil-Saturated Sample, g
LC7-1	Lower Chang-7	0	2.50	4.11	47.3395
LC7-2		4	2.50	4.19	47.7744
LC7-3		8	2.51	5.40	65.7224
LC7-4		12	2.50	5.54	65.4972
UC7-1	Upper Chang-7	10	2.49	4.08	48.4081
UC7-2		10	2.50	1.95	23.3551
UC7-3		10	2.50	6.44	81.14
UC7-4		10	2.49	3.10	36.2698
UC7-5		10	2.50	3.02	35.1438
QT-1	Quantou formation	10	2.50	3.5	44.9562
QT-2		10	2.52	3.60	45.3792
WEH-1	Wuerhe formation	10	2.51	5.02	57.4809
WEH-2		10	2.50	4.313	51.1658

XRD analysis was carried out on rock fragments using a Rigaku D/Max-2500/PC-type X-ray diffractometer (Co Kα radiation, 40 kV, 30 mA), which is made by the Rigaku Manufacture in the Woodlands, TX, USA, to study the effect of mineral composition on seepage results. The results of whole-rock mineral composition and clay mineral composition are shown in Figure 1. Sample WE is mainly composed of quartz and clay minerals, with 40% clay minerals. Sample QT is mainly composed of quartz, feldspar, and clay minerals at 26%, 34%, and 24%, respectively. Samples LC7 and UC7, on the other hand, are mainly composed of quartz and feldspar, both of which can reach 70%. However, their clay mineral content is less than 20%. According to Yang et al. [16], a moderate content of clay minerals (20.5–30.2%) can induce microfracture extension to mitigate reservoir damage, while a high content of clay minerals (36.8%) may disrupt matrix pore space and thus aggravate reservoir damage. The higher clay mineral content of WEH and QT can be used to analyze the effect of clay minerals on pore structure during percolation. The lower clay mineral content of samples LC7 and UC7 indicates that we can disregard the effect of their water absorption and swelling during percolation. Therefore, these samples can be used to analyze the pore size distribution and the effect of natural fractures on the percolation absorption.

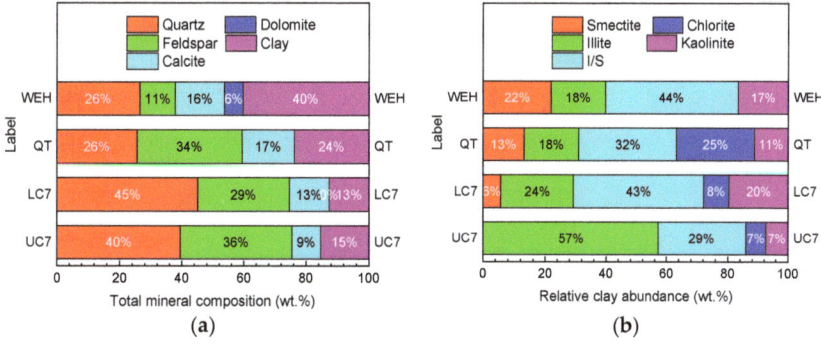

Figure 1. The results of mineral composition analysis: (a) percentage of whole-rock mineral composition; (b) percentage of clay mineral composition; I/S represents the illite/smectite mixed layer.

The experimental fluids consist of high-purity (≥99%) deuterium water and kerosene. The basic properties of density, viscosity and surface tension are shown in Table 2. In this case, the density and viscosity of paraffin and deuterium water were measured under the ASTM standard, and the surface tension of the liquids was measured using a surface tension meter under the standard [25]. Since the NMR equipment could not distinguish the signals of kerosene and water, deuterium water with similar properties to water was selected as the imbibition solution to reduce the impact on the reservoir. The NMR signals are all from kerosene in the core. In the percolation experiments, deuterium water is used as a wetting fluid to displace the non-wetting fluid kerosene.

Table 2. The basic properties of the experimental fluid (25 °C).

Fluid	Density, g/cm^3	Viscosity, cp	Surface Tension, mN/m
Deuterium water	1.107	0.91	72.2
Kerosene	0.81	1.32	29

2.2. Experimental Apparatus

The device for testing the core mass in the experiment is a Mettler balance (model ME204E) with an accuracy of 0.0001 g and a range of 220 g, as shown in Figure 2a. The change in sample mass and the difference in oil–water density can be used to calculate the volume of imbibed water and discharged oil during imbibition.

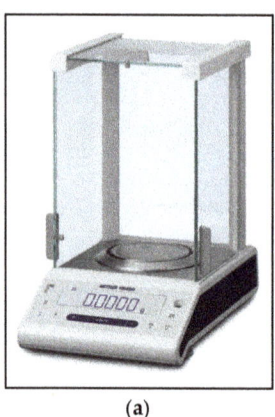
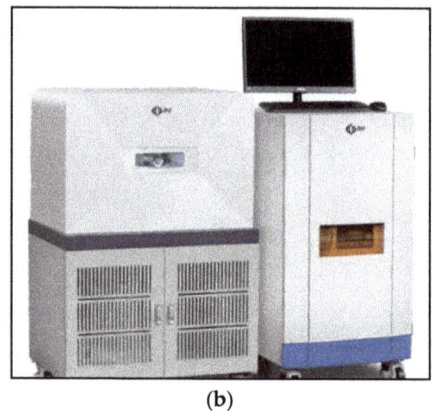

(a)   (b)

Figure 2. The apparatus for the experiment: (**a**) analytical balance, (**b**) nuclear magnetic resonance analyzer.

The NMR instrument is provided by Suzhou Niumai Analytical Instruments Corporation (Suzhou, China), with the model number MiniMR-VTP and a magnetic field strength of 0.5 T, as shown in Figure 2b. The test temperature is 25 °C, the humidity is 40%, and the pressure is atmospheric pressure. NMR is a nondestructive testing method to analyze the physical characteristics of rocks by measuring the content of hydrogen elements within the rocks. The NMR T_2 spectrum can reflect the pore structure and fluid distribution characteristics well. The higher the relaxation time T_2, the larger the pore size of the fugitive fluid. The more fluid is presented in a certain pore size of the rock, the larger the T_2 spectrum amplitude [26–29].

A forced imbibition experimental device consists of a pressurized pump, intermediate container, pressure gauge, iron pipeline, valve, etc. The upper part of the intermediate container piston is deuterium water, and the lower part of the water comes from the pressurized pump. The pressure applied in the vessel is read by a pressure gauge.

2.3. Experimental Procedures

In this study, the mass of core samples was measured under various pressurization conditions and correlated with NMR T_2 spectra over time. The experimental procedure comprised several key steps:

(1) Preparation and Cleaning: Initially, the tight reservoir samples were meticulously washed to remove residual oil. To effectively clean the cores and minimize the solvent's impact on rock wettability, a 1:4 alcohol/benzene mixture was used. After washing, the sample mass was recorded.

(2) Oil Saturation: The cleaned samples were then placed in a saturated oil device and evacuated for 2 to 3 h to eliminate any air. Subsequently, kerosene was injected at a pressure of 20 MPa, and the samples were left to saturate for 72 h.

(3) Mass and T_2 Measurement: Once the saturation period was complete, the mass and size of the oil-saturated samples were measured. These cores were then placed into deuterium water within the forced imbibition experimental apparatus. Pressure was applied using a pressurization pump. After a specified duration, the pressure was released, the core was removed, and its mass was measured. The T_2 spectrum was then analyzed using an NMR instrument, which is provided by Suzhou Niumai Analytical Instruments Corporation (Suzhou, China).

(4) Repetition and Analysis: This process (step 3) was repeated to observe changes in the T_2 spectrum over immersion time, enabling the creation of a detailed plot depicting the variation in the T_2 spectrum with time.

A schematic diagram illustrating the experimental procedure is provided in Figure 3. This methodical approach enabled a comprehensive understanding of fluid distribution and migration under applied pressure, shedding light on the imbibition mechanisms in tight reservoirs.

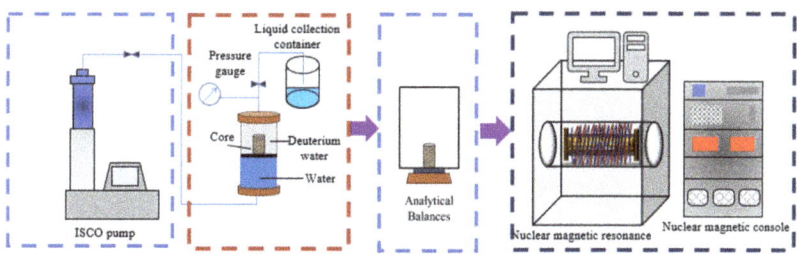

(a) Pressure loading system (b) Imbibition generation system (c) Quality monitoring (d) NMR measurement system

Figure 3. The schematic diagram of the experimental process. (**a**) Pressure loading system. (**b**) Imbibition generation system. (**c**) Qua-monitoring. (**d**) NMR measurement system.

3. Experimental Results and Discussion

3.1. Effect of Applied Pressure on Oil Transport during Imbibition

According to Meng et al. [22], tight rock pores can be divided into micropores (<1 ms), small pores (1–10 ms), large pores (10–100 ms), and the larger pores (>100 ms). The larger pores and macropores mainly include matrix pores and microcracks. Figure 4 shows the T_2 spectrum curves of samples LC7-1 to LC7-4 under different applied pressures. From the T_2 spectrum, it can be seen that the sample is mainly composed of micropores and small pores. The T_2 spectra in the first three groups all have bimodal characteristics, but the right bimodal characteristics are not obvious. The left peak T_2 value is within the range of 0.02~9 ms, and the right peak T_2 value is within the range of 9~714 ms. The relaxation time is between 0.1 and 10 ms, indicating that water absorption is mainly concentrated in nanopores and mesopores. The T_2 spectrum formed under an external pressure of 12 MPa is characterized by a single peak. Its range is between 0.02 and 821 ms. By comparing the results of spontaneous imbibition, the amplitude shows a decreasing trend as the fluid enters. Obviously, as the external pressure increases, the amplitude also increases, and the curve in the later stage of imbibition tends to be more consistent, which is similar to the T_2 spectrum characteristics of spontaneous imbibition. At 4 MPa, 8 MPa, and 12 MPa, the decrease amplitude is 18.6%, 28.3%, and 32.3%, respectively. And the reduction in micropores and small pores is significantly greater than that of large pores. This phenomenon may be attributed to the effect of external pressure, which reduces the contact angle between the core and the liquid. This reduction in contact angle leads to an increase in capillary pressure within the smaller pores, thereby enhancing their imbibition effect.

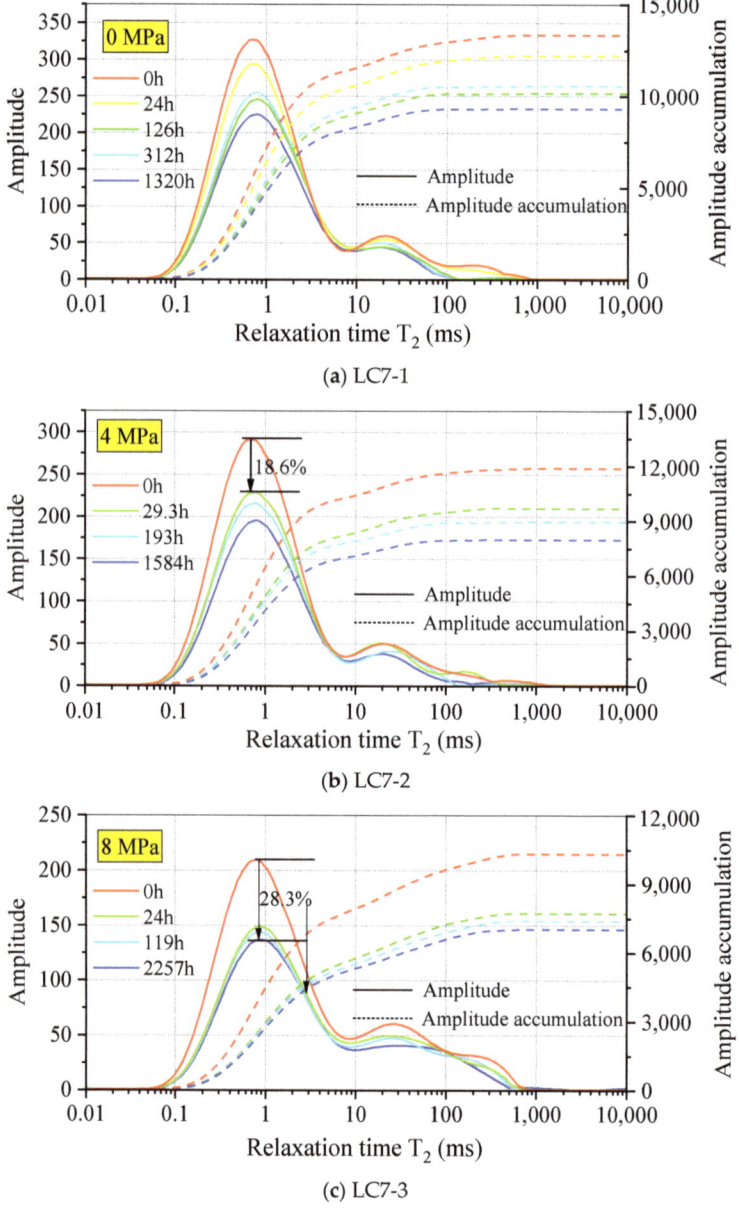

Figure 4. *Cont.*

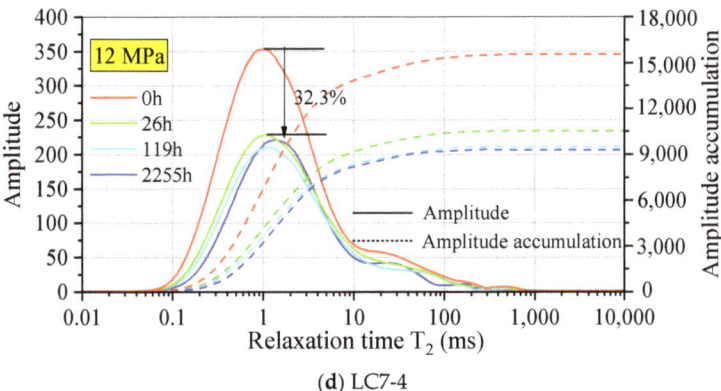

(d) LC7-4

Figure 4. The T_2 spectrum and recovery rates of imbibition of sample LC7-1, LC7-2, LC7-3, and LC7-4 at different applied pressures: (**a**) 0 MPa, (**b**) 4 MPa, (**c**) 8 MPa, (**d**) 12 MPa.

Figure 5 shows the imbibition recovery at different applied pressures. As the applied pressure increased, the overall imbibition recovery increased. The imbibition recovery at different pressures corresponds to 8.5%, 18.6%, 28.3%, and 32.3%, respectively. With increasing time, their corresponding recovery rates are 28.7%, 32.8%, 31.3%, and 39.8%. It is shown that with the increase in applied pressure, there is a positive correlation effect on the recovery rate, while for the late stage of imbibition, the effect is rather insignificant. In the initial stage of imbibition, the oil is mainly replaced by the water in the micropores and small pores for discharge, while the applied pressure mainly changes the imbibition rate by affecting the capillary pressure, which has the greatest effect on the micropores and small pores. Therefore, the overall imbibition rate can be improved. In the later stage of imbibition, the oil is mainly replaced through the large empty pores. The applied pressure has less influence on the large pores, and thus the change in imbibition recovery in the later stage is not obvious. It can be observed that the recovery rate at a pressure of 8 MPa is slightly lower than at 4 MPa. This variation can be attributed to the heterogeneity of the rock, as it is not possible to ensure uniform lithology across all samples. Despite this, the overall recovery rate exhibited an upward trend, indicating that increased pressure generally has a positive effect on recovery.

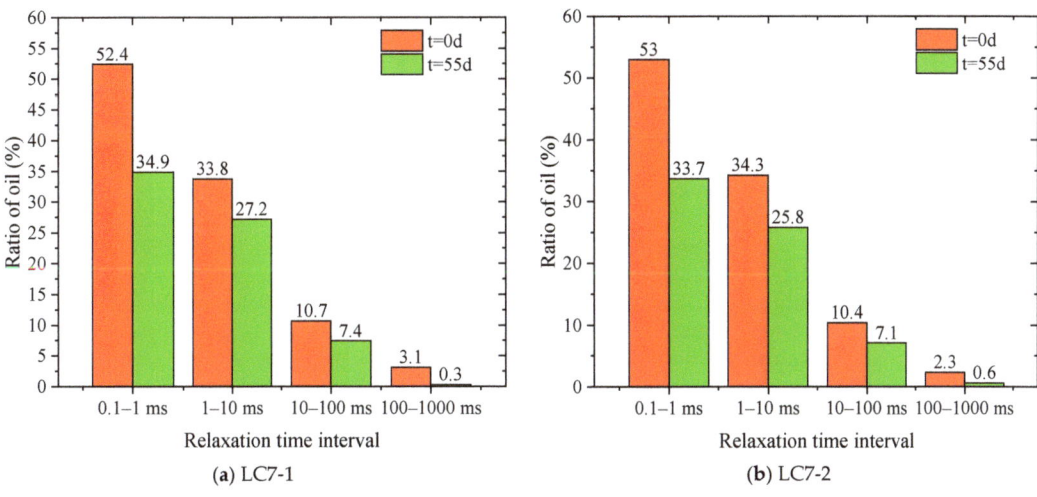

(**a**) LC7-1   (**b**) LC7-2

Figure 5. *Cont.*

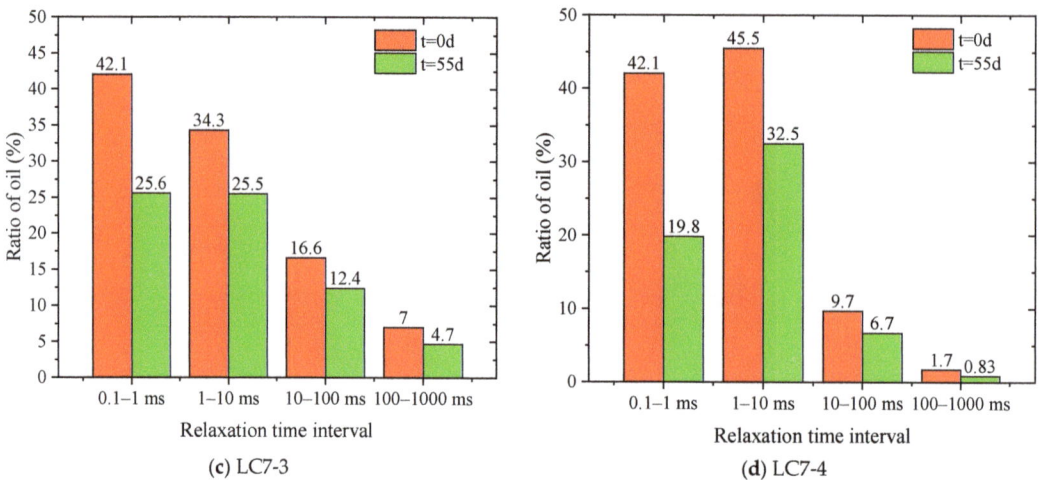

Figure 5. The imbibition recovery of different samples at different pressures on the first day and onwards.

3.2. Effect of Pore Size Distribution on Imbibition

In this section, we discuss the influence of different pore sizes and their distributions on the pressure imbibition effect. Figure 6 shows the T_2 spectral curves of UC7-1, UC7-2, and UC7-3 under an external pressure of 10 MPa. Both have the same T_2 spectral characteristics, and the amplitudes corresponding to different apertures show a significant downward trend. The T_2 spectrum of sample UC7-1 shows a three-peak distribution, with the left peak ranging from 0.07 to 50 ms, the middle peak ranging from 50 to 100 ms, and the right peak ranging from 100 to 700 ms. The T_2 spectrum of UC7-2 shows a multimodal distribution, with the left peak ranging from 0.07 to 50 ms and the right peak ranging from 50 to 400 ms. The T_2 spectrum of sample UC7-3 shows a single peak distribution, with a range of 0.07~10 ms. The amplitude accumulation curve also shows that the three samples have different pore size distribution characteristics. The pore distribution of UC7-1 is relatively uniform. In UC7-2 and UC7-3 samples, the oil in the pores of the former is mainly present in mesopores and larger pores, while the latter is mainly present in micropores and small pores. When the imbibition time exceeds 410 h, the cumulative curves of different T_2 spectra basically overlap and do not change, and the curves remain within a certain range. It indicates the presence of residual oil in the pores, and the use of external pressure and capillary force alone cannot completely drain the oil, which may be related to the trapping effect of the narrow pore throat channels inside.

Figure 7 shows the variation in oil content in different pores. In Figure 7a, due to the uniform distribution of pores, the overall imbibition recovery rate is relatively average, while the recovery rate of micropores is relatively low. In Figure 7b, after 52 days of imbibition, the oil content in all pores decreases, but that in micropores and small pores is more significant. In Figure 7c, after 52 days of imbibition, the oil content in micropores and small pores significantly decreases, while the oil content in large pores increases. It indicates that under the action of capillary force, water preferentially displaces oil from micropores and flows into larger pores. The capillary force of micropores and small pores is greater, so the permeability of pores is stronger. Due to the larger micropores and smaller pore sizes of sample UC7-3, its capillary pressure is stronger and its permeability is stronger.

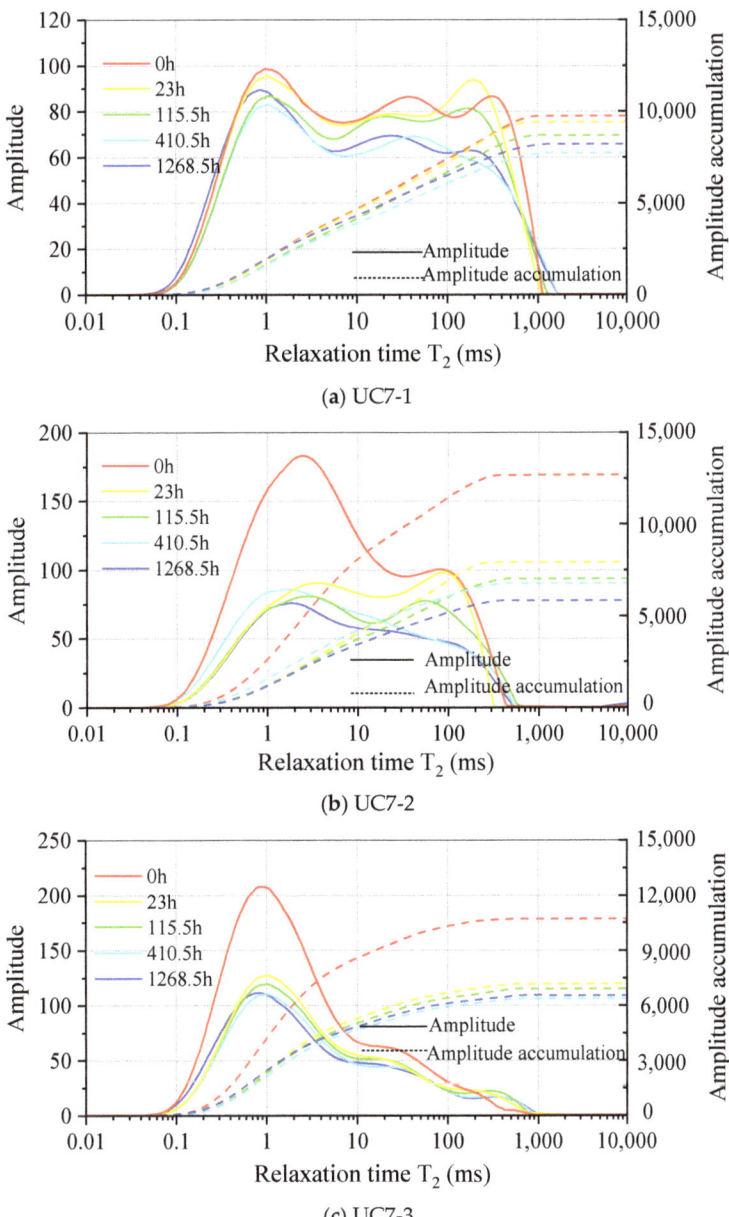

Figure 6. The T_2 spectra of different samples under 10 MPa forced imbibition: (**a**) UC7-1, (**b**) UC7-2, (**c**) UC7-3.

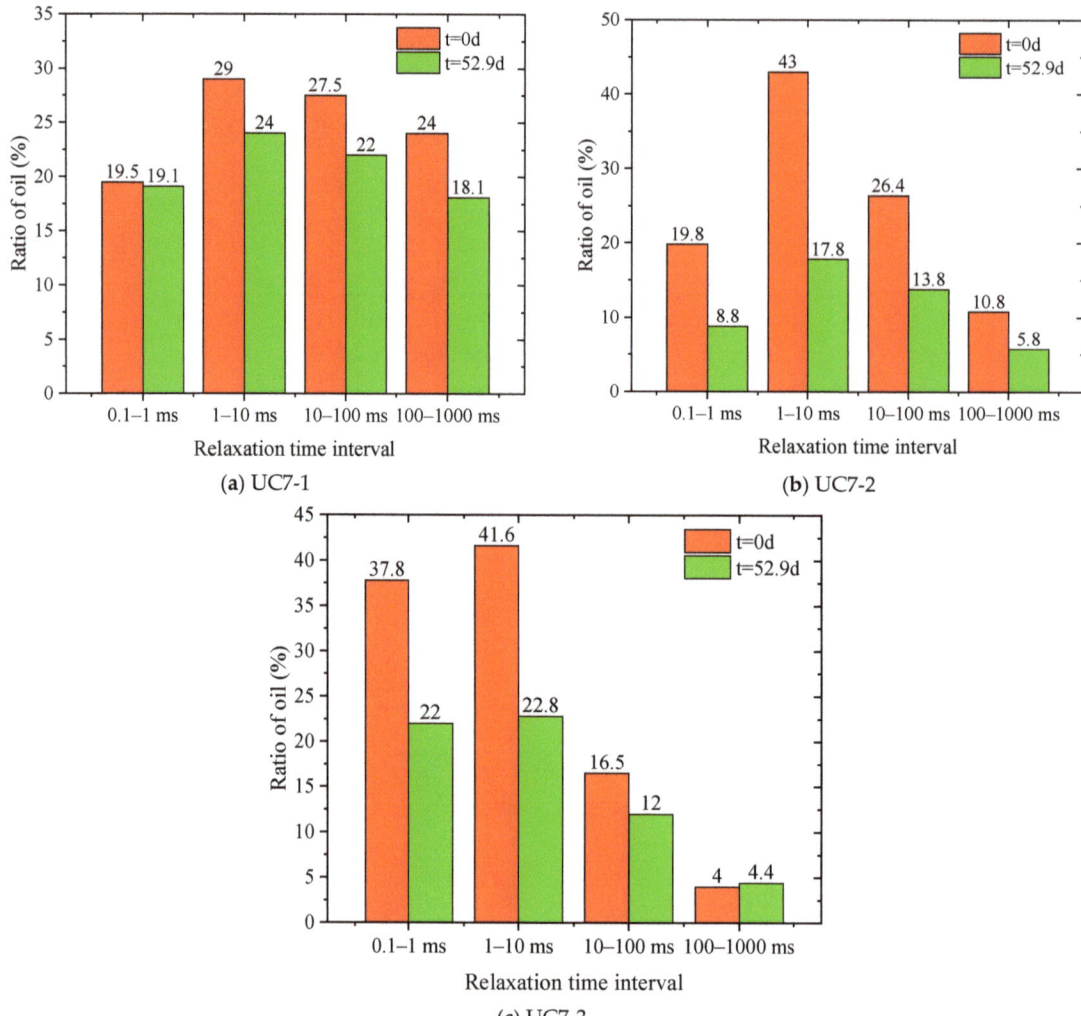

Figure 7. The changes in oil content in different pores of the sample: (**a**) UC7-1, (**b**) UC7-2, (**c**) UC7-3.

3.3. Effect of Natural Cracks on Imbibition Absorption

Figure 8 shows the T_2 spectra of spontaneous and pressurized imbibition of samples UC7-5 and LC7-5. According to Yang [17], spontaneous imbibition samples are enveloped by microcracks and exhibit different characteristics of oil migration. The amplitude decrease rate of macropores and macropores is greater than that of micropores. Compared with micropores, microfractures are the main channels for oil and gas migration. The T_2 spectrum distribution of UC7-5 sample under pressure imbibition is the same as that of spontaneous imbibition, with oil mainly distributed in larger pores (>10 ms) and also containing many microcracks, ranging from 0.05 to 1000 ms overall. LC7-5 contains fewer microcracks, and oil mainly exists in micropores. As the imbibition time increases, the amplitude of large pores in forced imbibition decreases significantly compared to spontaneous imbibition, and tends to stabilize in the later stage. From the T_2 spectrum curve at the beginning of imbibition, it can be seen that the cracks in sample UC7-5 are mainly macropores, with more parts >10 ms. Perhaps due to the presence of more natural fractures, the residual oil saturation in the later stage of imbibition is higher.

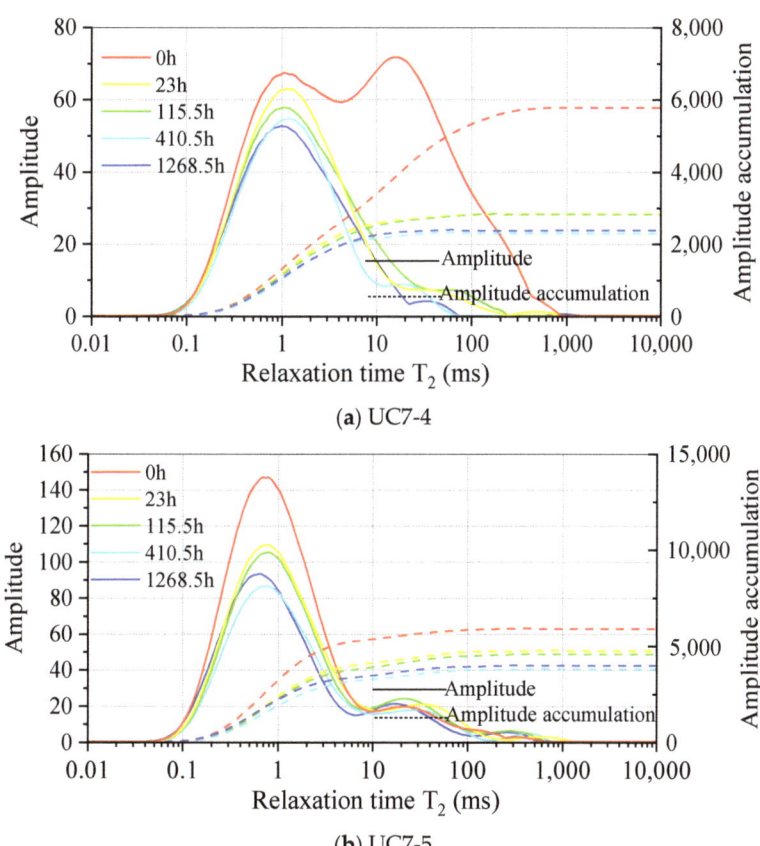

Figure 8. The T_2 spectra of different samples under forced and spontaneous imbibition at 10 MPa: (**a**) UC7-4, (**b**) UC7-5.

Figure 9 shows the spontaneous imbibition of sample UC7-2 and the changes in oil content in different pores of samples UC7-4 and UC7-5 under the pressure of 10 MPa. Before imbibition, oil in sample UC7-5 mainly exists in micropores, small pores, and larger pores, accounting for 21.9%, 36%, and 33.8%, respectively. After 52.9 days of imbibition, the oil content in larger pores and macropores significantly decreases, while the oil content in micropores changes less. The oil recovery rate is significantly higher than that of micropores. In sample LC7-5, oil mainly exists in micropores and small pores, accounting for 53% and 37%, respectively. The obvious difference is that the recovery rates of larger and larger pores are significantly lower than those of sample UC7-5, and the recovery rates of micropores are higher. Perhaps due to the presence of natural microcracks, water preferentially enters the natural fractures to replace oil, reducing the oil recovery rate of the pores.

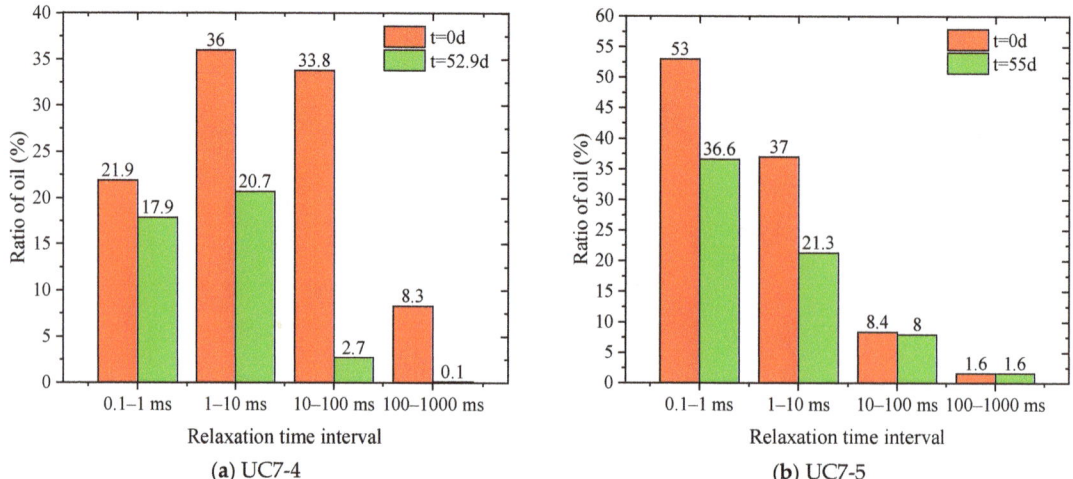

Figure 9. The changes in oil content in different pores of the sample: (**a**) UC7-4, (**b**) UC7-5.

3.4. Effect of Crack Expansion on Imbibition Absorption

Figure 10 shows the T_2 spectral curves of sample QT-1 spontaneous imbibition and QT-2 forced imbibition at 10 MPa. The T_2 spectra formed by both have unimodal characteristics, with T_2 values ranging from 0.07 to 900 ms. As the imbibition time increases, the amplitude shows a significant downward trend. Unlike the previous T_2 spectrum curve, QT-1 shows a certain upward trend in the later stage of imbibition, and the cumulative curve distribution is relatively uniform. In the early stage of imbibition, the amplitude curve decreases, indicating that water enters the micropores and replaces oil. In the later stage of imbibition, cracks expand due to the expansion of clay minerals. The oil discharged from micropores enters the expanding fractures, causing the amplitude curves of large and larger pores to show an upward trend. The QT-2 amplitude curve also shows a certain upward trend, and the pressure causes the amplitude curve to move downwards to the right, which has a significant impact on the middle and late stages of imbibition.

Figure 11 shows the oil content changes in different pores of samples QT-1 and QT-2. Before imbibition, the oil content in micropores, small pores, large pores, and larger pores of sample QT-1 are 22.3%, 50.5%, 22%, and 5.3%, respectively. After 42 days of imbibition, the content is 4.1%, 32%, 17.6%, and 1.9%, respectively. Due to the upward trend of the amplitude curve in the later stage, the recovery rate in larger pores is relatively low. Before imbibition, the oil content in micropores, small pores, large pores, and larger pores of sample QT-2 is 7.7%, 43.1%, 44.1%, and 5.6%, respectively. After 52.9 days of imbibition, the contents are 4.4%, 7.7%, 24.6%, and 12.1%, respectively. It can be observed that the oil recovery rate is highest in small pores, while the oil recovery rate in large pores shows a negative value. During the imbibition process, cracks expand due to the expansion of clay minerals, causing the oil displaced from micropores and small pores to enter the newly generated cracks. Pressurization causes the overall recovery rate of the pores to shift towards the direction of the larger pores.

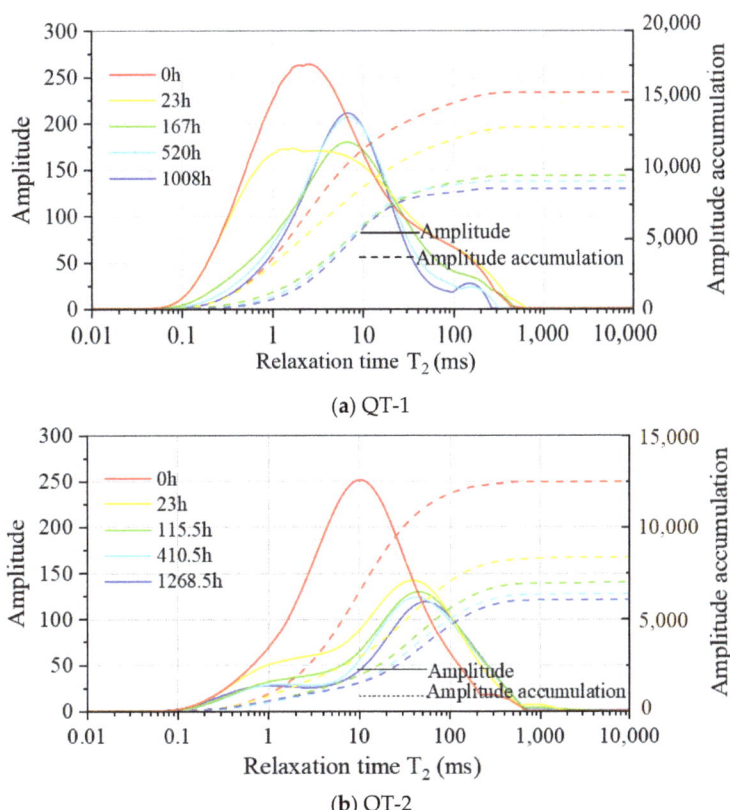

Figure 10. The T_2 spectra of different samples under 10 MPa forced imbibition: (**a**) QT-1, (**b**) QT-2.

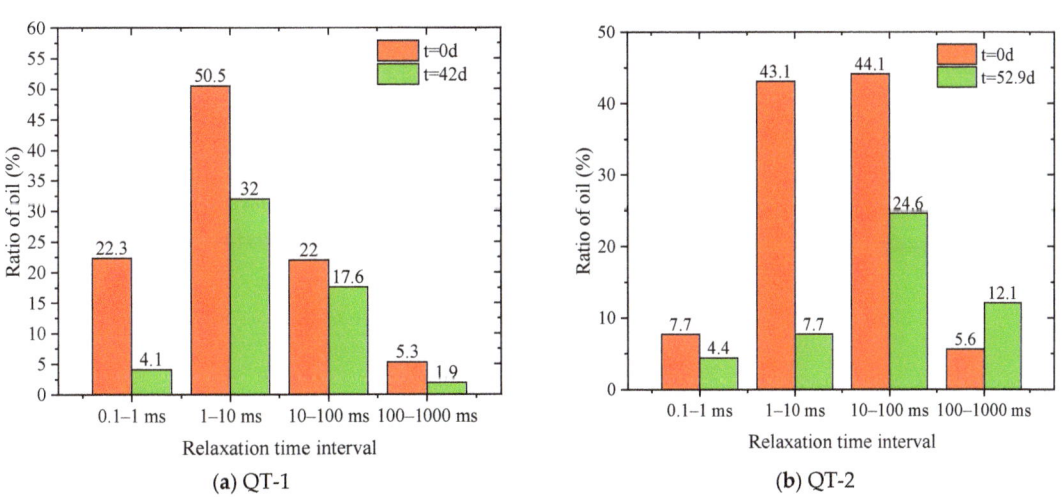

Figure 11. The changes in oil content in different pores of the sample: (**a**) QT-1, (**b**) QT-2.

3.5. Effect of Clay Minerals on Imbibition

Figure 12 shows the T_2 spectral curves of spontaneous imbibition of sample WEH-1 and forced imbibition of sample WEH-2. From the figure, it can be seen that the T_2 spectra of WEH-1 and WEH-2 both exhibit bimodal characteristics. The left peak range of WEH-1 is 0.7~80 ms, the right peak range is 80~900 ms, the left peak range of WEH-2 is 0.7~80 ms, and the right peak range is 80~1000 ms. As the imbibition time increases, the amplitude curve shows a downward trend. The initial amplitude curve of sample WEH-1 is larger than that of sample WEH-2, and the oil content in micropores and small pores is higher.

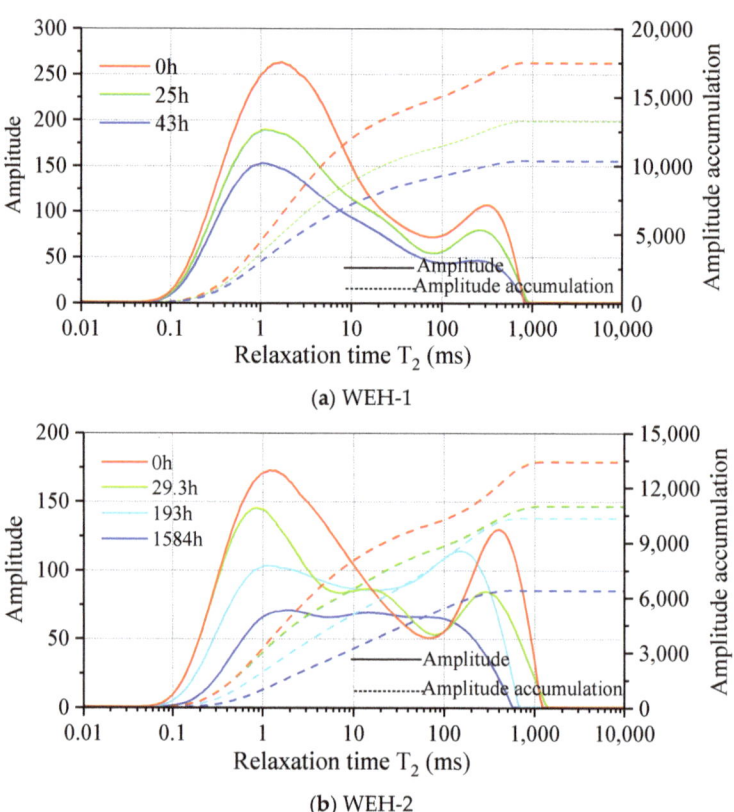

Figure 12. The T_2 spectra of different samples under 10 MPa forced imbibition: (**a**) QWEH-1, (**b**) WEH-2.

Figure 13 shows the changes in oil content of WEH-1 and WEH-2 samples in different pore sizes. Before imbibition, the oil content in micropores, small pores, large pores, and larger pores of WEH-1 is 24.6%, 43.2%, 17.9%, and 14.3%, respectively. After 1.8 days of imbibition, its content is 16.4%, 24%, 13.2%, and 6.6%, respectively. In WEH-2, before imbibition, the oil content in micropores, small pores, large pores, and larger pores is 23.3%, 35.9%, 16.8%, and 24%, respectively. After 17.1 days of imbibition, its contents are 6.8%, 16.7%, 16.4%, and 7.8%, respectively. The oil content in micropores, small pores, and large pores decreased to varying degrees after seepage. However, compared to the QT, LC7, and UC7 samples, the reduction in saturated oil content was less pronounced, which can be attributed to the clay content in the core. Clay particles can migrate and block pore throats, thereby reducing the overall permeability of the rock. Additionally, clay minerals can alter the wettability of the rock surface. For instance, certain clays can render the rock surface more oleophilic, thereby diminishing the efficiency of water imbibition.

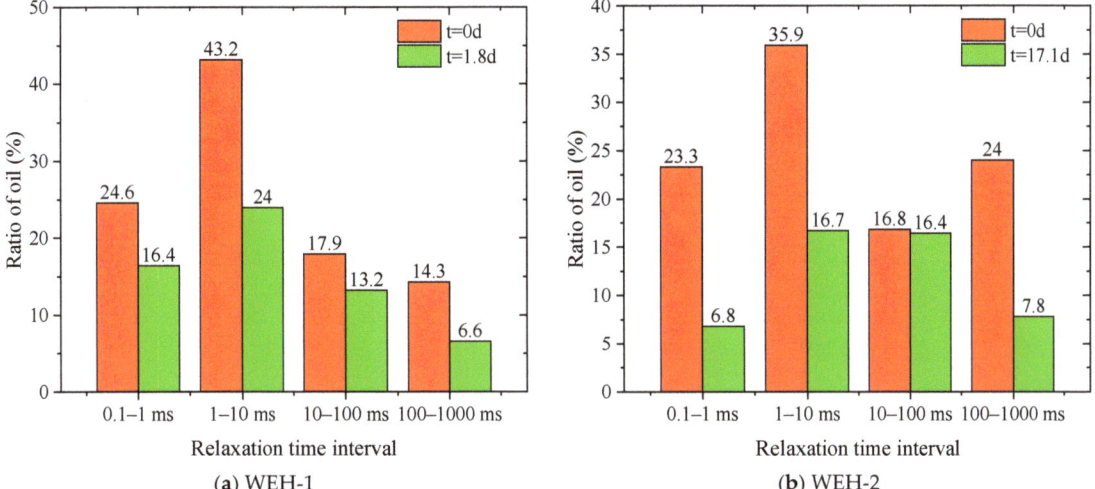

Figure 13. The changes in oil content in different pores of the sample: (**a**) WEH-1, (**b**) WEH-2.

3.6. The Implications and Future Potential Studies

Studying the effects of pressure, pore size distribution, microfractures, fracture propagation, and clay minerals on fluid distribution and transport during imbibition is significant for enhanced oil recovery (EOR) for several reasons:

(1) Improving oil recovery efficiency

A detailed understanding of pressure, pore size distribution, fracture expansion, and their effects on fluid transport can help optimize the design and operation of the EOR process. By combining the effects of these factors, EOR strategies can be tailored to specific reservoir conditions.

(2) Reduced environmental impact

A better understanding of fluid distribution and transport can lead to more targeted EOR interventions, reducing the amount of water and chemicals required. This not only reduces operating costs, but also minimizes environmental impact.

(3) Sustainability and economic efficiency

By optimizing the EOR process through a better understanding of these factors, more sustainable extraction methods can be found, extending the productive life of the field and improving its economic efficiency.

Future potential research may include the following:

(1) Advanced Characterization Techniques:

Develop advanced imaging and characterization methods to better understand pore structures, microfractures, and mineral compositions at micro- and nanoscales.

(2) Multiscale Modeling and Simulation:

Integrate data across different scales (micro, meso, macro) to create comprehensive models of fluid transport and distribution. Utilize machine learning and artificial intelligence to enhance predictive modeling capabilities.

(3) Integration with Other Technologies:

Combine enhanced oil recovery (EOR) techniques with technologies such as carbon capture and storage (CCS) to improve overall efficiency and environmental outcomes. Explore the use of renewable energy sources to power EOR operations, thereby reducing the carbon footprint of oil recovery processes.

In summary, understanding the effects of pressure, pore size distribution, microfractures, fracture propagation, and clay minerals on fluid distribution and transport during imbibition is vital for optimizing oil recovery, improving reservoir management, minimizing environmental impact, and enhancing economic viability. Future studies aligned with this work will likely focus on advanced characterization, modeling, field trials, environmental sustainability, integration with other technologies, and economic assessments to further enhance the effectiveness and sustainability of EOR techniques.

4. Conclusions

In this study, we conducted forced imbibition experiments on tight reservoir samples. And by comparing the T_2 spectrum curves of different samples and the changes in oil content in different pores, the distribution and migration patterns of fluids during the process of pressurized imbibition were studied. By studying the effects of external pressure, pore size distribution, microcracks, crack propagation, and clay minerals, the conclusions are as follows:

(1) With the increase in external forces, the imbibition rate of micropores and small pores is positively correlated, accelerating the displacement rate of oil in them. The external pressure may increase the capillary pressure at the smaller pores, increasing the imbibition effect of the smaller pores, but the impact on the larger pores is not significant. Overall, with the increase in external pressure, the overall imbibition recovery rate will increase.

(2) In tight reservoirs, without the influence of other factors, the smaller the pore size, the greater the capillary pressure, and the stronger the imbibition. Under capillary pressure, water will preferentially enter smaller pores to displace oil. The presence of residual oil in pores cannot be completely discharged by only applying external pressure and capillary force, which may be related to the trapping effect of the narrow pore throat channels inside.

(3) Due to the presence of natural fractures, the residual oil saturation in the later stage of imbibition is relatively high. And it prioritizes the water entering the pores to replace the oil, reducing the oil recovery rate of the micropores.

(4) During the imbibition process, the expansion of clay minerals in the sample causes cracks to expand, allowing the displaced oil to enter the newly generated cracks. And applying pressure causes the overall recovery rate of the pores to move towards the direction of the larger pores.

(5) More clay minerals are more prone to induce fracture after imbibition and contact with water, causing damage to their microscopic pore structure. This in turn affects the subsequent imbibition results.

Author Contributions: Conceptualization, X.L.; Methodology, L.Y.; Software, D.S.; Investigation, B.L.; Resources, S.W. All authors have read and agreed to the published version of the manuscript.

Funding: This work is supported by the National Key Research and Development Program of China under grant (2022YFE0206700).

Data Availability Statement: The original contributions presented in the study are included in the article, further inquiries can be directed to the corresponding author.

Conflicts of Interest: Author Xiaoshan Li was employed by Xinjiang Oilfield Company. Authors Dezhi Sun and Bingjian Ling were employed by China Railway Tianjin Metro. The remaining authors declare that the research was conducted in the absence of any commercial or financial relationships that could be construed as a potential conflict of interest.

References

1. Soeder, D.J. The successful development of gas and oil resources from shales in North America. *J. Pet. Sci. Eng.* **2018**, *163*, 399–420. [CrossRef]
2. Guo, C.; Wei, M.; Liu, H. Study of gas production from shale reservoirs with multi-stage hydraulic fracturing horizontal well considering multiple transport mechanisms. *PLoS ONE* **2018**, *13*, e0188480. [CrossRef]
3. Tian, W.; Wu, K.; Gao, Y.; Chen, Z.; Gao, Y.; Li, J. A Critical Review of Enhanced Oil Recovery by Imbibition: Theory and Practice. *Energy Fuels* **2021**, *35*, 5643–5670. [CrossRef]
4. Andersen, P.Ø.; Evje, S.; Kleppe, H. A Model for Spontaneous Imbibition as a Mechanism for Oil Recovery in Fractured Reservoirs. *Transp. Porous Media* **2014**, *101*, 299–331. [CrossRef]
5. Javaheri, A.; Habibi, A.; Dehghanpour, H.; Wood, J.M. Imbibition oil recovery from tight rocks with dual-wettability behavior. *J. Pet. Sci. Eng.* **2018**, *167*, 180–191. [CrossRef]
6. Cheng, Z.; Ning, Z.; Yu, X.; Wang, Q.; Zhang, W. New insights into spontaneous imbibition in tight oil sandstones with NMR. *J. Pet. Sci. Eng.* **2019**, *179*, 455–464. [CrossRef]
7. Wang, Y.; Liu, H.; Li, Y.; Wang, Q. Numerical Simulation of Spontaneous Imbibition Under Different Boundary Conditions in Tight Reservoirs. *ACS Omega* **2021**, *6*, 21294–21303. [CrossRef] [PubMed]
8. Binazadeh, M.; Xu, M.; Zolfaghari, A.; Dehghanpour, H. Effect of Electrostatic Interactions on Water Uptake of Gas Shales: The Interplay of Solution Ionic Strength and Electrostatic Double Layer. *Energy Fuels* **2016**, *30*, 992–1001. [CrossRef]
9. Yu, S. Post-frac evaluation of multi-stage fracturing on horizontal wells based on early flowback history. *Pet. Drill. Tech.* **2021**, *49*, 1–7.
10. Chen, Z.; Liu, H.; Li, Y.; Shen, Z.; Xu, G. The current status and development suggestions for shale oil reservoir stimulation at home and abroad. *Pet. Drill. Tech.* **2021**, *49*, 1–7.
11. Cai, J.; Li, C.; Song, K.; Zou, S.; Yang, Z.; Shen, Y.; Meng, Q.; Liu, Y. The influence of salinity and mineral components on spontaneous imbibition in tight sandstone. *Fuel* **2020**, *269*, 117087. [CrossRef]
12. Zhou, H.; Zhang, Q.; Dai, C.; Li, Y.; Lv, W.; Wu, Y.; Cheng, R.; Zhao, M. Experimental investigation of spontaneous imbibition process of nanofluid in the ultralow permeable reservoir with nuclear magnetic resonance. *Chem. Eng. Sci.* **2019**, *201*, 212–221. [CrossRef]
13. Wei, B.; Liu, J.; Zhang, X.; Xiang, H.; Zou, P.; Cao, J.; Bai, M. Nuclear Magnetic Resonance (NMR) mapping of remaining oil distribution during sequential rate waterflooding processes for improving oil recovery. *J. Pet. Sci. Eng.* **2020**, *190*, 107102. [CrossRef]
14. Chen, T.; Yang, Z.; Ding, Y.; Luo, Y.; Qi, D.; Lin, W.; Zhao, X. Waterflooding Huff-n-puff in Tight Oil Cores Using Online Nuclear Magnetic Resonance. *Energies* **2018**, *11*, 1524. [CrossRef]
15. Li, T.; Wang, Y.; Li, M.; Ji, J.; Chang, L.; Wang, Z. Study on the IMPacts of Capillary Number and Initial Water Saturation on the Residual Gas Distribution by NMR. *Energies* **2019**, *12*, 2714. [CrossRef]
16. Yang, Z.; Liu, X.; Li, H.; Lei, Q.; Luo, Y.; Wang, X. Analysis on the influencing factors of imbibition and the effect evaluation of imbibition in tight reservoirs. *Pet. Explor. Dev. Online* **2019**, *46*, 779–785. [CrossRef]
17. Yang, L.; Wang, S.; Tao, Z.; Leng, R.; Yang, J. The Characteristics of Oil Migration due to Water Imbibition in Tight Oil Reservoirs. *Energies* **2019**, *12*, 4199. [CrossRef]
18. Farahani, M.V.; Guo, X.W.; Zhang, L.X.; Yang, M.Z. Effect of thermal formation/dissociation cycles on the kinetics of formation and pore-scale distribution of methane hydrates in porous media: A magnetic resonance imaging study. *Sustain. Energy Fuels* **2021**, *5*, 1567–1583. [CrossRef]
19. Zhu, X.; Pan, R.; Zhu, S.; Wei, W. Research progress and core issues in tight reservoir exploration. *Earth Sci. Front.* **2018**, *25*, 141.
20. Hassanpouryouzband, A.; Joonaki, E.; Farahani, M.V.; Takeya, S.; Ruppel, C.; Yang, J.H. Gas hydrates in sustainable chemistry. *Chem. Soc. Rev.* **2020**, *49*, 5225–5309.
21. Cui, X.; Bustin, A.M.M.; Bustin, R.M. Measurements of gas permeability and diffusivity of tight reservoir rocks: Different approaches and their applications. *Geofluids* **2009**, *9*, 208–223. [CrossRef]
22. Xu, G.; Jiang, Y.; Shi, Y.; Han, Y.; Wang, M.; Zeng, X. Experimental investigations of fracturing fluid flow back and retention under forced imbibition in fossil hydrogen energy development of tight oil based on nuclear magnetic resonance. *Int. J. Hydrogen Energy* **2020**, *45*, 13256–13271. [CrossRef]
23. Wang, C.; Gao, H.; Gao, Y.; Fan, H. Influence of Pressure on Spontaneous Imbibition in Tight Sandstone Reservoirs. *Energy Fuels* **2020**, *34*, 9275–9282. [CrossRef]
24. Jiang, Y.; Shi, Y.; Xu, G.; Jia, C.; Meng, Z.; Yang, X.; Zhu, H.; Ding, B. Experimental Study on Spontaneous Imbibition under Confining Pressure in Tight Sandstone Cores Based on Low-Field Nuclear Magnetic Resonance Measurements. *Energy Fuels* **2018**, *32*, 3152–3162. [CrossRef]
25. SY/T 5370-1999; The Method for Measurement of Surface Tension & Interfacial Tension. State Bureau of Petroleum and Chemical Industry: Beijing, China, 1999.
26. Li, C.; Li, C.; Hou, Y.; Shi, Y.; Wang, C.; Hu, F.; Liu, M. Well logging evaluation of Triassic Chang 7 Member tight reservoirs, Yanchang Formation, Ordos Basin, NW China. *Pet. Explor. Dev.* **2015**, *42*, 667–673. [CrossRef]
27. Meng, M.; Ge, H.; Ji, W.; Wang, X. Research on the auto-removal mechanism of shale aqueous phase trapping using low field nuclear magnetic resonance technique. *J. Pet. Sci. Eng.* **2016**, *137*, 63–73. [CrossRef]

28. Yang, L.; Wang, H.; Xu, H.; Guo, D.; Li, M. Experimental study on characteristics of water imbibition and ion diffusion in shale reservoirs. *Geoenergy Sci. Eng.* **2023**, *229*, 212167. [CrossRef]
29. Yang, L.; Yang, D.; Zhang, M.-y.; Wang, S.; Su, Y.; Long, X. Application of nano-scratch technology to identify continental shale mineral composition and distribution length of bedding interfacial transition zone—A case study of Cretaceous Qingshankou formation in Gulong Depression, Songliao Basin, NE China. *Geoenergy Sci. Eng.* **2024**, *234*, 212674. [CrossRef]

Disclaimer/Publisher's Note: The statements, opinions and data contained in all publications are solely those of the individual author(s) and contributor(s) and not of MDPI and/or the editor(s). MDPI and/or the editor(s) disclaim responsibility for any injury to people or property resulting from any ideas, methods, instructions or products referred to in the content.

Article

Experimental Evaluation of Enhanced Oil Recovery in Shale Reservoirs Using Different Media

Jiaping Tao [1,2], Siwei Meng [1,2,*], Dongxu Li [3], Lihao Liang [1] and He Liu [1,2]

[1] PetroChina Research Institute of Petroleum Exploration & Development, Beijing 100083, China; taojiaping93@sina.com (J.T.)
[2] State Key Laboratory of Continental Shale Oil, Daqing 163002, China
[3] PetroChina Daqing Oilfield Co., Ltd., Daqing 163002, China
* Correspondence: mengsw@petrochina.com.cn

Abstract: The presence of highly developed micro-nano pores and poor pore connectivity constrains the development of shale oil. Given the rapid decline in oil production, enhanced oil recovery (EOR) technologies are necessary for shale oil development. The shale oil reservoirs in China are mainly continental and characterized by high heterogeneity, low overall maturity, and inferior crude oil quality. Therefore, it is more challenging to achieve a desirably high recovery factor. The Qingshankou Formation is a typical continental shale oil reservoir, with high clay content and well-developed bedding. This paper introduced high-precision non-destructive nuclear magnetic resonance technology to carry out a systematic and targeted study. The EOR performances and oil recovery factors related to different pore sizes were quantified to identify the most suitable method. The results show that surfactant, CH_4, and CO_2 can recover oil effectively in the first cycle. As the huff-and-puff process continues, the oil saturated in the shale gradually decreases, and the EOR performance of the surfactant and CH_4 is considerably degraded. Meanwhile, CO_2 can efficiently recover oil in small pores (<50 nm) and maintain good EOR performance in the second and third cycles. After four huff-and-puff cycles, the average oil recovery of CO_2 is 38.22%, which is much higher than that of surfactant (29.82%) and CH_4 (19.36%). CO_2 is the most applicable medium of the three to enhance shale oil recovery in the Qingshankou Formation. Additionally, the injection pressure of surfactant increased the fastest in the injection process, showing a low flowability in nano-pores. Thus, in the actual shale oil formations, the swept volume of surfactant will be suppressed, and the actual EOR performance of the surfactant may be limited. The findings of this paper can provide theoretical support for the efficient development of continental shale oil reservoirs.

Keywords: high clay content; huff-and-puff; nuclear magnetic resonance; T_2 spectrum; quantitative evaluation

1. Introduction

Unconventional oil and gas resources are playing an increasingly significant role in the development of the oil and gas industry, with their abundant reserves and great exploration and development potential. Shale oil is considered the most valuable unconventional oil and gas resource for development [1–3]. The shale oil revolution in the US has rapidly increased oil production in America and profoundly changes the global energy landscape. China has rich deposits of shale oil; the geological reserves of shale oil in China are about 28.3 billion tons, showing great potential for development [4–6]. The scaled-up cost-effective recovery of shale oil is of critical strategic value for the sustained development of China's petroleum industry.

Shale oil reservoirs possess highly developed micro-nano pores, with poor pore connectivity and extremely low permeability [7–9]. They require the use of horizontal wells and multi-stage fracturing technologies to achieve beneficial development [10–12]. However, the production of shale oil wells suffers from rapid production decline. The production of a fractured shale oil well in the US declines by about 70% after one year of production, and the recovery factor is typically less than 10% [13,14]. Compared with the largely marine shale oil reservoirs of the US, shale oil reservoirs in China are generally formed in continental sedimentary environments. They are characterized by small distribution area, high heterogeneity, low overall maturity, and inferior crude oil quality, and therefore, it is challenging to achieve a desirably high recovery factor [15–18]. Given the above-stated obstacles, enhanced oil recovery (EOR) treatments are necessary for improving shale oil recovery and mitigating well production decline.

The ultra-low porosity and permeability of shale oil reservoirs severely reduce the applicability of conventional EOR technologies to effectively enhance shale oil recovery. Therefore, strengthening the research and development of shale-appropriate EOR technologies is an important direction for future development. Previous research demonstrates that the gas huff-and-puff process and surfactant imbibition are currently more feasible technologies for the EOR of shale oil [19–23]. With the appropriate schemes of post-frac well soaking and production, the gas huff-and-puff method can efficiently supplement energy formation and enhance oil recovery [24,25]. Through the mechanisms of wettability alteration and interfacial tension reduction, surfactant imbibition can also effectively enhance oil recovery [26,27].

The Songliao Basin is a typical large continental sedimentary basin, possessing abundant hydrocarbon resources. The Qingshankou and Nenjiang Formations, which were formed in the Late Cretaceous period, are the main source rocks for the basin, with their high-quality organic-matter-rich shale [28,29]. Preliminary exploration shows that the shale oil geological reserves of the Qingshankou Formation reach about 5.46 billion tons, and it is this gargantuan development potential that makes this layer the main field for shale oil development in the Daqing Oil Field. In recent years, breakthroughs have been made in the production practices of Qingshankou shale oil. In order to support exploration and development, researchers have carried out a large number of basic studies on the Qingshankou Formation. Liu et al. clarified the shale oil exploration "sweet spots", based on the detailed description and analysis of the cores and shale lithofacies characteristics [30]. Yuan et al. proposed the key theoretical and technical issues and countermeasures for effective development, targeting reservoir characteristics [31]. Yang et al. employed nano-scratch technology to continuously investigate the damage and failure mechanisms of the Qingshankou Formation, which is one of the most important measurement methods for studying water–CO_2–shale interaction [32,33].

Related studies have stimulated the development of Qingshankou shale oil. However, due to the high clay content and well-developed bedding of the Qingshankou Formation [30,34,35], the shale oil wells still suffer from the rapid production decline in the production stage. In this situation, the recovery of shale oil is very low, and it is difficult to achieve beneficial development. Therefore, targeted EOR studies are urgently needed to identify the most suitable EOR technology for recovering Qingshankou shale oil. Targeting the characteristics of the Qingshankou Formation, this paper introduced high-precision non-destructive nuclear magnetic resonance (NMR) technology to carry out a systematic study. Multicycle huff-and-puff experiments using surfactant, CH_4, and CO_2 were performed under the reservoir conditions using cores and crude oil collected from the Qingshankou Formation. The oil recovery factor for different pore sizes was quantified during the multicycle huff-and-puff process; the EOR performances of different EOR technologies in the Qingshankou Formation were compared to identify the most suitable method. The findings of this paper can provide theoretical support for the efficient development of the Qingshankou Formation shale oil.

2. Experimental Section

2.1. Sample Preparation

The shale samples used in this paper were collected from the Qingshankou Formation in the Songliao Basin. The Qingshankou Formation is mostly shale, with local interbeds of sandstone, limestone, and limy mudstone. The shale, predominantly composed of clay minerals, quartz, feldspar, and carbonates, possesses well-developed nano pores and micro fractures. The crude oil used in this paper was collected from the production well. According to the slim-tube test, the minimum miscibility pressure between CO_2 and crude oil was 23.9 MPa. The viscosity of this oil is 1.51 mPa·s at 90 °C.

Collected core samples were formed into cylindrical specimens (5 cm × Φ2.5 cm) via wire-cutter machining. Then, the specimens with an intact appearance and no considerable fracture were selected and polished using abrasive paper to obtain a smooth surface and flattened ends. Finally, the polished specimens were treated with petroleum ether and benzene to remove crude oil in these cores. The specimens were inspected to ensure that they were intact and undamaged during oil removal.

2.2. Experimental Apparatus and Methods

The AP-608 Automated Permeameter-Porosimeter (Figure 1), from Coretest systems, was used to measure the porosity and permeability of the Qingshankou Formation shale. The available confining pressure of the apparatus is 0–9000 psi, and the measurement ranges of porosity and permeability are 0.1–40.0% and 0.001 mD–10,000 mD, respectively. During the experiments, the oil-removed shale specimens were dried at 60 °C in the oven for 24 h before being placed in the core holder. The confining pressure was set at 3000 psi for the porosity and permeability measurements. The measured initial porosity and permeability of the selected shale specimens are listed in Table 1.

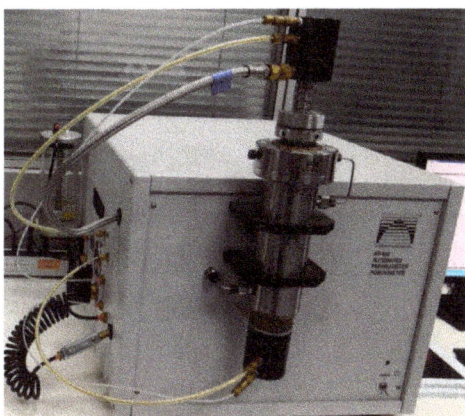

Figure 1. Experimental equipment used for porosity and permeability analysis.

Table 1. Core parameters of shale samples.

Sample	Diameter (cm)	Length (cm)	Porosity (%)	Permeability (mD)
1	2.50	5.00	5.31	0.0254
2	2.49	5.01	4.58	0.0428

Multicycle huff-and-puff physical simulation experiments: The schematic depiction of the experimental apparatus is shown in Figure 2. The apparatus mainly consists of an ISCO pump, an intermediate container, a core holder, a hand pump, and two pressure gauges. In these experiments, the following assumptions were adopted: the samples can represent the geological characteristics of the target formation, and the damage to the specimens during processing can be ignored. After saturation, the oil distribution in the cores can represent the actual occurrence of crude oil in the formation. The physical simulation experiment meets the requirements of the main influencing factors, based on the similarity criteria. The experimental procedures are briefly described as follows. (1) Use the wire cutter to cut the specimen in half radially. (2) Place the specimen in the intermediate container and hold under vacuum for 24 h. Then, saturate the specimen with crude oil, and record the volume of oil saturated into the specimen. (3) Place the specimen in the NMR apparatus to obtain the T_2 spectrum distribution to analyze the oil distribution under the saturated state. (4) Place the specimen in the core holder and constantly maintain the confining pressure at 2 MPa higher than the injection pressure. Continuously inject surfactant/CH_4/CO_2 into the specimen until the injection pressure reaches 30 MPa. Then, close the inlet valve and soak the specimen for 24 h. (5) After soaking, open the inlet valve and record the oil production. Then, obtain the T_2 spectrum distribution using the NMR apparatus to analyze the oil distribution after the huff-and-puff process. (6) Repeat Steps (4) and (5) to perform four huff-and-puff cycles to analyze the EOR performance of surfactant/CH_4/CO_2 in the shale cores. The surfactant used in this experiment is a kind of petroleum sulfonate, with a concentration of 0.3 wt%. According to the conditions of the Qingshankou Formation, the pressure and temperature during these experiments were set at 30 MPa and 90 °C, respectively.

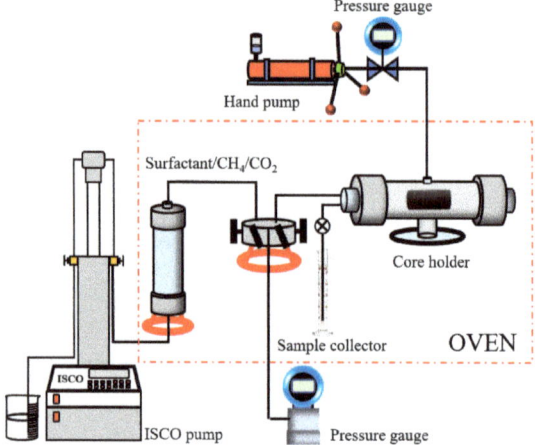

Figure 2. Schematic depiction of the experimental setup for huff-and-puff.

3. Results and Discussion

3.1. Analysis of EOR Performance of Surfactant in Shale Oil Reservoirs

Surfactant was continuously injected into the shale core until the injection pressure reached 30 MPa. Then, the core holder was sealed for 24 h, after which the inlet valve was opened for oil production. The experimental temperature was kept at 90 °C throughout the experiment. The T_2 spectrum was measured after each cycle of surfactant huff-and-puff to analyze the oil distribution in pores of different sizes in the shale cores.

The photos of Specimen 1-1 at different stages of the huff-and-puff process are shown in Figure 3. The surface of the oil-saturated shale core is dark black, with notable oil traces. With the progress of the surfactant huff-and-puff process, the core surface becomes drier. After three cycles of huff-and-puff, there were no notable oil traces on the core surface. The color of the shale core is also fading during the experiment. The change in the shale core is attributed to the oil production from the shale, and the mentioned phenomenon is significant, as noted in the comparison between the core photos at 0 H and 96 H. It is demonstrated that surfactant can diffuse into the micro pores of shale during soaking, helping to produce oil in such pores. This finding is also supported by the T_2 spectrums.

Figure 3. Shale core sample after surfactant huff-and-puff.

The T_2 spectrums of Specimen 1-1 at different stages of the huff-and-puff process are shown in Figure 4. The T_2 spectrum of oil-saturated Specimen 1-1 has two peaks, with the left peak being higher than the right one. This indicates that the nano-scale pores are well developed in the shale. The radii of the pores are predominantly smaller than 50 nm, and larger pores are relatively limited.

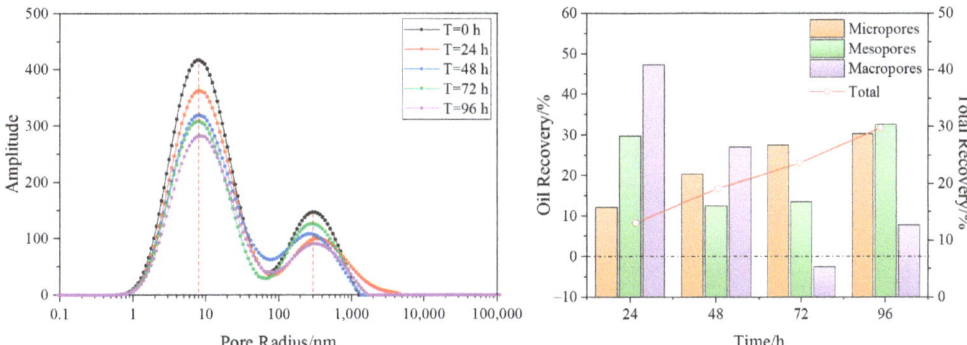

Figure 4. T_2 spectrums and oil recovery after surfactant huff-and-puff.

During the surfactant huff-and-puff experiment, the oil in the core is effectively recovered. After the first cycle of huff-and-puff, the oil recovery factor reaches 13.03%. The amount of oil produced in the small pores is significant—the amount of oil produced in pores with radii smaller than 50 nm (the small scale) is 12.12%; for those 50–500 nm (the medium scale), it is 29.76%; for those larger than 500 nm (the large scale), it is 47.31%. The oil recovery of the second cycle is 19.14%. Specifically, the amount of oil produced in the small pores is 20.36%, with a significant increase. However, as a part of the oil migrates from smaller pores to larger ones, the oil in the larger pores increases, and the amount of oil produced in medium (50–500 nm) and large (>500 nm) pores drop to 12.48% and 26.97%; both values are lower than those of the first cycle. The recovery factor grows to 23.62% and 29.82% after the third and fourth cycles, respectively. The amount of oil produced in the small pores continues to increase to 27.45% and 30.30%, respectively. That of the medium pores grows to 13.50% and 32.56% in the third and fourth cycles, while that of the large pores falls to −2.53% and 7.76%, respectively, due to the migration of the oil in the small and medium pores to the large pores.

The above results show that the surfactant exhibits good EOR performance for the Qingshankou Formation shale. During the surfactant huff-and-puff process, the oil in the small pores (with radii below 50 nm) of the shale is gradually recovered, with the corresponding amounts produced climbing constantly. For medium and large pores (with radii of 50–500 nm and >500 nm, respectively), the produced amounts are generally positive. Occasionally, there is a phenomenon of an increase in oil occurrence in larger pores, as a part of the oil in the small pores and throats migrates to the larger pores during the huff-and-puff process.

3.2. Analysis of EOR Performance of CH$_4$ in Shale Oil Reservoirs

Similarly, CH$_4$ was continuously injected into the shale core until the pressure reached 30 MPa, and flowback and production were executed after 24 h of soaking. The experimental temperature throughout the experiment was also 90 °C. The oil distribution at different stages of CH$_4$ huff-and-puff was captured using NMR to reveal the EOR performance of CH$_4$ in the shale oil reservoirs.

The photos of Specimen 2-1 at different stages of the huff-and-puff process are shown in Figure 5. The surface of Specimen 2-1 is dark black, with notable oil traces. As huff-and-puff proceeds, the specimen becomes drier, and the oil traces fade away. The color of the shale core also changes; it gradually becomes lighter in some zones. The stated phenomenon is significant when compared to that noted in the photos at 0 H and 96 H. It can be concluded that the injected CH$_4$ effectively recovered oil in the pores of the shale during soaking.

Figure 5. Shale core sample after CH$_4$ huff-and-puff.

The T$_2$ spectrums of Specimen 2-1 at different stages of CH$_4$ huff-and-puff are shown in Figure 6. After oil-saturation, the T$_2$ spectrum of the shale core exhibits three peaks, a high peak on the left, and two small peaks in the middle and right, which demonstrates the significant development of nano-scale pores in shale, with the majority of pores smaller than 50 nm. For such shale, there are fewer larger pores, and it is difficult to recover the oil.

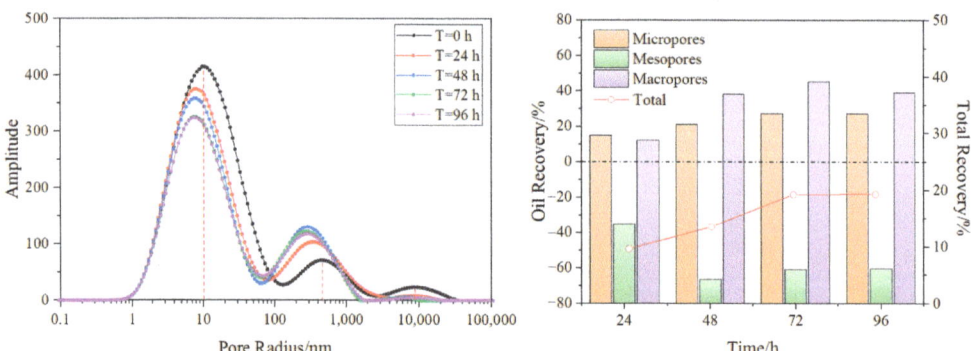

Figure 6. T$_2$ spectrums and oil recovery after CH$_4$ huff-and-puff.

The EOR effect of the CH$_4$ huff-and-puff process is relatively poor in the shale core. The oil recovery is 9.65% after the first cycle of huff-and-puff, and the oil in the small pores (<50 nm) is effectively produced (at 14.90%). Due to the migration of oil from the smaller pores to the larger pores, the oil content in the medium pores (50–500 nm) grows, and the production rate is −35.15%. The amount of oil produced in the large pores (>500 nm) is 12.1%. After the second and third cycles, the oil recovery rates are 13.58% and 19.30%, respectively. The oil production rate of the small pores (<50 nm) rises to 21.04% and 27.24%, respectively. As a large quantity of oil transfers from the small to medium (50–500 nm) pores, the oil in the latter increases considerably, and the production levels drop to −66.39% and −60.83%, respectively. In these two cycles, the oil in the large pores (>500 nm) is effectively recovered (at 38.11% and 45.23%). However, the EOR performance of CH_4 is degraded rapidly in the fourth cycle. The oil recovery factor after the fourth cycle is only 19.36%, nearly equal to that after the 3rd cycle. The oil production rates exhibit no notable changes, regardless of pore sizes.

3.3. Analysis of EOR Performance of CO_2 in Shale Oil Reservoirs

Similarly, the CO_2 multicycle huff-and-puff physical simulation experiments were performed at 90 °C and 30 MPa. During the four cycles of CO_2 huff-and-puff, the differentiation of oil production from different-sized pores was clarified via the T_2 spectrums, and the EOR performance of CO_2 in the shale oil reservoirs was assessed.

The photos of Specimen 1-2 and 2-2 at different stages of CO_2 huff-and-puff are shown in Figures 7 and 8. The cores are also dark black, with notable oil traces on their surface after oil saturation. With the ongoing CO_2 huff-and-puff process, the oil inside the cores is recovered, and the surface oil traces are reduced. The color of these cores gradually fades, and some zones even change from black to light grey. The above results demonstrate the high EOR capacity of CO_2 in shale, as CO_2 can diffuse into the tiny pores of the shale and help to effectively recover oil in such pores.

Figure 7. Shale core sample 1-2 after CO_2 huff-and-puff.

Figure 8. Shale core sample 2-2 after CO_2 huff-and-puff.

The T_2 spectrums of Specimen 1-2 at different stages of CO_2 huff-and-puff are shown in Figure 9. The T_2 spectrum of Specimen 1-2 is bimodal after oil saturation. A high left peak associated with a low right peak shows that the shale possesses highly developed nano-scale pores. The majority of pores have radii smaller than 50 nm, and the quantity of larger pores is limited. It is difficult to recover oil from such shale.

During CO_2 huff-and-puff, oil is effectively produced in the shale core. After the first cycle of CO_2 huff-and-puff, the oil recovery reaches 8.98%. Moreover, the oil in the small pores (<50 nm) is efficiently recovered at a rate of 11.99%. The amount of oil produced in the medium pores (50–500 nm) reaches 4.69%, and for large pores (>500 nm), it is −35.71%, due to oil migration from the smaller to the larger pores. The second cycle also effectively recovers oil from the shale. The oil recovery increases to 18.02%, and the amount of oil produced in the small pores rises considerably (to 20.43%). The amount of oil produced in the medium pores (50–500 nm) is 16.69%, and for the large pores (>500 nm), it is −30.29%. CO_2 maintains excellent EOR performance in the third cycle, i.e., the oil recovery rate rises to 30.98%, which is much higher than that after three cycles of surfactant huff-and-puff. The amount of oil produced in the small (<50 nm) and medium (50–500 nm) pores continue to grow (25.84% and 64.77%, respectively). As oil flows from the small and medium pores to the large pores, the oil produced by the latter (>500 nm) falls to −52.05%. In the fourth cycle, the EOR performance of CO_2 is slightly degraded. The recovery factor grows to 34.04%, and is associated with small incremental increases in the amount of oil produced by the small and medium pores (28.13% and 65.79%, respectively). Nevertheless, the oil in the large pores (>500 nm) is effectively recovered during this period, and the production rate increases from −52.05% to −18.66% (this negative value means that the amount of oil in the large pores after the fourth cycle is still greater than the initial amount observed in the large pores). During the CO_2 huff-and-puff, the oil recovery factor is 4.22% higher than that noted for the surfactant huff-and-puff process.

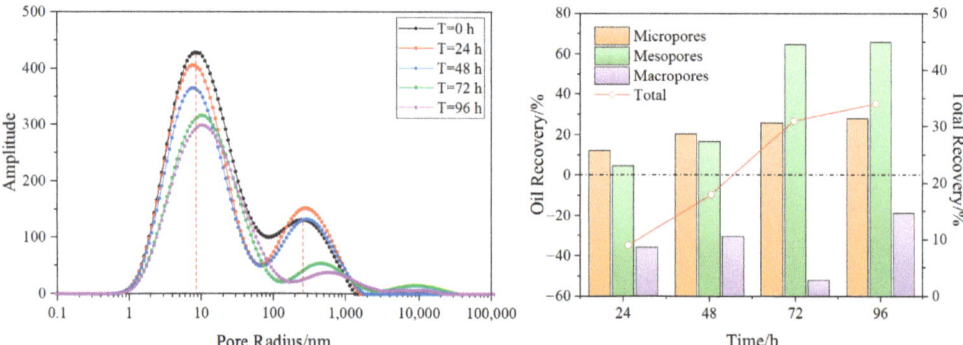

Figure 9. T_2 spectrums and oil recovery of core 1-2 after CO_2 huff-and-puff.

The T_2 spectrums of Specimen 2-2 after different stages of CO_2 huff-and-puff are shown in Figure 10. The T_2 spectrum of the oil-saturated core exhibits three peaks, with the higher left peak and lower middle and right peaks indicating the significant development of nano-scale pores in the shale. The oil is mostly stored in pores with radii smaller than 50 nm, which leads to high difficulties in oil recovery.

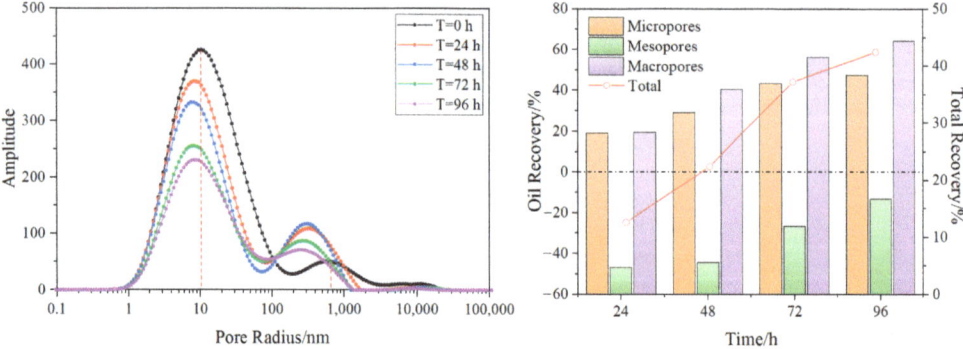

Figure 10. T_2 spectrums and oil recovery of core 2-2 after CO_2 huff-and-puff.

Similar to the case of Specimen 1-2, the oil in Specimen 2-2 is effectively recovered by CO_2 huff-and-puff. After the first cycle of CO_2 huff-and-puff, the oil recovery factor reaches 12.57%. The amount of oil produced in the small pores (<50 nm) is 19.15%, suggesting an effective recovery of saturated oil. The oil production rate of the medium pores (50–500 nm) is −46.72% because of the additional oil in the medium pores flowing from the small pores. As for large pores (>500 nm), the oil production rate is 19.57%. In the second and third cycles, CO_2 huff-and-puff maintains excellent EOR performance and delivers recovery factors of 22.40% and 37.21%, respectively, which are far higher than those for the CH_4 huff-and-puff process. Considerable increases are seen in the oil production rates of the small pores (29.17% and 43.43%, respectively). As for medium pores (50–500 nm), the oil is effectively produced; however, the production rates are still negative (−44.39% and −26.71%, respectively) due to the remaining excess of oil, compared with the oil initially observed in the core after saturation. The oil production rates of the large pores (>500 nm) are 40.53% and 56.26%, respectively, also showing an effective recovery level. The EOR performance of CO_2 huff-and-puff is sightly degraded in the fourth cycle. The resulting recovery factor is 42.4%, and the changes in the oil production rates in the pores of different sizes are similar to those noted in the second and third cycles (47.52% for small pores, −13.31% for medium pores, and 64.27% for large pores), which are all higher than those for the last cycle. Compared with CH_4 huff-and-puff, CO_2 huff-and-puff significantly increases the oil recovery factor by 23.04%.

3.4. Comparative Analysis of EOR Effect of Different Media

With the above-stated experimental results, the oil recovery effect for shale, with different media used for the huff-and-puff process, is investigated to clarify the differences in EOR performance obtained using surfactant, CH_4, and CO_2 in the Qingshankou Formation.

During the huff-and-puff process, all media can effectively recover oil from shale in the first cycle. As the huff-and-puff process proceeds, the oil saturated in the shale gradually decreases, and the oil recovery factor grows. The EOR performance of surfactant and CH_4 is considerably degraded, but CO_2 maintains good EOR performance in the second and third cycles. Therefore, after four cycles of huff-and-puff, the ultimate oil recovery rate of CO_2 is much higher than that of surfactant or CH_4, and the ultimate oil recovery rate of CH_4 is the lowest.

The oil production in pores of different sizes is shown in Figure 11. For small pores (<50 nm), all of the three media show similar EOR performances. The oil production rates of CO_2 are slightly higher, while those of surfactant and CH_4 are close to each other. For medium pores (50–500 nm), the oil production rates are significantly different in various cores. For Core 1-1 and 1-2, the oil in the medium pores is effectively produced, and the oil production rate for CO_2 is higher than that of surfactant after multiple cycles of huff-and-puff. Meanwhile, due to the migration of some oil from the smaller pores to the larger pores during the huff-and-puff process, the ultimate oil production rates for the medium pores are negative in Core 2-1 and 2-2.

However, it should be noted that those of CO_2 are still higher than those of CH_4. In the large pores (>500 nm), the oil production is considerably differentiated for the different cores. In Core 1-1, surfactant can effectively recover oil from the large pores in the first huff-and-puff cycle. But the production rates decline in the subsequent cycles, as oil in the smaller pores flows to the larger pores, while in Core 1-2, the oil production rates in the large pores remain negative throughout the CO_2 huff-and-puff experiment, also due to oil migration from the smaller pores to the larger pores. For Core 2-1 and 2-2, both CH_4 and CO_2 can effectively recover oil in the large pores. The EOR performance of CO_2 is always higher than that of CH_4. Moreover, the excellent EOR performance of CO_2 is maintained throughout the four cycles, while CH_4 only exhibits good EOR performance in the first two cycles.

Since the experiments used drilled core samples and crude oil from the production well, there are three main limitations of this study that must be considered. Firstly, China's shale oil reservoirs are mostly formed in a continental sedimentary environment and present high heterogeneity. Although the porosity and permeability are similar between the cores, there is significant variation in pore connectivity and pore size distribution. As a result, the oil production from pores of different sizes is greatly varied for different cores during CO_2 huff-and-puff. Therefore, targeted analysis for each shale oil formation is necessary in order to clarify the actual EOR performance of the huff-and-puff process. Secondly, the shale specimens for the huff-and-puff experiments used in this paper are limited in size, and the huff-and-puff media can sweep the entire specimen. In the actual shale oil formations, the swept volume is usually limited by the permeability of the formation and the viscosity of the injected fluid. Thus, the recovery rates obtained in experiments are usually higher than the actual recovery rates in the field. Additionally, compared with CO_2 and CH_4, the swept volume of the surfactant will be suppressed due to its poorer liquidity. Accordingly, the actual EOR performance of the surfactant may be limited. Thirdly, the experiments were carried out using produced oil. Under actual formation conditions, crude oil contains a large amount of dissolved gas, so the physical properties are different from those of the produced oil. During long-term production, the physical properties of crude oil will gradually change, which may also have an impact on the recovery rates.

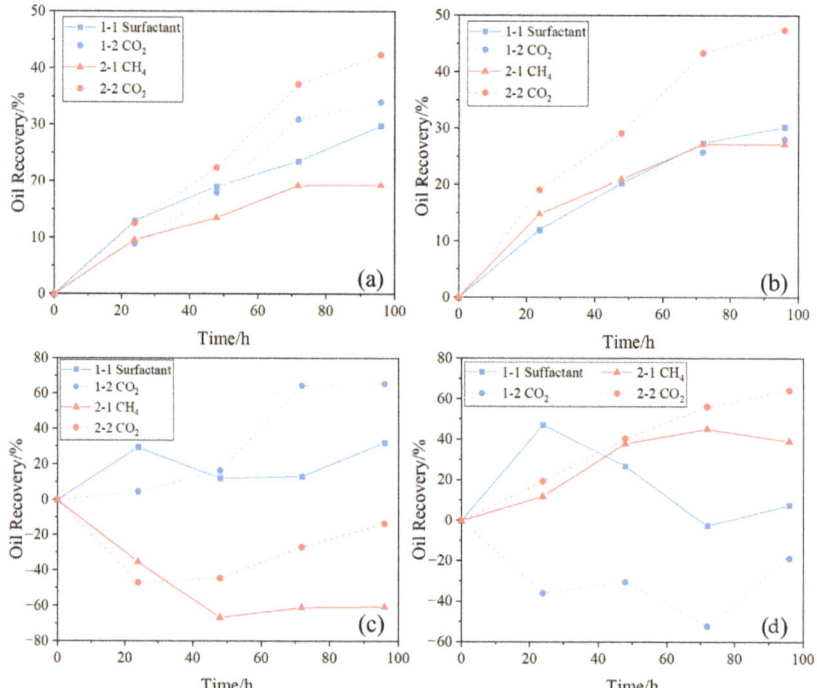

Figure 11. Comparative analysis of EOR effect during huff-and-puff of different media: (**a**) oil recovery of shale cores; (**b**) oil recovery of small pores; (**c**) oil recovery of medium pores; (**d**) oil recovery of large pores.

This study was carried out using the high clay-content shale samples in the Qingshankou Formation. In the future, the EOR performance of different media can be subsequently investigated using a wider range of shale formations with different pore structures and mineralogical compositions to clarify the most suitable EOR technology for different shale reservoirs and to provide theoretical support for the development of shale reservoirs.

4. Conclusions

The EOR performance of surfactant, CH_4, and CO_2 in a shale oil reservoir was systematically analyzed through physical simulation experiments. The conclusions are as follows: CO_2 is the most applicable medium of the three to enhance shale oil recovery in the Qingshankou Formation. It maintains good EOR performance in the second and third cycles. Therefore, after four cycles of huff-and-puff, the ultimate oil recovery of CO_2 is much higher than that of either surfactant or CH_4. Although the shale cores exhibit similar porosity and permeability, the pore connectivity is different. Thus, the oil production from pores of different sizes is greatly varied. This means that more microscopic research is necessary regarding shale reservoirs to clarify the influence of nano-pore characteristics on EOR performance. The injection pressure of the surfactant increased the fastest, which showed a low flowability in the nano-pores. This will greatly affect the sweep efficiency in shale reservoirs. Although the EOR performance of the surfactant is significant in the core scale experiment, the actual EOR performance of the surfactant may be even lower at the field scale.

Author Contributions: Conceptualization, S.M. and H.L.; funding acquisition, S.M.; investigation, J.T., D.L. and L.L.; methodology, J.T. and S.M.; project administration, S.M.; supervision, S.M.; visualization, J.T.; writing—original draft, J.T.; writing—review and editing, S.M. All authors have read and agreed to the published version of the manuscript.

Funding: This research was funded by the "Enlisting and Leading" Science and Technology Project of Heilongjiang Province (No. RIPED-2022-JS-1740), and the Technology Project of CNPC (No. 2023ZZ08).

Data Availability Statement: The original contributions presented in the study are included in the article, further inquiries can be directed to the corresponding author.

Conflicts of Interest: Author Dongxu Li was employed by the company PetroChina Daqing Oilfield Co., Ltd. The remaining authors declare that the research was conducted in the absence of any commercial or financial relationships that could be construed as a potential conflict of interest.

References

1. Sheng, J.; Chen, K. Evaluation of the EOR potential of gas and water injection in shale oil reservoirs. *J. Unconv. Oil Gas Resour.* **2014**, *5*, 1–9. [CrossRef]
2. Song, Z.; Song, Y.; Li, Y.; Bai, B.; Song, K.; Hou, J. A critical review of CO_2 enhanced oil recovery in tight oil reservoirs of North America and China. *Fuel* **2020**, *276*, 118006. [CrossRef]
3. Fakher, S.; Imqam, A. Application of carbon dioxide injection in shale oil reservoirs for increasing oil recovery and carbon dioxide storage. *Fuel* **2020**, *265*, 116944. [CrossRef]
4. Wang, J.; Feng, L.; Steve, M.; Tang, X.; Gail, T.; Mikael, H. China's unconventional oil: A review of its resources and out-look for long-term production. *Energy* **2015**, *82*, 31–42. [CrossRef]
5. Hu, S.; Zhao, W.; Hou, L.; Yang, Z.; Zhu, R.; Wu, S.; Bai, B.; Jin, X. Development potential and technical strategy of continental shale oil in China. *Pet. Explor. Dev.* **2020**, *47*, 819–828. [CrossRef]
6. Li, G.; Zhu, R. Progress, challenges and key issues of unconventional oil and gas development of CNPC. *China Petrol. Explor.* **2020**, *25*, 1–13.
7. Tang, J.; Wang, X.; Du, X.; Ma, B.; Zhang, F. Optimization of integrated geological-engineering design of volume fracturing with fan-shaped well pattern. *Pet. Explor. Dev.* **2023**, *50*, 971–978. [CrossRef]
8. Tao, J.; Meng, S.; Li, D.; Rui, Z.; Liu, H.; Xu, J. Analysis of CO_2 effects on porosity and permeability of shale reservoirs under different water content conditions. *Geoenergy Sci. Eng.* **2023**, *226*, 211774. [CrossRef]
9. Zhou, D.; Zhang, G.; Huang, Z.; Zhao, J.; Wang, L.; Qiu, R. Changes in microstructure and mechanical properties of shales exposed to supercritical CO_2 and brine. *Int. J. Rock Mech. Min.* **2022**, *160*, 105228. [CrossRef]
10. Mojid, M.; Negash, B.; Abdulelah, H.; Jufar, S.; Adewumi, B. A state-of-art review on waterless gas shale fracturing technologies. *J. Pet. Sci. Eng.* **2021**, *196*, 108048. [CrossRef]
11. Yu, J.; Li, N.; Hui, B.; Zhao, W.; Li, Y.; Kang, J.; Hu, P.; Chen, Y. Experimental simulation of fracture propagation and extension in hydraulic fracturing: A state-of-the-art review. *Fuel* **2024**, *363*, 131021. [CrossRef]
12. Tang, J.; Zhang, M.; Guo, X.; Geng, J.; Li, Y. Investigation of creep and transport mechanisms of CO_2 fracturing within natural gas hydrates. *Energy* **2024**, *300*, 131214. [CrossRef]
13. Hu, Y.; Weijermars, R.; Zuo, L.; Yu, W. Benchmarking EUR estimates for hydraulically fractured wells with and without fracture hits using various DCA methods. *J. Pet. Sci. Eng.* **2018**, *162*, 617–632. [CrossRef]
14. Jin, L.; Hawthorne, S.; Sorensen, J.; Pekot, L.; Kurz, B.; Smith, S.; Heebink, L.; Herdegen, V.; Bosshart, N.; Torres, J.; et al. Advancing CO_2 enhanced oil recovery and storage in unconventional oil play—Experimental studies on Bakken shales. *Appl. Energy* **2017**, *208*, 171–183. [CrossRef]
15. Wu, J.; Wang, H.; Shi, Z.; Wang, Q.; Zhao, Q.; Dong, D.; Li, S.; Liu, D.; Sun, S.; Qiu, Z. Favorable lithofacies types and genesis of marine-continental transitional black shale: A case study of Permian Shanxi Formation in the eastern margin of Ordos Basin, NW China. *Pet. Explor. Dev.* **2021**, *48*, 1137–1149. [CrossRef]
16. Tian, H.; He, K.; Huangfu, Y.; Liao, F.; Wang, X.; Zhang, S. Oil content and mobility in a shale reservoir in Songliao Basin, Northeast China: Insights from combined solvent extraction and NMR methods. *Fuel* **2024**, *357*, 129678. [CrossRef]
17. Gong, D.; Bai, L.; Gao, Z.; Qin, Z.; Wang, Z.; Wei, W.; Yang, A.; Wang, R. Occurrence mechanisms of laminated-type and sandwich-type shale oil in the Fengcheng Formation of Mahu Sag, Junggar Basin. *Energy Fuels* **2023**, *37*, 13960–13975. [CrossRef]

18. Zhao, X.; Jin, F.; Liu, X.; Zhang, Z.; Cong, Z.; Li, Z.; Tang, J. Numerical study of fracture dynamics in different shale fabric facies by integrating machine learning and 3-D lattice method: A case from Cangdong Sag, Bohai Bay basin, China. *J. Pet. Sci. Eng.* **2022**, *218*, 110861. [CrossRef]
19. Dai, J.; Wang, T.; Tian, K.; Weng, J.; Li, J.; Li, G. CO_2 huff-n-puff combined with radial borehole fracturing to enhance oil recovery and store CO_2 in a shale oil reservoir. *Geoenergy Sci. Eng.* **2023**, *228*, 212012. [CrossRef]
20. Jia, B.; Tsau, J.; Barati, R. A review of the current progress of CO_2 injection EOR and carbon storage in shale oil reservoirs. *Fuel* **2019**, *236*, 404–427. [CrossRef]
21. Alvarez, J.; Saputra, I.; Schechter, D. The Impact of surfactant imbibition and adsorption for improving oil recovery in the Wolfcamp and Eagle Ford Reservoirs. *SPE J.* **2018**, *23*, 2103–2117. [CrossRef]
22. Lu, M.; Qian, Q.; Zhong, A.; Zhang, Z.; Zhang, L. Investigation on the flow behavior and mechanisms of water flooding and CO_2 immiscible/miscible flooding in shale oil reservoirs. *J. CO2 Util.* **2024**, *80*, 102660. [CrossRef]
23. Burrows, L.; Haeri, F.; Cvetic, P.; Sanguinito, S.; Shi, F.; Tapriyal, D.; Goodman, A.; Enick, R. A Literature review of CO_2, natural gas, and water-based fluids for enhanced oil recovery in unconventional reservoirs. *Energy Fuels* **2020**, *34*, 5331–5380. [CrossRef]
24. Huang, S.; Jiang, G.; Guo, C.; Feng, Q.; Yang, J.; Dong, T.; He, Y.; Yang, L. Experimental study of adsorption/desorption and enhanced recovery of shale oil and gas by zwitterionic surfactants. *Chem. Eng. J.* **2024**, *487*, 150628. [CrossRef]
25. Alvarez, J.; Saputra, I.; Schechter, D. Potential of improving oil recovery with surfactant additives to completion fluids for the Bakken. *Energy Fuels* **2017**, *31*, 5982–5994. [CrossRef]
26. Cheng, C.; Ming, G. Investigation of cyclic CO_2 huff-and-puff recovery in shale oil reservoirs using reservoir simulation and sensitivity analysis. *Fuel* **2017**, *188*, 102–111. [CrossRef]
27. Wan, T.; Zhang, J.; Jing, Z. Experimental evaluation of enhanced shale oil recovery in pore scale by CO_2 in Jimusar reservoir. *J. Pet. Sci. Eng.* **2022**, *208*, 109730. [CrossRef]
28. Li, L.; Bao, Z.; Li, L.; Li, Z.; Ban, S.; Li, Z.; Wang, T.; Li, Y.; Zheng, N.; Zhao, C.; et al. The source and preservation of lacustrine shale organic matter: Insights from the Qingshankou Formation in the Changling Sag, Southern Songliao Basin, China. *Sediment. Geol.* **2024**, *466*, 106649. [CrossRef]
29. Sun, L.; Cui, B.; Zhu, R.; Wang, R.; Feng, Z.; Li, B.; Zhang, J.; Gao, B.; Wang, Q.; Zeng, H.; et al. Shale oil enrichment evaluation and production law in Gulong Sag, Songliao Basin, NE China. *Pet. Explor. Dev.* **2023**, *50*, 505–519. [CrossRef]
30. Liu, B.; Wang, H.; Fu, X.; Bai, Y.; Bai, L.; Jia, M.; He, B. Lithofacies and depositional setting of a highly prospective lacustrine shale oil succession from the Upper Cretaceous Qingshankou Formation in the Gulong sag, northern Songliao Basin, northeast China. *AAPG Bull.* **2019**, *103*, 405–432. [CrossRef]
31. Yuan, S.; Lei, Z.; Li, J.; Mo, Z.; Li, B.; Wang, R.; Liu, Y.; Wang, Q. Key theoretical and technical issues and countermeasures for effective development of Gulong shale oil, Daqing Oilfield, NE China. *Pet. Explor. Dev.* **2023**, *50*, 562–572. [CrossRef]
32. Yang, L.; Yang, D.; Zhang, M.; Meng, S.; Wang, S.; Su, Y.; Long, X. Application of nano-scratch technology to identify continental shale mineral composition and distribution length of bedding interfacial transition zone—A case study of Cretaceous Qingshankou Formation in Gulong Depression, Songliao Basin, NE China. *Geoenergy Sci. Eng.* **2024**, *234*, 212674. [CrossRef]
33. Yang, L.; Wang, H.; Xu, H.; Guo, D.; Li, M. Experimental study on characteristics of water imbition and ion diffusion in shale reservoirs. *Geoenergy Sci. Eng.* **2023**, *229*, 212167. [CrossRef]
34. Zhang, J.; Zhu, R.; Wu, S.; Jiang, X.; Liu, C.; Cai, Y.; Zhang, S.; Zhang, T. Microscopic oil occurrence in high-maturity lacustrine shales: Qingshankou Formation, Gulong Sag, Songliao Basin. *Pet. Sci.* **2023**, *20*, 2726–2746. [CrossRef]
35. Wang, X.; Cui, B.; Feng, Z.; Shao, H.; Huo, L.; Zhang, B.; Gao, B.; Zeng, H. In-situ hydrocarbon formation and accumulation mechanisms of micro- and nano- scale pore-fracture in Gulong shale, Songliao Basin, NE China. *Pet. Explor. Dev.* **2023**, *50*, 1105–1115. [CrossRef]

Disclaimer/Publisher's Note: The statements, opinions and data contained in all publications are solely those of the individual author(s) and contributor(s) and not of MDPI and/or the editor(s). MDPI and/or the editor(s) disclaim responsibility for any injury to people or property resulting from any ideas, methods, instructions or products referred to in the content.

Article

Evaluation of Favorable Fracture Area of Deep Coal Reservoirs Using a Combination of Field Joint Observation and Paleostress Numerical Simulation: A Case Study in the Linxing Area

Shihu Zhao [1,2,3,*], Yanbin Wang [4], Yali Liu [1,2,3], Zengqin Liu [1,2,3], Xiang Wu [5], Xinjun Chen [1,2,3] and Jiaqi Zhang [1,2,3]

1. State Key Laboratory of Shale Oil and Gas Enrichment Mechanism and Efficient Development, Beijing 102206, China
2. Sinopec Key Laboratory of Shale Oil and Gas Exploration and Production, Beijing 102206, China
3. Sinopec Petroleum Exploration and Production Research Institute, Beijing 102206, China
4. College of Geoscience and Surveying Engineering, China University of Mining and Technology-Beijing, Beijing 100083, China
5. China United Coalbed Methane Co., Ltd., Beijing 100015, China
* Correspondence: zhaoshh0310.syky@sinopec.com; Tel.: +86-133-9159-9957

Abstract: The development of fractures under multiple geological tectonic movements affects the occurrence and efficient production of free gas in deep coal reservoirs. Taking the No.8 deep coal seam of the Benxi formation in the Linxing area as the object, a method for evaluating favorable fracture areas is established based on the combination of field joint staging, paleogeological model reconstruction under structural leveling, finite element numerical simulation, and fracture development criteria. The results show that a large number of shear fractures and fewer tensile joints are developed in the Benxi formation in the field and mainly formed in the Yanshanian and Himalayan periods. The dominant strikes of conjugate joints in the Yanshanian period are NWW (100°~140°) and NNW (150°~175°), with the maximum principal stress magnitude being 160 MPa along the NW orientation. Those in the Himalayan period are in the NNE direction (0°~40°) and the EW direction (80°~110°), with the maximum principal stress magnitude being 100 MPa along the NE orientation. The magnitudes of the maximum principal stress of the No. 8 deep coal seam in the Yanshanian period are between −55 and −82 MPa, indicative of compression; those in the Himalayan period are from −34 to −70 MPa in the compressive stress form. Areas with high shear stress values are mainly distributed in the central magmatic rock uplift, indicating the influence of magmatic rock uplift on in situ stress distribution and fracture development. Based on the comprehensive evaluation factors of fractures, the reservoir is divided into five classes and 24 favorable fracture areas. Fractures in Class I areas and Class II areas are relatively well developed and were formed under two periods of tectonic movements. The method for evaluating favorable fracture areas is not only significant for the prediction of fractures and free gas contents in this deep coal reservoir but also has certain reference value for other reservoirs.

Keywords: deep coal reservoir; field joint observation; joint staging; finite element method; favorable fracture area

Citation: Zhao, S.; Wang, Y.; Liu, Y.; Liu, Z.; Wu, X.; Chen, X.; Zhang, J. Evaluation of Favorable Fracture Area of Deep Coal Reservoirs Using a Combination of Field Joint Observation and Paleostress Numerical Simulation: A Case Study in the Linxing Area. *Energies* **2024**, *17*, 3424. https://doi.org/10.3390/en17143424

Academic Editors: Reza Rezaee and Ákos Török

Received: 6 May 2024
Revised: 27 June 2024
Accepted: 9 July 2024
Published: 11 July 2024

Copyright: © 2024 by the authors. Licensee MDPI, Basel, Switzerland. This article is an open access article distributed under the terms and conditions of the Creative Commons Attribution (CC BY) license (https://creativecommons.org/licenses/by/4.0/).

1. Introduction

Deep coalbed methane (deeper than 1500 m) has enormous resource potential and has achieved multiple breakthroughs in basins including Piceance basin in the U.S., the Alberta basin in Canada, the Cooper basin in Australia, the Junggar basin, the Ordos basin, and the Sichuan basin in China [1–3]. Exploration practice has shown that most deep coal reservoirs belong to dry coal systems with low water content and relatively high free gas content [4–8]. As the main spaces for the occurrence of free gas, the formation and development of fractures are directly influenced by the stress field of different geological periods, and clarification of paleostress field distribution characteristics is of great significance for the prediction of favorable fracture areas [9–11]. Previous studies have mainly focused on the formation mechanism, controlling factors, characterization methods and prediction methods of fractures [12–16], and the numerical simulation method for 2D or 3D models combined with field fracture data, acoustic emission and rock mechanics experiments, imaging logging, and fracture criteria is widely used to predict paleostress [17–19]. However, geological models used in the numerical simulation are based on the current burial depth of reservoirs, which is a superimposed product of multiple tectonic movements. Additionally, the influence of subsequent tectonic movements on the stress numerical simulation of the early period cannot be eliminated. Therefore, two key problems still need to be examined in depth: (1) the formation and distribution characteristics of fractures under the superposition of multiple structure movements of deep coal reservoirs; and (2) techniques for evaluating favorable fracture areas constrained by multi-stage structural movements of deep coal reservoirs.

In this study, two periods of tectonic fractures are identified based on field joint observations; the maximum principal stress direction and magnitude are obtained using a joint staging and conjugate joint angle estimation method; 3D heterogeneous geological models of deep coal reservoirs in two periods are established using the tectonic trace recovery method; the distributions of stress fields of deep coal reservoirs during different tectonic periods are obtained based on the finite element method; and the fracture development characteristics of different periods in deep coal reservoirs are predicted based on Mohr–Coulomb and Griffith criteria. The results will be beneficial for the effective exploration and development of deep coalbed methane.

2. Geological Setting

The Linxing area is located in the northern part of Jinxi Fold at the eastern margin of the Ordos Basin, with a generally southwest oriented monocline structure that dips westward at 1–5° (Figure 1a,b). It has Cenozoic, Mesozoic, and Paleozoic strata from top to bottom, and the Upper Carboniferous and Lower Permian strata are exposed in the area (Figure 1c). The No. 8 + 9 coal seam in the Benxi formation is the main production layer of coalbed methane, with a thickness varying between 2.5 and 15.5 m and a depth varying between 1093 and 2114 m. The No. 8 + 9 coal seam has undergone four uplifts under multi-stage tectonic movements, resulting in the development of multi-directional folds and faults, and uplift amplitudes in the latter two periods are relatively larger (Figure 1d), reflecting stronger tectonic movements during the Yanshanian and Himalayan periods [20–22]. In addition, due to the magmatic activity during Yanshanian period, the Zijinshan magmatic pluton formed an uplift (magmatic rock uplift) in the central part of the study area [23,24].

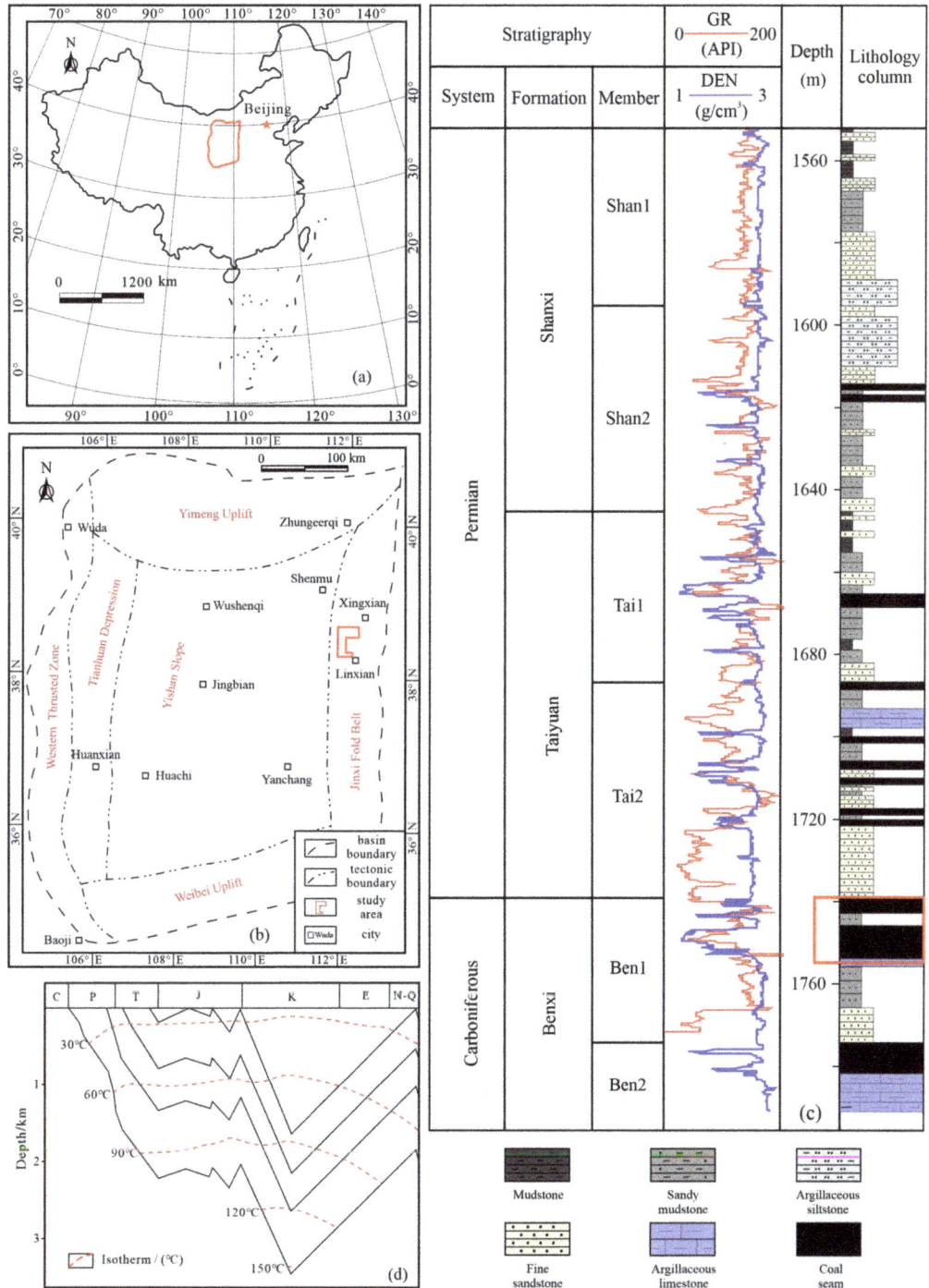

Figure 1. Location and stratigraphic column of the study area: (**a**) location of the Ordos basin (red line); (**b**) location of the Linxing area; (**c**) lithology column of the Linxing area; (**d**) buried history of the Linxing area.

3. Methodology

3.1. Field Joint Observations

Field joint observation is performed to obtain the fracture characteristics (strike, dip direction, and dip angle) of Carboniferous–Permian rocks in the study area and its surrounding areas, including the Baode–Palougou Section, Fugu–Sunjiagou Section, Xingxian–Guanjiaya Section, and Liulin–Chengjiazhuang Section. The observed fractures mainly include shear and tension joints from sandstone, mudstone, limestone, and the coal seam. The GPS is used for positioning, and the compass is used to measure the strike, dip direction, and dip angle of joints.

3.2. Staging of Joints

The tectonic movements of different stages form different types of joint combinations and the staging of joints is essential for the determination of the paleostress direction. In the article, the regional geological setting, fracture occurrence, and the intersection relationship (staggering, limiting, intercutting, tracking, utilizing, transforming) of joints are combined to determine the formation sequence of joints and clarify the joints combination in the same tectonic period.

3.3. Numerical Simulation of Paleostress Fields

The paleostress field is the stress field of the paleogeological period, which is influenced by the geological body, mechanical properties, and boundary conditions (direction and magnitude of principal stress) [25–27]. Based on the restoration of the geological model, mechanical parameters and boundary conditions of the paleogeological period, the finite element method, and ANSYS software (18.0 version) are used to produce 3D simulations of the paleostress fields.

3.3.1. Paleogeological Model

The geological body went through a process of sedimentation, compaction, folding, and fracture during the sedimentary evolution. To study the paleostress state of a geological body, the restoration of the paleogeological model is necessary, including defaultization, defolding, and decompaction. The construction-flattening method [20], based on the superposition theory of waves, is used to flatten the folds of geological bodies and restore the geological bodies of different geological periods, which is effective in paleostress studies.

In this study, the structural flattening method is used to restore paleogeological models based on the contour lines of the 8 + 9 # coal seam floor of current period. Additionally, a combination strata of roof–coal seam–floor is adopted in the model to consider the influence of the roof and floor on the in situ stress of the coal seam. Additionally, the geological model is discretized into hexahedron elements, with 70,529 elements and 24,430 nodes.

3.3.2. Mechanical Parameters

The mechanical parameters of the 8 + 9 # coal seam are calculated by logging interpretation under the constraints of triaxial compression tests, including Young's modulus, Poisson's ratio, compressive strength, tensile strength, shear strength, cohesion, and internal friction angle of the coal seam (Formulas (1)–(6)) [28–30].

$$E_d = \frac{10^3 \rho v_s^2 \left[3(v_p/v_s)^2 - 4 \right]}{\left[(v_p/v_s)^2 - 1 \right]} \quad (1)$$

$$\mu_d = \frac{(v_p/v_s)^2 - 2}{2\left[(v_p/v_s)^2 - 1 \right]} \quad (2)$$

$$\sigma_c = 12\sigma_t = 0.0045 E_d (1 - V_{sh}) + 0.008 E_d V_{sh} \quad (3)$$

$$k = 0.026\sigma_c / \left[\frac{3(1-2\mu_d)}{E_d} \times \frac{\psi}{1-\psi} \times 10^6\right] \quad (4)$$

$$C = 5.44 \times 10^{-3}\rho^2 v_p^4 \left(\frac{1+\mu_d}{1-\mu_d}\right)^2 (1-2\mu_d)(1+0.78V_{sh}) \quad (5)$$

$$\varphi = 90 - \frac{360}{\pi}\arctan\left(1/\sqrt{4.73 - 0.098\psi}\right) \quad (6)$$

where E_d represents Young's modulus, MPa; ρ represents logging density, g/cm³; V_p represents primary wave velocity of logging, km/s; vs. represents secondary wave velocity of logging, km/s; μ_d represents Poisson's ratio; σ_c represents uniaxial compressive strength, MPa; σ_t represents uniaxial tensile strength, MPa; k represents shear strength, MPa; V_{sh} represents volume percent of shale, %; ψ represents porosity, %; C represents cohesion, MPa; and φ represents internal friction angle, °.

3.3.3. Boundary Conditions

Boundary conditions include the determination of stress direction, stress magnitude, and displacement constraints on the boundary of geological model. The stress direction is determined by conjugate shear joint strike of different tectonic stages. The stress magnitude is calculated based on the relationship expression between the conjugate shear joint angle and the stress magnitude (Formulas (7) and (8)) [31,32].

$$\sigma_1 = \sigma_t - \frac{k^2}{4\sigma_t} + \frac{k^2}{2\sigma_t}\left(\frac{1}{\cos\theta} - \frac{1}{2\cos^2\theta}\right) \quad (7)$$

$$\sigma_3 = \sigma_t - \frac{k^2}{4\sigma_t} - \frac{k^2}{2\sigma_t}\left(\frac{1}{\cos\theta} + \frac{1}{2\cos^2\theta}\right) \quad (8)$$

where σ_1 represents the maximum principal stress, MPa; σ_3 represents the minimum principal stress, MPa; and θ represents the conjugate angle of conjugate shear joint, °.

3.4. Fracture Development Criteria

The fracture is formed in coal reservoirs when the paleostress field reaches the initiate threshold [33–35]. Based on the Mohr–Coulomb and Griffith criteria, the shear fracture coefficient (C_S, Formulas (9)–(11)) and the tensile fracture coefficient (C_T, Formula (2)) are established to evaluate the development degree of shear fracture and tensile fracture, respectively.

$$C_S = \left\{\sigma_1 - \left[\sigma_3 tan^2\left(45° + \frac{\varphi}{2}\right) + 2C \cdot tan\left(45° + \frac{\varphi}{2}\right)\right]\right\}/\sigma_1 \quad (9)$$

$$\text{When } \sigma_1 + 3\sigma_3 \leq 0 \ C_T = (\sigma_3 - \sigma_t)/\sigma_3 \quad (10)$$

$$\text{When } \sigma_1 + 3\sigma_3 > 0 \ C_T = \left(\frac{(\sigma_1 - \sigma_3)^2}{8(\sigma_1 + \sigma_3)} + \sigma_t\right)/\sigma_3 \quad (11)$$

where C_S represents the shear fracture coefficient, and C_T represents the tensile fracture coefficient.

It is obvious that when C_S is less than 0, the rock has not undergone shear fracture. When C_S is greater than or equal to 0, the rock has undergone shear fracture, and the degree of fracture development increases with the increase in the C_S value; Similarly, when C_T is less than 0, the rock has not undergone tensile fracture. When C_T is greater than or equal to 0, the rock has undergone tensile fracture, and the fracture development degree increases with the increase in the C_T value.

4. Results and Discussions

4.1. Field Joints' Characteristics

Field joint observations are conducted on 81 geological points, and a total of 230 sets of joint orientation observation data are obtained (Appendix A Table A1), showing that sub-vertical joints are developed in different geological points. Additionally, based on the strike rose diagram of field joints (Figure 2), the dominant strikes of joints are NNE (10°~25°), NEE (50°~80°), near-EW (85°~110°), NW (130°~150°), and NNW (160°~175°) trending. Most of the triangular rock blocks sandwiched between the joint surfaces have fallen off.

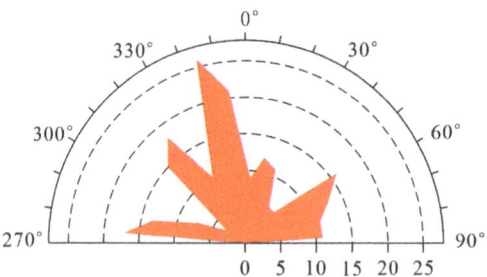

Figure 2. Strike rose diagram of field joints.

4.2. Paleo Tectonic Stress Characteristics

Based on the development characteristics of conjugate joints and the tectonic evolution history in the study area, four combinations of different types and periods are classified using the stereographic projection method [36] (Figure 3): the type I combination has a set of conjugated joints with NWW (50°~80°) and NNW (150°~175°) strikes, with joint surfaces approximately perpendicular to the geological strata formed in the Yanshanian period and the maximum stress direction in the near-NW orientation (Figure 3a,b).

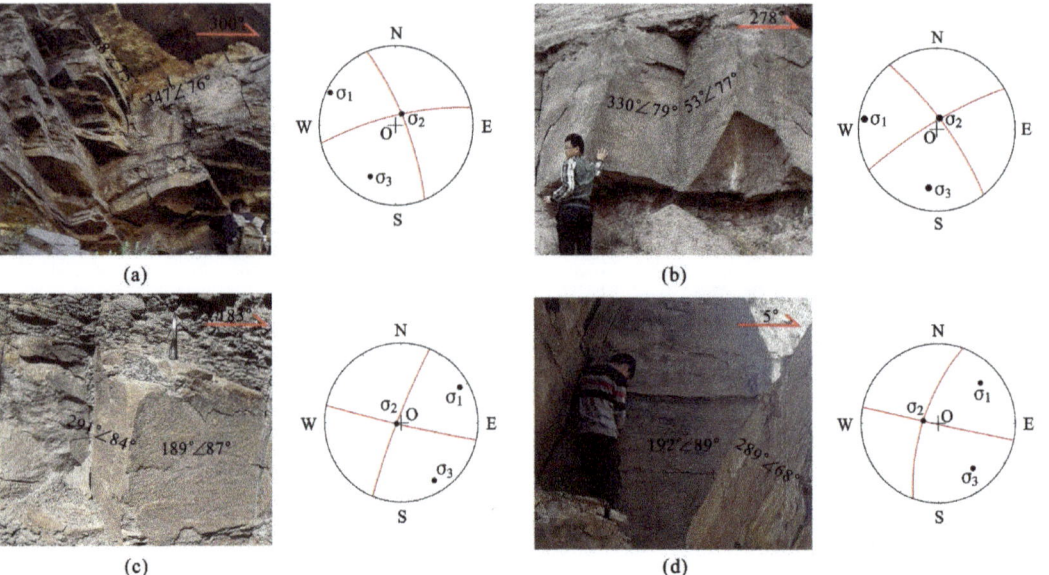

Figure 3. Combination types and stereographic projection of conjugate joints in the field: (**a,b**) represent the type I combination; (**c,d**) represent the type II combination.

The conjugated joint strikes of the type II combination are in the NNE direction (0°~40°) and the near-EW direction (80°~110°), corresponding to the Himalayan period, with the maximum stress direction being in the near-NE orientation (Figure 3c,d).

It is obvious that the joints in the study area are mainly formed in the Yanshanian period and the Himalayan period. Additionally, the paleostress magnitude is calculated using the conjugate shear angle estimation method, which shows the maximum horizontal principal stress in the Yanshanian period is 160 MPa, and the minimum principal stress is 10 MPa. The maximum horizontal principal stress in the Himalayan period is 100 MPa, and the minimum principal stress is 20 MPa.

4.3. Paleogeological Models

The compression with NW orientation in the Yanshanian period causes the coal reservoir to produce NW oriented folds. After the compression with NE orientation in the Himalayan period, which is nearly perpendicular to the Yanshanian period, the NW oriented folds are superimposed on the NE oriented folds, forming a superimposed fold structure. The 8 + 9 # coal reservoir in the study area forms a large number of synclines, anticlines, and saddle-shaped structures. For the restoration of the ancient geological model in the study area, the core is the products of these superimposed structures. Firstly, taking the Benxi formation 8 + 9 # coal reservoir in the Linxing area as the research object, the structural traces of the current coal seam floor contour lines are analyzed, and structures including anticlines, synclines, and folds are categorized. The structural superposition method [20] is applied to flatten the superimposed products of anticlines, synclines, and saddle structures and obtain the contour map of the coal seam floor in paleo periods (Figure 4).

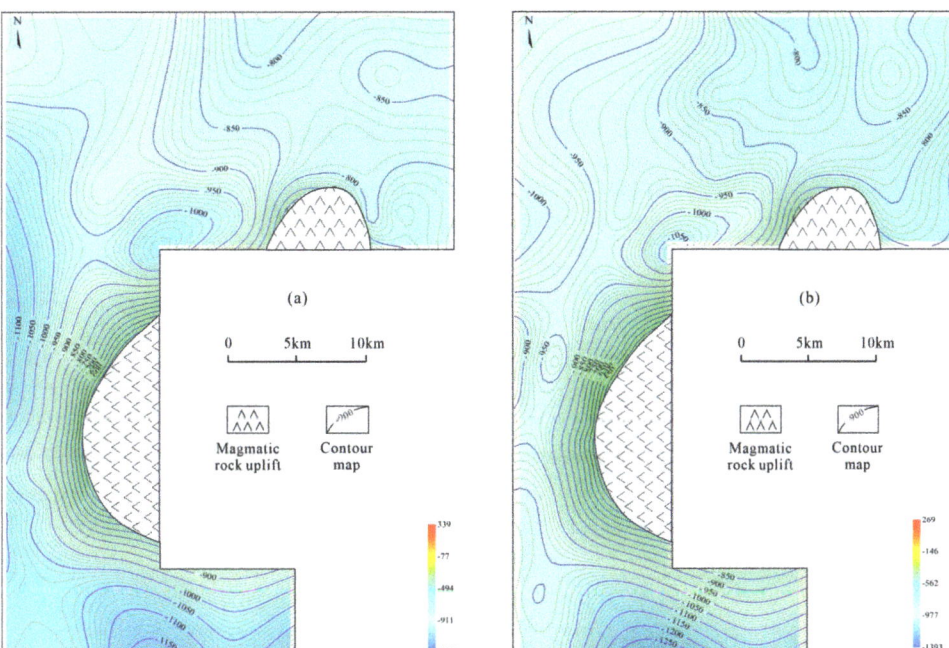

Figure 4. Floor contour map of the No. 8 + 9 coal seam during paleogeological periods: (**a**) Yanshanian period; (**b**) Himalayan period.

Based on the contour map of coal thickness and ancient coal seam floor, an isotropic idealized geological model is established using triangular meshes in ANSYS software (18.0 version). In addition, the stress of the reservoir is greatly affected by the roof and floor rock layers, so the geological model is constructed as a roof–coal–floor combination type (Figure 5), and a cube is considered to surround the roof–coal–floor combination model to simulate the stress condition of the surrounding rock.

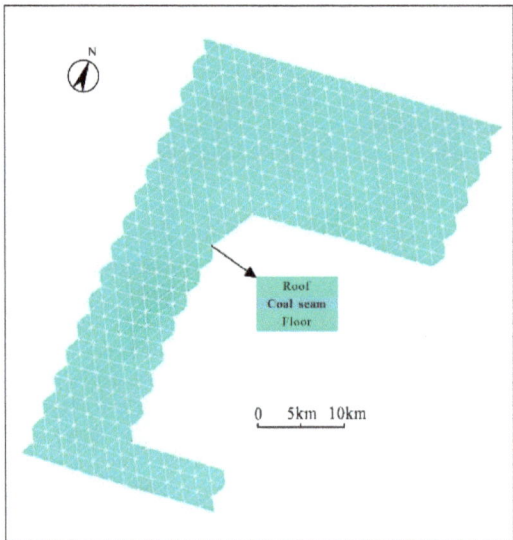

Figure 5. Roof–coal–floor combination type of paleogeological model in the study area.

4.4. Paleostress Field

Based on the tectonic stress direction and magnitude, overburden stress, Young's modulus, Poisson's ratio, and density of geological model (Table 1), geomechanical models with boundary conditions for different paleogeological periods are applied, and the finite element method and ANSYS software (18.0 version) are used to obtain the stress distribution of the 8 + 9 # coal reservoir in different geological periods.

Table 1. Mechanical parameters for numerical simulation of the paleostress field in the Linxing area.

Geologic Bodies	Poisson's Ratio	Young's Modulus (GPa)	Density (g/cm^3)
Roof	0.22	21.33	2.730
Coal seam	0.36	6.2	1.480
Floor	0.21	21.55	2.750
Others	0.23	20	1.655

4.4.1. Yanshanian Period

The stress distribution of the 8 + 9 # coal reservoir in the Yanshanian period is shown in Figure 6, where the values of maximum principal stress are between −55 and −82 MPa, with an average of −67 MPa, which is indicative of compression (Figure 6a). Excluding the influence of model boundaries on the results, the overall maximum principal stress gradually increases from north to south, and high-value zones are mainly distributed in the southern syncline (mostly −76 to −82 MPa) and central magmatic rock uplift (mainly between −73 and −80 MPa). The minimum principal stress values are distributed between −8 and −35 MPa, with an average of −20 MPa, which is also indicative of compression (Figure 6b). The value of minimum principal stress gradually decreases from east to west, and the central magmatic rock uplift has a low minimum principal stress value, with values below −10 MPa.

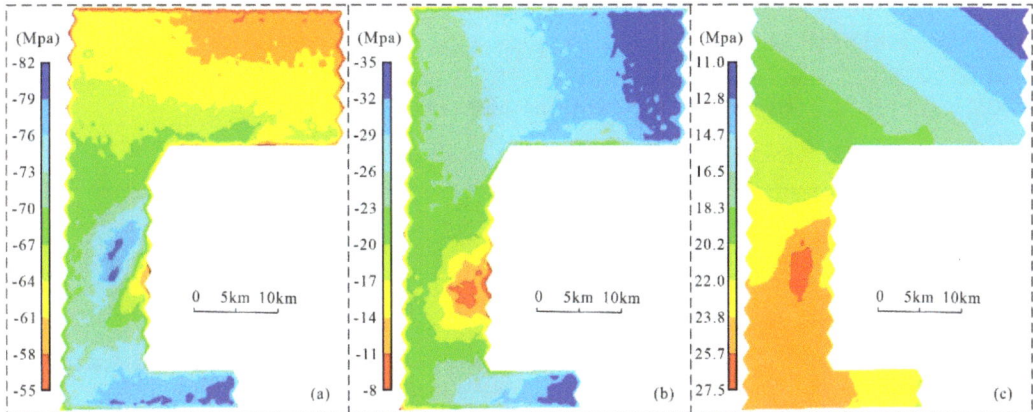

Figure 6. The stress distribution of the 8 + 9 # coal reservoir in the Yanshanian period: (**a**) maximum principal stress; (**b**) minimum principal stress; (**c**) shear stress.

Figure 6c indicates that shear stress values in the Yanshanian period are generally between 11 MPa and 27.5 MPa. The stress value in the northern zone of the study area gradually increases from northeast to southwest, exhibiting a strip-shaped distribution; the central magmatic rock uplift and southwest zones show high shear stress values, which means shear fractures are prone to occur in those zones.

4.4.2. Himalayan Period

Figure 7 shows the stress distribution of the 8 + 9 # coal reservoir in the Himalayan period, in which the maximum and minimum principal stress values are from −34 to −70 MPa and from −2 to −27 MPa, respectively, both in the form of compressive stress. The overall trends of high in south zones and low in the north zones of maximum and minimum principal stress are displayed in the study area (Figure 7a,b), and local low value zones of maximum and minimum principal stress are distributed in the central magmatic rock uplift.

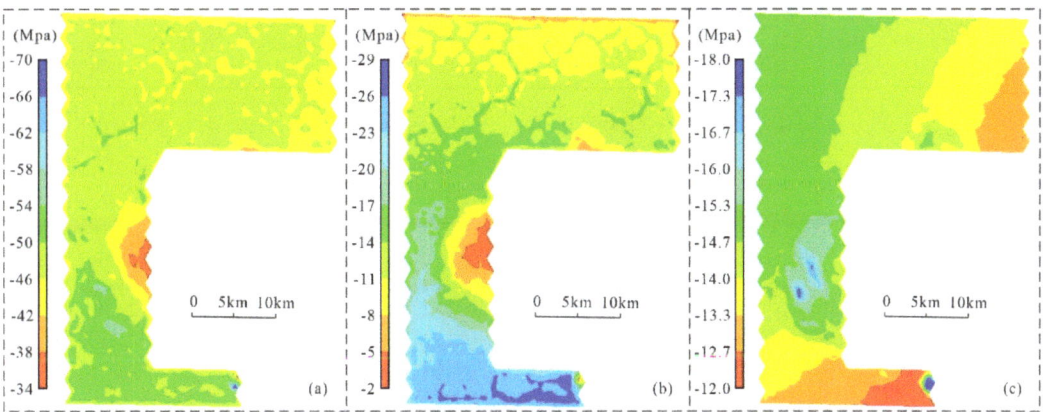

Figure 7. The stress distribution of the 8 + 9 # coal reservoir in the Himalayan period: (**a**) maximum principal stress; (**b**) minimum principal stress; (**c**) shear stress.

As is shown in Figure 7c, the values of shear stress range from −12 to −18 MPa, with an average of −15 MPa. The value of shear stress gradually increases from southeast to northwest zones, and high shear stress value occurs in the central magmatic rock uplift, indicating the influence of magmatic rock uplift on in situ stress distribution and fracture development.

4.5. Evaluation of Favorable Fracture Area

Based on the numerical simulation results, combined with the tensile strength, cohesion, and internal friction angle calculated from logging, the tensile and shear fracture coefficients are calculated. The results are shown in Table 2.

Table 2. Fracture coefficients in two paleogeological periods.

Nodes	C_{T1}	C_{T2}	C_{S1}	C_{S2}	F	Nodes	C_{T1}	C_{T2}	C_{S1}	C_{S2}	F
1149	−0.06	−0.60	−0.06	0.04	0.55	11,651	−0.05	−0.51	−0.09	−0.12	0.45
1409	−0.07	−0.83	−0.05	0.02	0.54	11,916	−0.04	−0.53	−0.09	−0.30	0.41
1565	−0.07	−0.95	−0.06	0.04	0.53	12,124	−0.04	−0.44	−0.09	−0.30	0.41
1669	−0.07	−1.01	−0.06	0.05	0.53	12,332	−0.04	−0.33	−0.09	−0.14	0.48
1773	−0.08	−1.09	−0.04	0.19	0.59	12,701	−0.02	−0.32	−0.09	−0.32	0.41
1877	−0.07	−1.14	−0.06	0.06	0.52	12,805	−0.03	−0.42	−0.09	−0.40	0.39
1981	−0.07	−1.19	−0.06	−0.01	0.49	12,909	−0.02	−0.35	−0.08	0.00	0.54
2033	−0.08	−1.24	−0.05	0.15	0.56	13,387	−0.02	−0.33	−0.10	−0.38	0.39
2298	−0.06	−0.69	−0.06	0.01	0.53	13,491	0.01	−0.21	−0.06	0.15	0.65
2402	−0.07	−0.76	−0.05	0.14	0.59	14,021	−0.01	−0.30	−0.09	−0.48	0.37
3291	−0.06	−0.62	−0.06	0.01	0.54	14,125	0.07	0.07	0.02	0.37	0.92
3499	−0.07	−0.80	−0.07	0.00	0.52	14,230	0.01	0.00	−0.12	0.28	0.59
3707	−0.07	−0.95	−0.07	−0.09	0.47	14,707	0.03	−0.08	−0.04	0.05	0.66
3915	−0.07	−1.08	−0.07	−0.14	0.44	14,811	0.05	0.12	−0.07	0.07	0.65
4123	−0.07	−1.15	−0.06	0.11	0.53	15,237	−0.01	−0.22	−0.10	−0.63	0.32
5225	−0.05	−0.52	−0.07	−0.05	0.52	15,341	0.02	−0.09	−0.07	−0.33	0.48
5433	−0.06	−0.67	−0.07	−0.06	0.50	15,445	0.05	0.11	−0.08	−0.26	0.51
5641	−0.07	−0.84	−0.08	−0.09	0.47	15,923	−0.01	−0.24	−0.09	−0.75	0.29
5849	−0.07	−0.98	−0.08	−0.08	0.45	16,027	0.00	−0.15	−0.07	−0.56	0.40
6057	−0.07	−1.10	−0.08	−0.02	0.46	16,132	0.00	−0.18	−0.10	−0.48	0.37
7367	−0.05	−0.53	−0.08	−0.12	0.48	16,505	−0.01	−0.25	−0.10	−0.76	0.27
7575	−0.06	−0.71	−0.08	−0.13	0.46	16,609	−0.01	−0.30	−0.10	−0.82	0.25
7783	−0.06	−0.86	−0.08	−0.18	0.42	16,713	−0.01	−0.34	−0.10	−0.85	0.24
7991	−0.07	−0.93	−0.08	−0.25	0.39	17,295	−0.02	−0.39	−0.10	−0.89	0.22
8199	−0.07	−1.09	−0.09	−0.16	0.40	17,400	−0.02	−0.41	−0.10	−0.92	0.21
9301	−0.04	−0.46	−0.08	−0.21	0.45	17,959	−0.01	−0.34	−0.10	−0.87	0.23
9509	−0.05	−0.59	−0.08	−0.11	0.47	18,115	−0.02	−0.47	−0.09	−1.00	0.18
9717	−0.06	−0.80	−0.09	−0.25	0.40	18,271	−0.03	−0.62	−0.09	−1.05	0.16
9925	−0.05	−0.70	−0.05	0.15	0.60	18,428	−0.04	−0.67	−0.09	−0.84	0.23
10,133	−0.07	−0.97	−0.10	−0.17	0.39	18,848	−0.03	−0.46	−0.11	−0.88	0.20
11,443	−0.04	−0.49	−0.09	−0.24	0.43	18,952	−0.03	−0.51	−0.11	−0.91	0.19

Notes: C_{T1} and C_{T2} represent tensile fracture coefficients of Yanshanian and Himalayan periods, respectively; C_{S1} and C_{S2} represent shear fracture coefficients of Yanshanian and Himalayan periods, respectively; and F represents the comprehensive evaluation factor.

It is obvious that tensile fracture coefficients are far lower than shear fracture coefficients in the coal reservoir, which indicates that shear behaviors are more likely to occur than tensile behaviors, and that shear fractures are more developed than tensile fractures. In addition, the shear fracture coefficients of the Himalayan period are higher than those of the Yanshanian period, reflecting that fractures are prone to developing during the Himalayan period. Thus the comprehensive evaluation factor F was established and calculated to show the development degree of fracture under multiple structural movements.

$$F = aN_{T1} + bN_{T2} + cN_{S1} + dN_{S2} \tag{12}$$

$$N = \frac{C_i - C_{min}}{C_{max} - C_{min}} \tag{13}$$

where F represents the comprehensive evaluation factor; N represents the normalization value of the fracture coefficient; C_i, C_{max}, and C_{min}, represent fracture coefficients of node i, the maximum fracture coefficient, and the minimum fracture coefficient, respectively; N_{T1}

and N_{T2} represent normalization values of the tensile fracture coefficients of the Yanshanian and Himalayan periods, respectively; N_{S1} and N_{S2} represent normalization values of the shear fracture coefficients of the Yanshanian and Himalayan periods, respectively; and a, b, c, and d represent weights of normalization values of fracture coefficients, with values of 0.1, 0.1, 0.3, and 0.5, respectively.

Additionally, the greater the value of F, the greater the development degree of fracture. According to the comprehensive evaluation results, the study area is divided into five classes in the degree of fracture development: Class I ($F > 0.50$), Class II ($0.45 < F \leq 0.50$), Class III $0.40 < F \leq 0.45$), Class IV ($0.35 < F \leq 0.40$), and Class V ($F \leq 0.35$), thereby performing a quantitative evaluation of the favorable fracture area.

As shown in Figure 8, the study area is divided into 24 fracture development areas, of which Class I areas are mainly distributed in the northwest and surround the magmatic rock uplift. Fractures in Class I areas are well developed and can be formed under two periods of tectonic movements. Class II areas are mainly distributed in the central and western regions of the research area, and fractures are relatively well developed. Class III areas are mainly distributed in the central, western, and southwestern parts of the study area with moderately developed fractures. Class IV and Class V areas have less-developed fractures, and fractures can only be formed through one tectonic movement period in some areas. Generally, fractures in the study area are controlled by tectonic stress, magmatic rock uplift, and buried depth, and favorable fracture areas are located surrounding magmatic rock uplift. Additionally, the division of reservoir fractures area can help to predict the development degree of reservoir fractures and provide a basis for coalbed methane extraction.

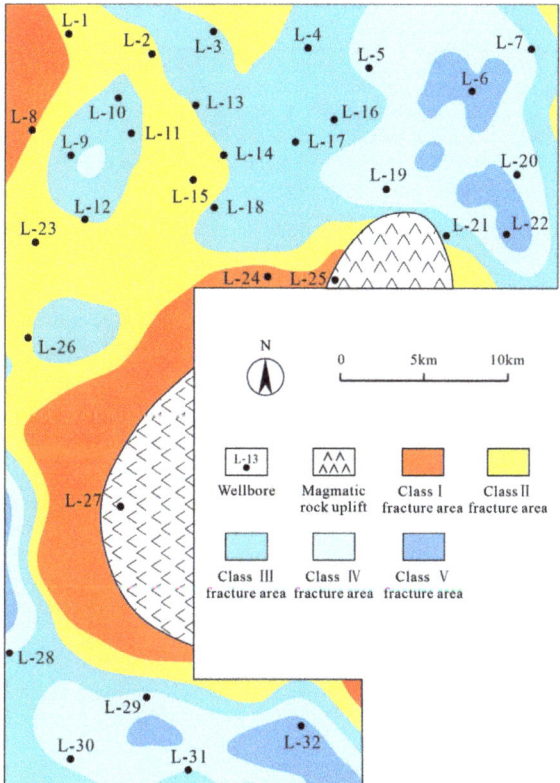

Figure 8. Evaluation of favorable fracture areas of the No. 8 + 9 coal reservoir in the study area.

5. Conclusions

Areas with high shear stress values are mainly distributed in the central magmatic rock uplift, indicating the influence of magmatic rock uplift on in situ stress distribution and fracture development. Based on the comprehensive evaluation factors of fractures, the reservoir is divided into five classes and 24 favorable fracture areas, and the following conclusions can be made:

(1) The 8+9 # coal reservoir in the Linxing area has mainly undergone two stages of tectonic movements, which are the compression in the Yanshanian period in the NW direction and the compression in the Himalayan period in the NE direction. The maximum horizontal principal stress during the Yanshanian period is 160 MPa, and the minimum principal stress is 10 MPa. The maximum horizontal principal stress during the Himalayan period is 110 MPa, and the minimum principal stress is 20 MPa.

(2) The degree of fracture development in deep coal reservoirs in the research area is directly influenced by the paleostress field, with the main fracturing periods being the Yanshanian and Himalayan periods. Based on the distribution of the paleostress field obtained from numerical simulation, the Mohr–Coulomb fracture criterion and Griffith fracture criterion are used to predict shear and tension fractures. It is found that the fracture threshold of shear fracture is smaller than that of tension fracture, and shear fractures are formed earlier than tensile fractures.

(3) Based on the comprehensive evaluation factors of fractures, the 8 + 9 # coal reservoir is divided into 24 favorable fracture areas from Class V to Class I. Fractures in Class I areas and Class II areas are relatively well developed and were formed under two periods of tectonic movements. Additionally, there are nine favorable zones in Class I and Class II, mainly distributed in the northwest of the study area and the magmatic rock uplift area.

Author Contributions: Conceptualization, S.Z., Y.W. and Y.L.; Methodology, S.Z. and Y.L.; Validation, S.Z. and Z.L.; Formal analysis, Y.W. and X.W.; Investigation, S.Z.; Writing—original draft, S.Z.; Writing—review & editing, S.Z. and X.C.; Visualization, S.Z. and J.Z.; Supervision, Y.L.; Project administration, J.Z.; Funding acquisition, Z.L. All authors have read and agreed to the published version of the manuscript.

Funding: This study was funded by the Sinopec Ministry of Science and Technology Project (No. P23208). The funders had no role in study design; the collection, analysis and interpretation of data; in the writing of the manuscript; and in the decision to submit the article for publication.

Data Availability Statement: The original contributions presented in the study are included in the article, further inquiries can be directed to the corresponding author.

Acknowledgments: We gratefully acknowledge the valuable discussions and feedback on the manuscript from our colleagues, the constructive comments by the anonymous reviewers, and the review and processing by the editors.

Conflicts of Interest: Author Xiang Wu was employed by the company China United Coalbed Methane Co., Ltd. The remaining authors declare that the research was conducted in the absence of any commercial or financial relationships that could be construed as a potential conflict of interest. The authors declare that this study received funding from Sinopec Ministry of Science and Technology Project. The funder was not involved in the study design, collection, analysis, interpretation of data, the writing of this article or the decision to submit it for publication.

Appendix A

Table A1. Field joint characteristics in the Linxing area and surroundings.

Number	Latitude	Longitude	Dip Direction and Angle	Number	Latitude	Longitude	Dip Direction and Angle
1	N 38°45'32.96"	E 111°08'13.97"	45°∠78°	116	N 39°03'31.40"	E 111°06'11.67"	325°∠69°
2	N 38°45'32.96"	E 111°08'13.97"	75°∠84°	117	N 39°02'56.54"	E 111°06'38.80"	235°∠70°
3	N 38°45'30.10"	E 111°08'9.74"	255°∠77°	118	N 39°02'56.54"	E 111°06'38.80"	151°∠86°
4	N 38°45'30.10"	E 111°08'9.74"	102°∠55°	119	N 39°02'56.54"	E 111°06'38.80"	316°∠73°
5	N 38°45'27.78"	E 111°08'6.06"	75°∠78°	120	N 39°02'56.54"	E 111°06'38.80"	225°∠69°
6	N 38°45'27.78"	E 111°08'6.06"	168°∠82°	121	N 39°02'25.24"	E 111°05'50.44"	187°∠84°
7	N 38°45'28.63"	E 111°08'3.04"	65°∠57°	122	N 39°02'25.24"	E 111°05'50.44"	86°∠81°
8	N 38°45'28.63"	E 111°08'3.04"	178°∠64°	123	N 39°02'21.39"	E 111°05'45.30"	184°∠76°
9	N 38°45'27.89"	E 111°07'51.18"	78°∠65°	124	N 39°02'21.39"	E 111°05'45.30"	77°∠54°
10	N 38°45'27.89"	E 111°07'51.18"	342°∠76°	125	N 39°02'21.39"	E 111°05'45.30"	189°∠71°
11	N 38°45'28.89"	E 111°07'47.44"	260°∠85°	126	N 39°02'21.39"	E 111°05'45.30"	76°∠67°
12	N 38°45'28.89"	E 111°07'47.44"	50°∠71°	127	N 39°02'57.69"	E 111°05'17.15"	102°∠81°
13	N 38°45'26.42"	E 111°07'43.21"	47°∠73°	128	N 39°02'57.69"	E 111°05'17.15"	182°∠85°
14	N 38°45'26.42"	E 111°07'43.21"	268°∠68°	129	N 39°02'57.69"	E 111°05'17.15"	98°∠80°
15	N 38°45'26.94"	E 111°07'41.98"	50°∠53°	130	N 39°02'57.69"	E 111°05'17.15"	65°∠81°
16	N 38°45'26.94"	E 111°07'41.98"	260°∠79°	131	N 39°02'49.33"	E 111°04'32.01"	324°∠82°
17	N 38°45'25.98"	E 111°07'40.60"	48°∠78°	132	N 39°02'49.33"	E 111°04'32.01"	345°∠78°
18	N 38°45'25.98"	E 111°07'40.60"	242°∠69°	133	N 39°02'49.33"	E 111°04'32.01"	60°∠79°
19	N 38°45'24.68"	E 111°07'34.20"	76°∠70°	134	N 39°02'49.33"	E 111°04'32.01"	68°∠75°
20	N 38°45'24.68"	E 111°07'34.20"	340°∠86°	135	N 39°02'49.33"	E 111°04'32.01"	347°∠76°
21	N 38°45'25.90"	E 111°07'29.62"	55°∠82°	136	N 39°02'21.47"	E 111°03'17.23"	335°∠83°
22	N 38°45'25.90"	E 111°07'29.62"	159°∠88°	137	N 39°02'21.47"	E 111°03'17.23"	62°∠78°
23	N 38°45'22.51"	E 111°07'24.62"	37°∠77°	138	N 39°02'21.47"	E 111°03'17.23"	325°∠87°
24	N 38°45'22.51"	E 111°07'24.62"	292°∠54°	139	N 39°02'21.47"	E 111°03'17.23"	84°∠77°
25	N 38°45'21.51"	E 111°07'22.61"	22°∠82°	140	N 39°02'11.29"	E 111°02'35.42"	335°∠82°
26	N 38°45'21.51"	E 111°07'22.61"	258°∠66°	141	N 39°02'11.29"	E 111°02'35.42"	54°∠51°
27	N 38°46'1.61"	E 111°04'9.33"	315°∠87°	142	N 39°02'11.29"	E 111°02'35.42"	63°∠54°
28	N 38°46'1.61"	E 111°04'9.33"	43°∠82°	143	N 39°02'11.29"	E 111°02'35.42"	358°∠85°
29	N 38°46'2.58"	E 111°04'11.09"	12°∠73°	144	N 38°30'14.85"	E 111°10'14.05"	97°∠56°
30	N 38°46'2.58"	E 111°04'11.09"	115°∠61°	145	N 38°30'14.85"	E 111°10'14.05"	23°∠82°
31	N 38°46'0.10"	E 111°04'11.78"	149°∠72°	146	N 38°30'14.85"	E 111°10'14.05"	352°∠87°
32	N 38°46'0.10"	E 111°04'11.78"	52°∠74°	147	N 38°30'14.85"	E 111°10'14.05"	94°∠73°
33	N 38°45'59.16"	E 111°04'12.44"	56°∠71°	148	N 38°30'15.85"	E 111°10'14.41"	22°∠85°
34	N 38°45'59.16"	E 111°04'12.44"	313°∠79°	149	N 38°30'15.85"	E 111°10'14.41"	87°∠81°
35	N 38°45'58.49"	E 111°04'20.51"	343°∠76°	150	N 38°30'12.00"	E 111°10'11.72"	70°∠82°
36	N 38°45'58.49"	E 111°04'20.51"	65°∠78°	151	N 38°30'12.00"	E 111°10'11.72"	142°∠83°
37	N 38°45'58.10"	E 111°04'22.05"	337°∠89°	152	N 38°30'1.18"	E 111°10'11.06"	78°∠75°
38	N 38°45'58.10"	E 111°04'22.05"	75°∠83°	153	N 38°30'1.18"	E 111°10'11.06"	20°∠82°
39	N 38°45'58.41"	E 111°04'23.18"	52°∠63°	154	N 38°30'1.02"	E 111°10'8.31"	84°∠82°
40	N 38°45'58.41"	E 111°04'23.18"	73°∠76°	155	N 38°30'1.02"	E 111°10'8.31"	341°∠81°
41	N 38°45'58.41"	E 111°04'23.18"	352°∠51°	156	N 38°30'0.25"	E 111°10'7.68"	57°∠78°
42	N 38°45'58.41"	E 111°04'23.18"	70°∠81°	157	N 38°30'0.25"	E 111°10'7.68"	120°∠71°
43	N 38°45'58.41"	E 111°04'23.18"	353°∠76°	158	N 38°29'57.27"	E 111°10'4.44"	43°∠75°
44	N 38°45'58.41"	E 111°04'23.18"	74°∠85°	159	N 38°29'57.27"	E 111°10'4.44"	86°∠86°
45	N 38°45'58.41"	E 111°04'23.18"	355°∠87°	160	N 38°29'57.27"	E 111°10'4.44"	43°∠83°
46	N 38°45'58.41"	E 111°04'23.18"	75°∠76°	161	N 38°29'57.27"	E 111°10'4.44"	335°∠81°
47	N 38°45'58.97"	E 111°04'40.51"	321°∠88°	162	N 38°29'55.65"	E 111°10'2.87"	64°∠86°
48	N 38°45'58.97"	E 111°04'40.51"	46°∠67°	163	N 38°29'55.65"	E 111°10'2.87"	26°∠82°
49	N 38°45'58.97"	E 111°04'40.51"	330°∠79°	164	N 37°26'27.77"	E 110°54'12.61"	47°∠79°
50	N 38°45'58.97"	E 111°04'40.51"	53°∠77°	165	N 37°26'27.77"	E 110°54'12.61"	125°∠64°
51	N 38°46'2.86"	E 111°05'14.02"	201°∠88°	166	N 37°26'39.75"	E 110°53'51.22"	145°∠84°
52	N 38°46'2.86"	E 111°05'14.02"	76°∠68°	167	N 37°26'39.75"	E 110°53'51.22"	62°∠89°
53	N 38°46'2.86"	E 111°05'14.02"	152°∠88°	168	N 37°26'41.60"	E 110°53'49.24"	16°∠55°
54	N 38°46'2.86"	E 111°05'14.02"	77°∠72°	169	N 37°26'41.60"	E 110°53'49.24"	107°∠80°
55	N 38°46'2.86"	E 111°05'14.02"	206°∠86°	170	N 37°26'41.60"	E 110°53'49.24"	117°∠83°
56	N 38°46'2.86"	E 111°05'14.02"	106°∠59°	171	N 37°26'41.60"	E 110°53'49.24"	44°∠75°
57	N 38°45'27.48"	E 111°06'44.02"	321°∠84°	172	N 37°26'41.60"	E 110°53'49.24"	86°∠88°
58	N 38°45'27.48"	E 111°06'44.02"	73°∠66°	173	N 37°33'30.36"	E 110°53'51.46"	298°∠80°
59	N 38°45'26.41"	E 111°06'46.27"	144°∠81°	174	N 37°33'30.36"	E 110°53'51.46"	195°∠84°
60	N 38°45'26.41"	E 111°06'46.27"	81°∠69°	175	N 37°33'35.10"	E 110°53'33.25"	290°∠87°
61	N 38°45'26.41"	E 111°06'46.27"	346°∠88°	176	N 37°33'35.10"	E 110°53'33.25"	75°∠75°
62	N 38°45'26.41"	E 111°06'46.27"	74°∠74°	177	N 37°33'35.07"	E 110°53'33.50"	297°∠86°
63	N 38°45'26.41"	E 111°06'46.27"	155°∠76°	178	N 37°33'35.07"	E 110°53'33.50"	194°∠89°

Table A1. Cont.

Number	Latitude	Longitude	Dip Direction and Angle	Number	Latitude	Longitude	Dip Direction and Angle
64	N 38°45′26.41″	E 111°06′46.27″	74°∠65°	179	N 37°33′8.90″	E 110°51′58.80″	52°∠83°
65	N 38°45′27.10″	E 111°06′47.15″	4°∠65°	180	N 37°33′8.90″	E 110°51′58.80″	141°∠68°
66	N 38°45′27.10″	E 111°06′47.15″	97°∠73°	181	N 37°33′9.01″	E 110°51′58.85″	183°∠89°
67	N 38°45′27.10″	E 111°06′47.15″	341°∠82°	182	N 37°33′9.01″	E 110°51′58.85″	81°∠66°
68	N 38°45′27.10″	E 111°06′47.15″	74°∠78°	183	N 37°33′10.65″	E 110°51′55.53″	75°∠70°
69	N 38°45′24.64″	E 111°06′56.57″	322°∠82°	184	N 37°33′10.65″	E 110°51′55.53″	155°∠78°
70	N 38°45′24.64″	E 111°06′56.57″	74°∠67°	185	N 37°33′9.98″	E 110°51′54.10″	196°∠89°
71	N 38°45′22.63″	E 111°07′0.53″	123°∠68°	186	N 37°33′9.98″	E 110°51′54.10″	81°∠75°
72	N 38°45′22.63″	E 111°07′0.53″	52°∠74°	187	N 37°33′10.09″	E 110°51′51.38″	85°∠79°
73	N 38°45′21.75″	E 111°07′3.08″	18°∠61°	188	N 37°33′10.09″	E 110°51′51.38″	184°∠83°
74	N 38°45′21.75″	E 111°07′3.08″	86°∠74°	189	N 37°33′9.76″	E 110°51′50.04″	183°∠84°
75	N 38°45′21.75″	E 111°07′3.08″	150°∠78°	190	N 37°33′9.76″	E 110°51′50.04″	285°∠81°
76	N 38°45′21.75″	E 111°07′3.08″	76°∠72°	191	N 37°33′9.76″	E 110°51′50.04″	78°∠89°
77	N 38°45′20.86″	E 111°07′14.95″	146°∠75°	192	N 37°33′9.76″	E 110°51′50.04″	152°∠73°
78	N 38°45′20.86″	E 111°07′14.95″	85°∠89°	193	N 37°32′55.11″	E 110°49′30.59″	57°∠81°
79	N 39°03′23.56″	E 111°07′6.74″	188°∠56°	194	N 37°32′55.11″	E 110°49′30.59″	129°∠84°
80	N 39°03′23.56″	E 111°07′6.74″	100°∠76°	195	N 37°32′55.11″	E 110°49′30.59″	51°∠79°
81	N 39°03′19.84″	E 111°07′4.13″	219°∠82°	196	N 37°32′55.11″	E 110°49′30.59″	142°∠80°
82	N 39°03′19.84″	E 111°07′4.13″	127°∠84°	197	N 37°35′33.50″	E 110°53′10.92″	290°∠74°
83	N 39°03′20.39″	E 111°07′3.14″	133°∠76°	198	N 37°35′33.50″	E 110°53′10.92″	185°∠86°
84	N 39°03′20.39″	E 111°07′3.14″	221°∠84°	199	N 37°35′33.50″	E 110°53′10.92″	190°∠87°
85	N 39°03′20.73″	E 111°07′2.48″	212°∠79°	200	N 37°35′33.50″	E 110°53′10.92″	290°∠82°
86	N 39°03′20.73″	E 111°07′2.48″	139°∠87°	201	N 37°35′30.47″	E 110°53′0.46″	294°∠82°
87	N 39°03′20.32″	E 111°07′2.01″	222°∠79°	202	N 37°35′30.47″	E 110°53′0.46″	193°∠82°
88	N 39°03′20.32″	E 111°07′2.01″	143°∠84°	203	N 37°35′30.47″	E 110°53′0.46″	190°∠89°
89	N 39°03′19.69″	E 111°07′1.11″	56°∠81°	204	N 37°35′30.47″	E 110°53′0.46″	290°∠84°
90	N 39°03′19.69″	E 111°07′1.11″	127°∠79°	205	N 37°35′30.47″	E 110°53′0.46″	191°∠87°
91	N 39°03′18.83″	E 111°07′0.56″	53°∠74°	206	N 37°35′30.47″	E 110°53′0.46″	275°∠88°
92	N 39°03′18.83″	E 111°07′0.56″	131°∠87°	207	N 37°35′26.86″	E 110°52′50.82″	190°∠89°
93	N 39°03′21.35″	E 111°07′4.26″	42°∠81°	208	N 37°35′26.86″	E 110°52′50.82″	285°∠87°
94	N 39°03′21.35″	E 111°07′4.26″	320°∠83°	209	N 37°35′26.86″	E 110°52′50.82″	198°∠88°
95	N 39°03′21.35″	E 111°07′4.26″	45°∠71°	210	N 37°35′26.86″	E 110°52′50.82″	281°∠71°
96	N 39°03′21.35″	E 111°07′4.26″	324°∠86°	211	N 37°35′25.39″	E 110°52′48.02″	80°∠73°
97	N 39°03′35.25″	E 111°06′19.22″	182°∠67°	212	N 37°35′25.39″	E 110°52′48.02″	193°∠88°
98	N 39°03′35.25″	E 111°06′19.22″	277°∠90°	213	N 37°35′25.28″	E 110°52′41.15″	291°∠84°
99	N 39°03′35.25″	E 111°06′19.22″	359°∠71°	214	N 37°35′25.28″	E 110°52′41.15″	189°∠87°
100	N 39°03′35.25″	E 111°06′19.22″	272°∠77°	215	N 37°35′22.00″	E 110°52′35.99″	192°∠89°
101	N 39°03′35.25″	E 111°06′19.22″	272°∠57°	216	N 37°35′22.00″	E 110°52′35.99″	289°∠68°
102	N 39°03′35.25″	E 111°06′19.22″	183°∠88°	217	N 37°35′20.02″	E 110°52′30.63″	189°∠89°
103	N 39°03′34.15″	E 111°06′20.15″	267°∠76°	218	N 37°35′20.02″	E 110°52′30.63″	282°∠79°
104	N 39°03′34.15″	E 111°06′20.15″	183°∠74°	219	N 37°35′20.02″	E 110°52′30.63″	186°∠85°
105	N 39°03′30.99″	E 111°06′19.14″	355°∠83°	220	N 37°35′20.02″	E 110°52′30.63″	82°∠66°
106	N 39°03′30.99″	E 111°06′19.14″	87°∠81°	221	N 37°35′19.18″	E 110°52′26.46″	82°∠81°
107	N 39°03′30.99″	E 111°06′19.14″	230°∠82°	222	N 37°35′19.18″	E 110°52′26.46″	193°∠86°
108	N 39°03′30.99″	E 111°06′19.14″	325°∠46°	223	N 37°35′13.14″	E 110°51′57.89″	179°∠86°
109	N 39°03′30.26″	E 111°06′18.75″	359°∠82°	224	N 37°35′13.14″	E 110°51′57.89″	270°∠79°
110	N 39°03′30.26″	E 111°06′18.75″	273°∠79°	225	N 37°35′13.14″	E 110°51′57.89″	175°∠84°
111	N 39°03′30.15″	E 111°06′13.40″	224°∠75°	226	N 37°35′13.14″	E 110°51′57.89″	283°∠84°
112	N 39°03′30.15″	E 111°06′13.40″	82°∠87°	227	N 37°35′13.70″	E 110°51′57.07″	184°∠82°
113	N 39°03′31.40″	E 111°06′11.67″	337°∠66°	228	N 37°35′13.70″	E 110°51′57.07″	272°∠85°
114	N 39°03′31.40″	E 111°06′11.67″	54°∠75°	229	N 37°35′14.15″	E 110°51′51.27″	177°∠72°
115	N 39°03′31.40″	E 111°06′11.67″	209°∠73°	230	N 37°35′14.15″	E 110°51′51.27″	280°∠69°

References

1. Johnson, R.D.; Flores, R.M. Developmental geology of coalbed methane from shallow to deep in Rocky Mountain basins and in Cook Inlet–Matanuska basin, Alaska, U.S.A. and Canada. *Int. J. Coal Geol.* **1998**, *35*, 241–282. [CrossRef]
2. Salmachi, A.; Rajabi, M.; Wainman, C.; Mackie, S.; McCabe, P.; Camac, B.; Clarkson, C. History, Geology, In Situ Stress Pattern, Gas Content and Permeability of Coal Seam Gas Basins in Australia: A Review. *Energies* **2021**, *14*, 2651. [CrossRef]
3. Li, S.; Qin, Y.; Tang, D.; Shen, J.; Wang, J.; Chen, S. A comprehensive review of deep coalbed methane and recent developments in China. *Int. J. Coal Geol.* **2023**, *279*, 104369. [CrossRef]
4. Kang, Y.; Huangfu, Y.; Zhang, B.; He, Z.; Jiang, S.; Ma, Y.Z. Gas oversaturation in deep coals and its implications for coal bed methane development: A case study in Linxing Block, Ordos Basin, China. *Front. Earth Sci.* **2023**, *10*, 1031493. [CrossRef]
5. Li, Y.; Wang, Z.; Tang, S.; Elsworth, D. Re-evaluating adsorbed and free methane content in coal and its ad- and desorption processes analysis. *Chem. Eng. J.* **2022**, *428*, 131946. [CrossRef]

6. Ouyang, Z.; Wang, H.; Sun, B.; Liu, Y.; Fu, X.; Dou, W.; Du, L.; Zhang, B.; Luo, B.; Yang, M.; et al. Quantitative Prediction of Deep Coalbed Methane Content in Daning-Jixian Block, Ordos Basin, China. *Processes* **2023**, *11*, 3093. [CrossRef]
7. Chen, B.; Stuart, F.M.; Xu, S.; Györe, D.; Liu, C. The effect of Cenozoic basin inversion on coal-bed methane in Liupanshui Coalfield, Southern China. *Int. J. Coal Geol.* **2022**, *250*, 103910. [CrossRef]
8. Sun, Y.; Lin, Q.; Zhu, S.; Han, C.; Wang, X.; Zhao, Y. NMR investigation on gas desorption characteristics in CBM recovery during dewatering in deep and shallow coals. *J. Geophys. Eng.* **2023**, *20*, 12–20. [CrossRef]
9. Ni, X.; Jia, Q.; Wang, Y. The Relationship between Current Ground Stress and Permeability of Coal in Superimposed Zones of Multistage Tectonic Movement. *Geofluids* **2019**, *2019*, 9021586. [CrossRef]
10. Li, Y.; Tang, D.; Xu, H.; Yu, T. In-situ stress distribution and its implication on coalbed methane development in Liulin area, eastern Ordos basin, China. *J. Petrol. Sci. Eng.* **2014**, *122*, 488–496. [CrossRef]
11. Zhao, J.; Tang, D.; Xu, H.; Li, Y.; Li, S.; Tao, S.; Lin, W.; Liu, Z. Characteristic of In Situ Stress and Its Control on the Coalbed Methane Reservoir Permeability in the Eastern Margin of the Ordos Basin, China. *Rock Mech. Rock Eng.* **2016**, *49*, 3307–3322. [CrossRef]
12. Reeher, L.J.; Hughes, A.N.; Davis, G.H.; Kemeny, J.M.; Ferrill, D.A. Finding the right place in Mohr circle space: Geologic evidence and implications for applying a non-linear failure criterion to fractured rock. *J. Struct. Geol.* **2023**, *166*, 104773. [CrossRef]
13. Maerten, L.; Maerten, F.; Lejri, M.; Gillespie, P. Geomechanical paleostress inversion using fracture data. *J. Struct. Geol.* **2016**, *89*, 197–213. [CrossRef]
14. Liu, J.; Luo, Y.; Tang, Z.; Lu, L.; Zhang, B.; Yang, H. Methodology for quantitative prediction of low-order faults in rift basins: Dongtai Depression, Subei Basin, China. *Mar. Petrol. Geol.* **2024**, *160*, 106618. [CrossRef]
15. Xie, Q.; Li, G.; Yang, X.; Peng, H. Evaluating the Degree of Tectonic Fracture Development in the Fourth Member of the Leikoupo Formation in Pengzhou, Western Sichuan, China. *Energies* **2023**, *16*, 1797. [CrossRef]
16. Liu, S.; Sang, S.; Pan, Z.; Tian, Z.; Yang, H.; Hu, Q.; Sang, G.; Qiao, M.; Liu, H.; Jia, J. Study of characteristics and formation stages of macroscopic natural fractures in coal seam #3 for CBM development in the east Qinnan block, Southern Quishui Basin, China. *J. Nat. Gas Sci. Eng.* **2016**, *34*, 1321–1332. [CrossRef]
17. Zamani, G.B. Geodynamics and tectonic stress model for the Zagros fold-thrust belt and classification of tectonic stress regimes. *Mar. Petrol. Geol.* **2023**, *155*, 106340. [CrossRef]
18. Li, J.; Qin, Q.; Li, H.; Zhou, J.; Wang, S.; Zhao, S.; Qin, Z. Numerical simulation of the palaeotectonic stress field and prediction of the natural fracture distribution in shale gas reservoirs: A case study in the Longmaxi Formation of the Luzhou area, southern Sichuan Basin, China. *Geol. J.* **2023**, *58*, 4165–4180. [CrossRef]
19. Jiu, K.; Ding, W.; Huang, W.; You, S.; Zhang, Y.; Zeng, W. Simulation of paleotectonic stress fields within Paleogene shale reservoirs and prediction of favorable zones for fracture development within the Zhanhua Depression, Bohai Bay Basin, east China. *J. Petrol. Sci. Eng.* **2013**, *110*, 119–131. [CrossRef]
20. Gao, X.; Wang, Y.; Ni, X.; Li, Y.; Wu, X.; Zhao, S.; Yu, Y. Recovery of tectonic traces and its influence on coalbed methane reservoirs: A case study in the Linxing area, eastern Ordos Basin, China. *J. Nat. Gas Sci. Eng.* **2018**, *56*, 414–427. [CrossRef]
21. Liu, J.; Cao, D.; Tan, J.; Zhang, Y. Gzhelian cyclothem development in the western North China cratonic basin and its glacioeustatic, tectonic, climatic and autogenic implications. *Mar. Petrol. Geol.* **2023**, *155*, 106355. [CrossRef]
22. Ju, W.; Shen, J.; Li, C.; Yu, K.; Yang, H. Natural fractures within unconventional reservoirs of Linxing Block, eastern Ordos Basin, central China. *Front. Earth Sci.* **2020**, *14*, 770–782. [CrossRef]
23. Pu, Y.; Li, S.; Tang, D.; Chen, S. Effect of Magmatic Intrusion on In Situ Stress Distribution in Deep Coal Measure Strata: A Case Study in Linxing Block, Eastern Margin of Ordos Basin, China. *Na. Resour. Res.* **2022**, *31*, 2919–2942. [CrossRef]
24. Shu, Y.; Lin, Y.; Liu, Y.; Yu, Z. Control of magmatism on gas accumulation in Linxing area, Ordos Basin, NW China: Evidence from fluid inclusions. *J. Petrol. Sci. Eng.* **2019**, *180*, 1077–1087. [CrossRef]
25. Paul, S.; Chatterjee, R. Mapping of cleats and fractures as an indicator of in-situ stress orientation, Jharia coalfield, India. *Int. J. Coal Geol.* **2011**, *88*, 113–122. [CrossRef]
26. Wang, J.; Wang, Y.; Zhou, X.; Xiang, W.; Chen, C. Paleotectonic Stress and Present Geostress Fields and Their Implications for Coalbed Methane Exploitation: A Case Study from Dahebian Block, Liupanshui Coalfield, Guizhou, China. *Energies* **2024**, *17*, 101. [CrossRef]
27. Han, W.; Wang, Y.; Li, Y.; Ni, X.; Wu, X.; Wu, P.; Zhao, S. Recognizing fracture distribution within the coalbed methane reservoir and its implication for hydraulic fracturing: A method combining field observation, well logging, and micro-seismic detection. *J. Nat. Gas Sci. Eng.* **2021**, *92*, 103986. [CrossRef]
28. Ning, F.; Wu, N.; Li, S.; Zhang, K.; Yu, Y.; Liu, L.; Sun, J.; Jiang, G.; Sun, C.; Chen, G. Estimation of in-situ mechanical properties of gas hydrate-bearing sediments from well logging. *Petrol. Explor. Develop.* **2013**, *40*, 542–547. [CrossRef]
29. Li, Z.; Chen, Z.; Yu, L.; Zhang, S.; Gai, K.; Fan, Y.; Huo, W. Geophysical logging in brittleness evaluation on the basis of rock mechanics parameters: A case study of sandstone of Shanxi formation in Yanchang gas feld, Ordos Basin. *J. Pet. Explor. Prod. Technol.* **2023**, *13*, 151–162. [CrossRef]
30. Wojtowicza, M.; Jarosiński, M. Reconstructing the mechanical parameters of a transversely-isotropic rock based on log and incomplete core data integration. *Int. J. Rock Mech. Min.* **2019**, *115*, 111–120. [CrossRef]

31. Xie, X.; Wang, W. Seismic Conjugate Ruptures and Limiting Principal Stresses Accompanying Variation of Depths in the Crust -Take 1975 Haicheng Earthquake with M7.3 as an Example. *Earthq. Res. China* **2002**, *18*, 166–174, (Chinese Journal with English Abstract).
32. Lin, Y. Relation Between Conjugate Shearing Angle and Values of Confining Pressure. *J. Prog. Geophys.* **1993**, *8*, 133–139, (Chinese Journal with English Abstract).
33. Mahetaji, M.; Brahma, J. A critical review of rock failure Criteria: A scope of Machine learning approach. *Eng. Fail. Anal.* **2024**, *159*, 107998. [CrossRef]
34. Feng, J.; Dai, J.; Lu, J.; Li, X. Quantitative Prediction of 3-D Multiple Parameters of Tectonic Fractures in Ti Sandstone Reservoirs Based on Geomechanical Method. *IEEE Access* **2018**, *6*, 39096–39116. [CrossRef]
35. Zhao, Y.; Mishra, B.; Shi, Q.; Zhao, G. Size-dependent Mohr-Coulomb failure criterion. *Bull. Eng. Geol. Environ.* **2023**, *82*, 218. [CrossRef]
36. Koca, M.Y.; Kincal, C.; Onur, A.H.; Koca, T.K. Determining the inclination angles of anchor bolts for sliding and toppling failures: A case study of Izmir, Turkiye. *Bull. Eng. Geol. Environ.* **2023**, *82*, 47. [CrossRef]

Disclaimer/Publisher's Note: The statements, opinions and data contained in all publications are solely those of the individual author(s) and contributor(s) and not of MDPI and/or the editor(s). MDPI and/or the editor(s) disclaim responsibility for any injury to people or property resulting from any ideas, methods, instructions or products referred to in the content.

MDPI AG
Grosspeteranlage 5
4052 Basel
Switzerland
Tel.: +41 61 683 77 34

Energies Editorial Office
E-mail: energies@mdpi.com
www.mdpi.com/journal/energies

Disclaimer/Publisher's Note: The statements, opinions and data contained in all publications are solely those of the individual author(s) and contributor(s) and not of MDPI and/or the editor(s). MDPI and/or the editor(s) disclaim responsibility for any injury to people or property resulting from any ideas, methods, instructions or products referred to in the content.

www.ingramcontent.com/pod-product-compliance
Lightning Source LLC
LaVergne TN
LVHW070414100526
838202LV00014B/1458